LTLGB 2012

Hosted by
**Low Carbon Research and Education Center,
Beijing Jiaotong University**

In Cooperation with
University of Liverpool, UK

Sponsored by
**NSFC—National Natural Science Foundation of China
K. C. Wong Education Foundation, Hong Kong**

Supported by
**The Shipping Research Centre (SRC)
The Hong Kong Polytechnic University**

**Center for Housing Innovations
The Chinese University of Hong Kong**

**Academy of Mathematics and Systems Science
Chinese Academy of Sciences**

**School of Economics and Management
Beijing University of Chemical Technology**

Feng Chen · Yisheng Liu · Guowei Hua
Editors

LTLGB 2012

Proceedings of International Conference on Low-carbon Transportation and Logistics, and Green Buildings

Beijing, China

October 12-13, 2012

Volume 1

 Springer

Editors
Feng Chen
School of Civil Engineering
Beijing Jiaotong University
Beijing
People's Republic of China

Yisheng Liu
Guowei Hua
School of Economics and Management
Beijing Jiaotong University
Beijing
People's Republic of China

ISBN 978-3-642-34650-7   ISBN 978-3-642-34651-4 (eBook)
DOI 10.1007/978-3-642-34651-4
Springer Heidelberg New York Dordrecht London

Library of Congress Control Number: 2012955044

© Springer-Verlag Berlin Heidelberg 2013
This work is subject to copyright. All rights are reserved by the Publisher, whether the whole or part of the material is concerned, specifically the rights of translation, reprinting, reuse of illustrations, recitation, broadcasting, reproduction on microfilms or in any other physical way, and transmission or information storage and retrieval, electronic adaptation, computer software, or by similar or dissimilar methodology now known or hereafter developed. Exempted from this legal reservation are brief excerpts in connection with reviews or scholarly analysis or material supplied specifically for the purpose of being entered and executed on a computer system, for exclusive use by the purchaser of the work. Duplication of this publication or parts thereof is permitted only under the provisions of the Copyright Law of the Publisher's location, in its current version, and permission for use must always be obtained from Springer. Permissions for use may be obtained through RightsLink at the Copyright Clearance Center. Violations are liable to prosecution under the respective Copyright Law.
The use of general descriptive names, registered names, trademarks, service marks, etc. in this publication does not imply, even in the absence of a specific statement, that such names are exempt from the relevant protective laws and regulations and therefore free for general use.
While the advice and information in this book are believed to be true and accurate at the date of publication, neither the authors nor the editors nor the publisher can accept any legal responsibility for any errors or omissions that may be made. The publisher makes no warranty, express or implied, with respect to the material contained herein.

Printed on acid-free paper

Springer is part of Springer Science+Business Media (www.springer.com)

Preface

This volume contains the *"Proceedings of the 2012 International Conference on Low-carbon Transportation and Logistics, and Green Buildings"* (LTLGB' 2012), held in Beijing, China, hosted by Beijing Jiaotong University, in cooperation with the University of Liverpool, supported by the Shipping Research Centre (SRC) of The Hong Kong Polytechnic University, Center for Housing Innovations of the Chinese University of Hong Kong, Academy of Mathematics and Systems Science of Chinese Academy of Sciences and School of Economics and Management of Beijing University of Chemical Technology, and sponsored by National Natural Science Foundation of China (NSFC) and K. C. Wong Education Foundation.

This conference is a prime international forum for both researchers and industry practitioners to exchange latest fundamental advances in the state of the art and practice of the fields of technology, policy, and management of carbon emission, with three simultaneous tracks covering different aspects, including: "Low Carbon Transportation", "Low Carbon Logistics" and "Green Building". It also had five special sessions and workshops, Low-carbon Technology and Low-carbon Policy, Low-carbon Project Management, Industrial Security under Low Carbon Development, Low-carbon Transportation and Low-carbon Tourism, and Green Supply Chain Management. Papers published in each track describe the state-of-the-art research work that is often oriented towards real-world applications and highlight the benefits of related methods and techniques for the emerging field of low carbon transportation, Low carbon logistics, and green building development.

LTLGB 2012 received 252 paper submissions from 11 countries and regions. One hundred and thirty-three papers were accepted and published after strict peer reviews. The total acceptance ratio is 52 %. Additionally, a number of invited talks, presented by internationally recognized specialists in different areas, have positively contributed to reinforce the overall quality of the conference and to provide a deeper understanding of related areas.

The program for this conference required the dedicated effort of many people. First, we must thank the authors, whose research and development efforts are recorded here. Second, we thank the members of the program committee and the additional reviewers for valuable help with their expert reviewing of all submitted papers. Third, we thank the invited speakers for their invaluable contribution and the time for preparing their talks. Fourth, we thank the special session chairs whose collaboration with LTLGB was much appreciated. Finally, many thanks are given to the colleagues from BJTU for their hard work in organizing this year's event.

A final selection of papers, from those presented at LTLGB 2012 in Beijing, will be done based on the classifications and comments provided by the Program Committee and on the assessment provided by session chairs. Extended and revised versions of the selected papers will be published in special issues of the four international journals.

We wish that you all enjoy an exciting conference and an unforgettable stay in Beijing, China.

Prof. Feng Chen
Prof. Yisheng Liu
Dr. Guowei Hua

Organizing Committees

Honorary Chairman

Shoubo Xu	Beijing Jiaotong University, China
T. C. Edwin Cheng	The Hong Kong Polytechnic University, China
Shouyang Wang	Chinese Academy of Sciences, China
Ziyou Gao	Beijing Jiaotong University, China

General Chair

Feng Chen — Beijing Jiaotong University, China

Program Chairs

Yisheng Liu	Beijing Jiaotong University, China
Yuanfeng Wang	Beijing Jiaotong University, China
Baohua Mao	Beijing Jiaotong University, China
Haishan Xia	Beijing Jiaotong University, China
Yugong Xu	Beijing Jiaotong University, China

Publication Chairs

Guowei Hua	Beijing Jiaotong University, China
Hongjie Lan	Beijing Jiaotong University, China

Financial Chairs

Bing Zhu	Beijing Jiaotong University, China
Liang Xu	Beijing Jiaotong University, China

Special Session/Workshop Chairs

Guowei Hua   Beijing Jiaotong University, China
Xuesong Feng   Beijing Jiaotong University, China

Publicity Chair

Yingqi Liu   Beijing Jiaotong University, China

Organization Chairs

Yisong Li   Beijing Jiaotong University, China
Xuesong Feng   Beijing Jiaotong University, China
Dan Chang   Beijing Jiaotong University, China

Secretary

Liang Xu   Beijing Jiaotong University, China
Zhongzhong Zeng   Beijing Jiaotong University, China

Program Committees

Program Chair

Feng Chen — Beijing Jiaotong University, China

Vice Chairs

Dong Li — University of Liverpool, UK
Jingyu Zou — The Chinese University of Hong Kong, Hong Kong, China

Executive Chairman

Yisheng Liu — Beijing Jiaotong University, China

Program Committees

Andre Brown — University of Liverpool, UK
Andy Plater — University of Liverpool, UK
Biao Liu — The Chinese University of Hong Kong, Hongkong, China
Christina W. Y. Wong — The Hong Kong Polytechnic University, Hongkong, China
C. T. Daniel Ng — The Hong Kong Polytechnic University, Hongkong, China
Dong-Wook Song — Heriot-Watt University, UK
Hossam Ismail — University of Liverpool, UK
Huajun Tang — Macau University of Science and Technology, Macau, China
Jian Li — Beijing University of Chemical Technology, China
Jianping Wu — Tsinghua University, China
Jorge Hernandez Hormazabal — University of Liverpool, UK

ix

Juliang Zhang	Beijing Jiaotong University, China
Kee-hung Lai	The Hong Kong Polytechnic University, Hongkong, China
Lincoln Wood	Curtin University, Australia
Mingjun Zhao	The Chinese University of Hong Kong, Hongkong, China
Mingyu Zhang	Beijing Jiaotong University, China
Paul Drake	University of Liverpool, UK
Quanxin Sun	Beijing Jiaotong University, China
Runtong Zhang	Beijing Jiaotong University, China
Shaochuan Fu	Beijing Jiaotong University, China
Shaofeng Liu	University of Plymouth, UK
Steven Sharples	University of Liverpool, UK
Xianliang Shi	Beijing Jiaotong University, China
Y. H. Venus Lun	The Hong Kong Polytechnic University, Hongkong, China
Yahaya Yusuf	Lancashire University, UK
Yanping Liu	Beijing Jiaotong University, China
Yongqing Shen	Beijing Jiaotong University, China
Zhenji Zhang	Beijing Jiaotong University, China
Blandine Ageron	University of Grenoble—IUT de Valence, France
Elie Awwad	American University of Beirut, Lebanon
Ping Yin	Beijing Jiaotong University, China
Shiwei He	Beijing Jiaotong University, China
Jia Bin	Beijing Jiaotong University, China
Justin Bishop	University of Oxford, UK
Peter Boelsterli	Berne University of Applied Sciences, Switzerland
Yann Bouchery	Laboratoire Génie Industriel, France
Chengxuan Cao	Beijing Jiaotong University, China
Amin Chaabane	Ecole de Technologie Superieure, Canada
Jian Chai	Shaanxi Normal University, China
Po-Han Chen	National Taiwan University, Taiwan, China
Baotian Dong	Beijing Jiaotong University, China
Robert Earley	The Innovation Center for Energy and Transportation
Tomas Echaveguren	Universidad de Concepcion
Gokhan Egilmez	Ohio University, USA
Raafat El-Hacha	University of Calgary, Canada
Yao Enjian	Beijing Jiaotong University, China
Wenbo Fan	Southwest Jiaotong University, China
Tao Feng	Eindhoven University of Technology, The Netherlands
Xuesong Feng	Beijing Jiaotong University, China
Terry Friesz	The Pennsylvania State University, USA

Program Committees

Bian Gavin	Beijing Jiaotong University, China
Vicente González	The University of Auckland, New Zealand
Li Haiying	North China University of Technology, China
Bruce Hartman	University of Saint Francis, USA
Xiaoming He	Beijing Jiaotong University, China
Yong He	Southeast University, China
Katsuya Hihara	University of Tokyo, Japan
Fangcheng Tang	Beijing Jiaotong University, China
Guowei Hua	Beijing Jiaotong University, China
Niu Huimin	Lanzhou Jiaotong University
Pawinee Iamtrakul	Thammasat University, Thailand
Umit Isikdag	Beykent University, Turkey
Chiappetta Jabbour	The Universidade Estadual Paulista, Brazil
Werner Jammernegg	Vienna University of Economics and Business, Austria
Ananda Jeeva	Curtin University of Technology, Australia
Li Jian	Beijing University of Chemical Technology, China
R. C. Jou	National Chi Nan University, Taiwan, China
Zhucui Jing	Beijing Jiaotong University, China
Wu Jun	Beijing University of Posts and Telecommunications, China
Liu Kai	Beijing Jiaotong University, China
Lenny Koh	The University of Sheffield Management School, UK
Susana Lagüela	University of Vigo
Gilbert Laporte	University of Montreal, Canada
Guoquan Li	Railway Technical Research Institute, Japan
Xiang Li	Nankai University, China
Xingang Li	Beijing Jiaotong University, China
Xuemei Li	Beijing Jiaotong University, China
Yanzhi Li	City University of Hong Kong, China
Yongjian Li	Nankai University, China
Tiezhu Li	Southeast University, China
Tianliang Liu	Beihang University, China
Yingqi Liu	Beijing Jiaotong University, China
Hector G. Lopez-Ruiz	University of Lyon, France
Dragana Macura	University of Belgrade
Qiang Meng	National University of Singapore, Singapore
Juan Moreno-Gutiérrez	University of Cadiz, Spain
Baozhuang Niu	Sun Yat-sen University, China
Makoto Okumura	Tohoku University, Japan
Ghim Ping Ong	National University of Singapore, Singapore
Tongyan Pan	The Catholic University of America
Kriengsak Panuwatwanich	Griffith University

Luiz Pereira-De-Oliveira	Universidade da Beira Interior, Portugal
Harilaos Psaraftis	National Technical University of Athens, Greece
Han Qiao	China University of Geosciences, China
Mansour Rahimi	University of Southern California, USA
Xiaofeng Shao	Shanghai Jiaotong University, China
Rafat Siddique	Thapar University, India
Stephen Skippon	Shell Global Solutions, UK
Huajun Tang	Macau University of Science and Technology, Macau, China
Humberto Varum	University of Aveiro
Vedat Verter	McGill University, Canada
Eleni Vlahogianni	The National Technical University of Athens, Greece
Chao Wang	Beijing Jiaotong University, China
Mingxi Wang	University of International Business and Economics, China
Xiaolei Wang	The Hong Kong University of Science and Technology, China
Yacan Wang	Beijing Jiaotong University, China
Ying-Wei Wang	National Penghu University of Science and Technology, Taiwan, China
Wang Wei	Beijing Jiaotong University, China
Wei Wei	Tsinghua University, China
Liu Weihua	Tianjin University, China
S. C. Wong	The University of Hong Kong
Lincoln Wood	Curtin University, Australia
Ning Wu	Ruhr University Bochum
Danqin Yang	Nanjing University of Science and Technology, China
Liu Yang	University of International Business and Economics, China
Sudong Ye	Beijing Jiaotong University, China
Bin Yu	Dalian Maritime University, China
Andrew Yuen	The Chinese University of Hong Kong, China
Yahaya Yusuf	University of Central Lancashire,
Massimiliano Zanin	INNAXIS Foundation & Research Institute
Guoxing Zhang	LanZhou University, China
Junyi Zhang	Hiroshima University, Japan
Yingxue Zhao	University of International Business and Economics, China
Leng Zhen	The Hong Kong Polytechnic University, China
Qian Zhen	Stanford University, USA
Ali Ülkü	Capital University, USA

Special Session Program Committee

Jianping Wu	Tshinghua University, China
Jie Cao	Nanjing University of Information Science & Technology, China
Zhucui Jing	Beijing Jiaotong University, China
Guoxing Zhang	Lanzhou University, China
Huajun Tang	Macau University of Science and Technology, Macau, China
Ping Yin	Beijing Jiaotong University, China

Contents

Part I Keynote Lectures

1 Environmental Material Flow Theory and System in China 3
  Shoubo Xu

2 Green Supply Chain Design and Management 5
  Zuo-Jun Max Shen

3 Integrating the Environment, Urban Planning, and Transport: Where Does Economics Fit in? 7
  Kenneth Button

4 The History and Challenges of Japan's Low-Carbon Transportation Systems 9
  Takayuki Morikawa

5 Low Carbon Urban Design 11
  Peter Boelsterli

6 Interdisciplinary Behavior Studies for Cross-Sector Energy Policies 13
  Junyi Zhang

7 International Journal of Shipping and Transport Logistics: An Insider's Perspective 15
  Y. H. Venus Lun

8 The Roles of Railway Freight Transport in Developing the Low-Carbon Society and Relevant Issues 17
  Guoquan Li

9	**Chinese Condition Must be Considered on Developing Green Building in China**......................... Youguo Qin	19

Part II Low Carbon Transportation

10	**Study on Traffic and Infrastructure Construction Performance Assessment Based on Sustainable Development**............. Jie Zhang, Huibing Xie, Minghui Liu and Kai Liu	23
11	**Sustainable Development of China's Road Transportation Infrastructure: Situation and Prospect**......... Jie Zhang, Kai Liu and Yurong Zhang	31
12	**Energy Demand and Emission from Transport Sector in China**...................................... Yin Huang and Mengjun Wang	39
13	**Exploring the Effect of Inter-Stop Transport Distances on Traction Energy Cost Intensities of Freight Trains**......... Xuesong Feng, Haidong Liu and Keqi Wu	45
14	**A Freeway/Expressway Shockwave Elimination Method Based on IoT**..................................... Ling Huang and Jianping Wu	51
15	**Allocating the Subsidy Among Urban Public Transport Enterprises for Good Performance and Low Carbon Transportation: An Application of DEA**................... Qianzhi Dai, Yongjun Li, Qiwei Xie and Liang Liang	59
16	**What Counts in the Bus Use for Commuting? A Probe Survey Based on Extended Theory of Planned Behavior**............. Wen Wu, Dong Ding and Ping Wu	67
17	**Evaluation Study on the City Bicycle Rental System**.......... Jianyou Zhao, Yunjiao Zhang and Cheng Zhang	75
18	**The Green Traffic Strategy in Low Carbon Community**........ Zesong Wei, Xia Wang and Xiaolong Pang	83

19	**Discussion on Countermeasures of China's Low-Carbon Tourism Development**............................ Xuefeng Wang and Hui Zhang	91
20	**A Study of Vehicle Tax Policy Adjustment Based on System Dynamics in the Background of Low-Carbon Transport** ..... Feifei Xie and Xuemei Li	101
21	**Economic Evaluation of Energy Saving and Emission Reduction for ETC**............................. Jia-hua Gan, Xiao-ming Zhang and Ze-bin Huang	111
22	**Model Calculating on Integrated Traffic Energy Consumption and Carbon Emissions in Beijing**...................... Ying-yue Hu, Feng Chen, Wei-ming Shen and Qi-bing Wu	119
23	**Evaluation Indexes of Public Bicycle System** .... Yue Ma and Xiao-ning Zhu	127
24	**Study on Urban ITS Architecture Based on the Internet of Things** .... Zinan Yang, Xifu Wang and Hongsheng Sun	139
25	**The Pedestrian and Cycling Planning in the Medium-Sized City: A Case Study of Xuancheng**..................... Yiling Deng, Xiucheng Guo, Yadan Yan and Xiaohong Jiang	145
26	**A Model to Evaluate the Modal Shift Potential of Subsidy Policy in Favor of Sea-Rail Intermodal Transport** .... Xuezong Tao	153
27	**Study of Training System Applying on Energy-Saving Driving**.. Haili Yuan, Bin Li and Wei Wang	161
28	**The Outlook of Low-Carbon Transport System: A Case Study of Jinan** .... Qiang Han and Yong Zhou	167
29	**The Logit Model in the Urban Low Carbon Transport and Its Application**............................ Zinan Yang and Xifu Wang	175

30	Comprehensive Evaluation of Highway Traffic Modernization Based on Low-Carbon Economy Perspective Linlin Zheng, Yongbo Lv, Li Chen and Le Huang	181
31	Traffic Congestion Measurement Method of Road Network in Large Passenger Hub Station Area Yu Han, Xi Zhang and Lu Yu	189
32	The Governance of Urban Traffic Jam Based on System Dynamics: In Case of Beijing, China Haoxiong Yang, Kaichun Lin, Yongsheng Zhou and Xinjian Du	197
33	The Design and Realization of Urban Mass Information Publishing System Kai Yan, Li-min Jia, Jie Xu and Jian-yuan Guo	209
34	Research on Multiple Attribute Decision Making of BRT System Considering Low Carbon Factors Jia-qing Wu, Rui Song and Li Zheng	217
35	Research on Time Cost of Urban Congestion in Beijing Qifu He	225
36	The Development and Application of Transport Energy Consumption and Greenhouse Gas Emission Calculation Software Based on the Beijing Low-Carbon Transport Research Weiming Shen, Feng Chen and Zijia Wang	237
37	Operational Planning of Electric Bus Considering Battery State of Charge Qian Qiu, Jun Li and Hongru Yu	243
38	Scheme Research of Urban Vehicle Restriction Measures According to Synthesis Criterion Long Chen, Ming-jiang Shen and Xing-yi Zhu	251
39	The Research of Low-Carbon Transportation Management System Wen-shuai Guo and Chao-he Rong	261
40	Urban Low-Carbon Transport System Peng Xing and Tianjun Hu	269

41	**Study on the Control of AC Dynamometer System for Hybrid Electrical Vehicle Test Bench**............ Ying Tian, Zhenhua Jing, Keli Wang, Shengfang Nie and Qingchun Lu	277
42	**Low-Carbon Transport System by Bicycle, in Malmö, Sweden**............................ Yingdong Hu and Xiaobei Li	283
43	**A Study on Low-Carbon Transportation Strategy Based on Urban Complex: Taking Shenzhen and Hong Kong as Examples**................ Yezi Dai	293
44	**Study on Data Storage Particle Size Optimization of Traffic Information Database for Floating Car Systems Based on Minimum Description Length Principle**........... Rui Zhao, Enjian Yao, Xin Li, Yuanyuan Song and Ting Zuo	301
45	**Design of Double Green Waves Scheme for Arterial Coordination Control**......................... Chengkun Liu, Qin Yong, Haijian Li, Yichao Liang, Yalong Zhao and Honghui Dong	309
46	**An Evaluation Indicator System of Low-Carbon Transport for Beijing**........................ Siyuan Zhu and Xuemei Li	317
47	**Design and Implementation of Regional Traffic Information Disseminating System Based on ZigBee and GPRS**..................... Weiran Li, Wei Guan, Jun Bi and Dongfusheng Liu	325
48	**Studying Electric Vehicle Batteries Consumption with Agent Based Modeling**................... Jinjin Fu and Xiaochun Lu	333
49	**Low-Carbon Scenario Analysis on Urban Transport of a Metropolitan of China in 2020**............. Xiaofei Chen and Zijia Wang	341
50	**Impact Study of Carbon Trading Market to Highway Freight Company in China**................... Li Chen, Boyu Zhang, Hanping Hou and Alfred Taudes	347

51	**The Importance and Construction Measures of Chinese Low-Carbon Transportation System** Xinyu Wang and Yurong Gong	355
52	**Planning Model of Optimal Modal-Mix in Intercity Passenger Transportation** Makoto Okumura, Huseyin Tirtom and Hiromichi Yamaguchi	361
53	**Research on the Optimization Scheme of Beijing Public Bicycle Rental System Life Cycle** Kaiyan Jiang and Hao Wu	367
54	**The Primary Condition of Bicycle Microcirculation System Benign Operation in Urban: Taking Hangzhou and Beijing for Example** Hao Wu and Xiao You	373
55	**The New Energy Buses in China: Policy and Development** Jingyu Wang, Yingqi Liu and Ari Kokko	379
56	**The Determinants of Public Acceptance of Electric Vehicles in Macau** Ivan Ka-Wai Lai, Donny Chi-Fai Lai and Weiwei Xu	387
57	**Synthetical Benefit Evaluation of High-Speed Rail, Take Beijing-Shanghai High-Speed Rail for Example** Han-bo Jin, Hua Feng and Fu-guang Cui	395
58	**Strategy Research on Planning and Construction of Low-Carbon Transport in Satellite Towns: The Case of Shanghai** Luwei Wang and Xinsheng Ke	403
59	**Study on Intensive Design of Urban Rail Transport Hub from the Perspective of Low-Carbon** Haishan Xia and Xiaobei Li	409
60	**The Roles of Railway Freight Transport in Developing Low-Carbon Society and Relevant Issues** Guoquan Li	417

Part III SS-Industrial Security Under Low Carbon Development

61 Preliminary Study on Coal Industrial Safety Evaluation Index System Under Low-Carbon Economy 427
Lei Zhang and Cheng Chen

62 China's Energy Economy from Low-Carbon Perspective 435
Xiaonan Qu

63 Analysis for Transformation and Development of China PV Industry 443
Shengzhen Ma

64 Non-decomposable Minimax Optimization on Distribution Center Location Selected 451
Zhucui Jing, Menggang Li and Chuanlong Wang

65 Green Finance and Development of Low Carbon Economy ... 457
Shuo Chen

Part IV Workshop on Green Supply Chain Management

66 Research on Network Optimization of Green Supply Chain: A Low-Carbon Economy Perspective 465
Cuizhen Cao and Guohao Zhao

67 The Research on Evolutionary Game of Remanufacturing Closed-Loop Supply Chain Under Asymmetric Situation 473
Jian Li, Weihao Du, Fengmei Yang and Guowei Hua

68 A Sequencing Problem for a Mixed-Model Assembly Line on Supply Chain Management 481
Hugejile, Shusaku Hiraki, Zhuqi Xu and Shaolan Yang

69 Price Competition in Tourism Supply Chain with Hotels and Travel Agency 489
Yun Huang

70 Evaluation on Bus Rapid Transit in Macau Based on Congestion and Emission Reduction 497
Huajun Tang, Xinlong Xu and Bo Huang

71	The Analysis and Strategy Research on Green Degree of Enterprise in Green Supply Chain .... Lijin Liu	503
72	The Ways for Improving the Operations of Hospital Industry: The Case in Macau................................ Yan Chen, Harry K. H. Chow and Ting Nie	511
73	The Social Costs of Rent-Seeking in the Regulation of Vehicle Exhaust Emission .... Yan Pu and Xia Liu	519

Part V Low Carbon Logistics

74	CO_2 Emissions Embodied in 42 Sectors' Exports of China .... Yufeng Wang, Shulin Liu and Changcai Qin	529
75	The Study on Risk Assess Model of Rail Transit Projects..... Xiangdong Zhu, Xiang Xiao and Chaoran Wu	539
76	Low Carbon Supply Chain Performance Evaluation Based On BSC-DEA Method......................... Yunlong Li and Xianliang Shi	547
77	Research on a Reverse Logistics of Waste Household Appliances Includes the Impact of Carbon Tax .... Youmei Gan and Xianliang Shi	553
78	Electric Power Enterprises Supply Relationships Integration: Achieve Low-Carbon Procurement .... Jingchen Gao, Jie Xu and Meiying Cheng	561
79	Coordination of Low Carbon Agricultural Supply Chain Under Contract Farming .... Guohua Sun and Shengyong Du	569
80	Logistics Financial Innovation Mode Analysis in the Low-Carbon Economy: Based on Comparative Analysis Between the Logistics Enterprise and the Professional Market.......................... ZeBin Wang	577

81	**Order Decision with Random Demand: A Research from the Perspective of Carbon Emission Cap and Carbon Trade Mechanism**	585
	Weihua Liu, Wenchen Xie and Guowei Hua	
82	**Evaluation of Low Carbon Inventory Control Policy for Creative Products in Hybrid Distributing Channels**	595
	Chun-rong Guo, Zhan-feng Zhu and Xiao-dong Zhang	
83	**Analysis of Cooperative Game in Low Carbon Supply Chain** ..	601
	Xiao-dong Zhang, Zhan-feng Zhu and Chun-rong Guo	
84	**Low-Carbon Economic Development Model on Road Freight Transport Industry in Beijing**	609
	Haoxiong Yang, Mengnan Zhang, Yongsheng Zhou and Zanbo Zhang	
85	**The Research of Carbon Footprint in the Manufacturing Supply Chain Management**	615
	Ruyan Hao and Shaochuan Fu	
86	**Research on Collaborative Pricing Decisions of Enterprises in Supply Chain Under Constraints of Carbon Emission**	623
	Qian Liu and Huiping Ding	
87	**Research on the Low Carbon District Development Mechanism of Beijing**	633
	Yingkui Zhang, Di Wu and Jia Liu	
88	**Current Trends for Development in the Aviation Industry World Integration Groups**	641
	Bo Wang and Shaolan Yang	
89	**Shipping Enterprise Develop Strategies Based on Low-Carbon Integrated Logistics**	647
	Lei Yang, Guilu Tu and Xiaocui Xiao	
90	**An Estimation Method of the Carbon Footprint in Manufacturing Logistics Systems**	657
	Xiaolong Qu and Bo Li	
91	**The Optimization Model and Algorithm of Reverse Logistics Network for Resource Recovery**	665
	Wei Cao, Xi Zhang, Te-lang Li and Ying-hui Liang	

92	Analysis of the Development of Low-Carbon Logistics Based on a Low-Carbon Economy................................ Xiu-Ying Liu	673
93	Eco-Efficient Based Logistics Network Design in Hybrid Manufacturing/Remanufacturing System Under Low-Carbon Restriction .... Yacan Wang, Xiaoxia Zhu and Tao Lu	681
94	Research on Household Electrical Appliances' Supply Chain Based on the LCA Method in the Situation of Low-Carbon Product Certification .... Honghao Gao and Xianliang Shi	691
95	Comparative Research on the Environmental Cost of Replacement and Maintenance of the Computer .... Jing Zhang and Yaoqiu Wang	699
96	Hoteling Price Competition Model Under the Carbon Emissions Constraints................................. Bin Zhang and Wenliang Bian	707
97	Reverse Logistics Practices: A Survey in Electronic Industry in Guangdong Province of China................ Yacan Wang, Junjun Yu and Yakun Wang	713
98	Study on the Legal System Development and Countermeasures of Low-Carbon Logistics in China...................... Chen Wang and Jia Jiang	721
99	Analysis of Warehouse Location in Low-Carbon Supply Chain Based on the Cost.............................. Zongxu Liu and Hongjie Lan	727
100	Research on Multi-Facility Weber Problem to Reduce Carbon Emissions............................ Sen Zheng and Jianqin Zhou	735
101	Impact of Carbon Emission Control Policies on Food Logistics Chain Speed and Cost Performance .... Zurina Hanafi and Dong Li	743

102 The Research on Driver Model of Sustainable
Supply Chain Management 751
Xiaohua Tang

Part VI Green Buildings

103 The Construction of Green Shipbuilding System 763
Hong-zhi Wang and Yang Zhao

104 Research on the Mahoney Tables Used in Shanghai
Building Energy Efficiency Design..................... 769
Bo Xia

105 Healthy Development of Green Real Estate a Report
on Current Status and Prospect of China's Green
Real Estate Development in 2012...................... 775
Xianming Huang, Junpeng Huang, Tao Li and Wei Gao

106 Research of Chinese Ancient Urban Morphologies
Based on Climate Adaptability 781
Zhongzhong Zeng, Haishan Xia and Haoxia Chen

107 Case Study of BIM-Based Building Energy Evaluation....... 787
Runmei Zhang, Changcheng Liu and Tao Xu

108 High Green Value with Low Resource Cost: Case Study
of Pearl Region Delta Greenway in China 799
Huibin Zhu

109 Research on Economic Incentive Policy to Promote
the Development of Green Buildings in China 805
Lei Fan, Dao-zhai Zhu and Yuan-feng Wang

110 A Study on the Measures in Multi-Angles for Developing
Green Building in Beijing............................ 815
Nana Zhang and Jing Liu

111 A Study on the Connotation and Evaluation System
of Green Railway Station 823
Gaiping Zhang and Chaohe Rong

112	Investigation of Application of Evaluation Standard for Green Building	829
	Ling Ye, Zhijun Cheng and Qingqin Wang	
113	Study and Application on China Railway Construction Project Scheduling Model Based on Resource Leveling	837
	Yuanjie Tang, Rengkui Liu and Quanxin Sun	
114	Study on Comprehensive Evaluation of External Thermal Insulation Composite Systems Based on Total Life Cycle of Building	847
	Yisheng Liu and Xiaowen Wang	
115	Analysis on Green Building's Technological Development and Economic Feasibility in China	855
	Jie Li	
116	Durability of Green Reactive Powder Concrete	863
	Yue Wang, Ming-zhe An, Zi-ruo Yu and Xin-tuo Hou	
117	Study on the Strategy of Green Buildings Development in China..................................	871
	Yisheng Liu and Mengyuan Hua	
118	The Green Building Materials Enterprises in the Management of Innovation and Production Technology Improvement	879
	Yunlu Li	
119	Building Life Cycle Energy Consumption Estimation Based on the Work Breakdown Structure	887
	Jian Xiao and Xueqing Zhang	
120	Research on Railway Tunnel Construction Scheduling Technique Based on LSM............................	895
	Liqiang Liu, Yisheng Liu, Yuanjie Tang and Qing Li	
121	Solar Design in the Application of the City Planning	905
	Xia Wang, Ze-Song Wei and Xiaolong Pang	
122	Discussions on Integration Designs of Solar Collectors and Building Envelopes...............................	913
	Lan Chen, Ya-Fei Zhang, Wen-Jing Liu and Jia-Huan Yin	

123 **Study on Collaborative Design of Green Building Based on BIM Technology** 921
Haishan Xia and Kuangyi Yi

124 **Research on the Structural Design of Real Estate Green Supply Chain** 929
Jingjuan Guo, Ting Xie and Aibo Hao

Part VII SS-Low-Carbon Technology and Low-Carbon Policy

125 **The Decoupling of Carbon Emissions from Economic Growth in Jiangsu, China: 2000–2010** 937
Hui Zhou and Jie Cao

126 **Research of the Criteria of Choosing Leading Industry in Under developed Areas: Guangxi Province** 945
Tong- Li, Shouji- Tu, Yin- Peng and Liqing- Li

127 **Analysis on China's Power Industry Development and Countermeasures in Low Carbon Economy Environment** 951
Ze-min Yan, Zhan-feng Zhu, Wen Qiao and Xiao-dong Zhang

128 **System Dynamics Analysis of Port City Development Under the Low-Carbon Economy-A Case Study of Ningbo** 959
Sen Yan and Fangchu Liang

129 **Empirical Analysis on Technical Factors Impacting Energy Consumption Efficiency** 967
Feixue Zhou and Zaiwu Gong

130 **Construction of Changsha-Zhuzhou-Xiangtan Low-Carbon Urban Agglomerations: Major Progress and Basic Experience** 973
Xinsha Peng and Dalun Tian

131 **Constructive Research of Carbon Accounting Information Disclosure of Listed Companies** 989
Bohan Wang, Xuemeng Guo and Dongfang Gao

132 An Empirical Study on the Factors Affecting Carbon
    Accounting Information Disclosure of Chinese
    Listed Companies.................................... 995
    Zengjun Gu, Xuemeng Guo and Lixia Jian

133 Study of Jiangsu Manufacturing Energy Consumption
    Structure Under Low Carbon Economy 1001
    Xiaodong Zhu, Chuhui Hua and Yingcui Sun

134 The Framework of Security Mechanism on the Internet
    of Things Based on RFID Boosting Low-Carbon Economy.... 1007
    Zhongyun Li and Xindi Wang

135 The Impact Brought by Global Warming
    and Countermeasures................................ 1015
    Cuifeng Huo, Menghan Xu and Xuan Ding

Part VIII SS-Low-Carbon Project Management

136 Evolutionary Analysis of Cooperative Behavior
    of the Countries in Cancun Climate Summit 1027
    Lei Zhao, Guorong Chai, Haizhou Wang and Guoping Li

137 How Does the Carbon Emission of China's Transportation
    Industry Change with the Fluctuation of GDP
    and International Oil Price?.......................... 1035
    Guoxing Zhang, Sujie Cheng, Peng Liu, Xutao Zhang
    and Guorong Chai

138 Cluster Analysis for Study Ecological Landscape Sustainability:
    An Empirical Study in Xi'an of China................. 1041
    Liyun Liu and Hongzhen Lei

139 The Construction and Empirical Study of Low-Carbon
    City Comprehensive Evaluation 1049
    Chungui Liu, Zhongxing Guo, Bin Han, Huting Yuan
    and Shaoyin Zhu

Part IX Workshop on Low-Carbon Transportation and Low-Carbon Tourism

140 SLP Method Based on Low-Carbon Logistics in Professional Agricultural Logistics Park Layout 1063
Yong Chen

141 Low-Carbon Tourism Planning Study: A Theoretical Framework 1069
Ping Yin

142 Measuring the Ecological Embeddedness of Tourism Industrial Chains 1077
Yan Wang and Hui Zhang

Part I
Keynote Lectures

Chapter 1
Environmental Material Flow Theory and System in China

Shoubo Xu

1.1 Introduction

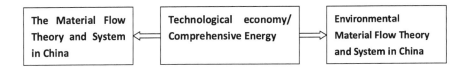

1.2 The Material Flow Theory and System in China

- The Material flow theory
- Comprehensive MF theory
- The MF element theory
- The MF nature theory
- The MF science & technology theory
- The MF engineering theory
- The MF Industry Theory

1.3 The hexa-structure theory for the MF engineering

- 6 MF forces

  Labor forces; Financial forces; Physical forces; Natural forces; Transport forces; Time forces

S. Xu (✉)
Beijing Jiaotong University, Beijing, People's Republic of China
e-mail: shbxu@bjtu.edu.cn

- The 6 MF elements

MF laborers; Objects to be worked on in the MF; Means of labor for the MF; MF work environments; MF labor space; MF labor time

1.4 Environmental Material Flow Theory and System in China

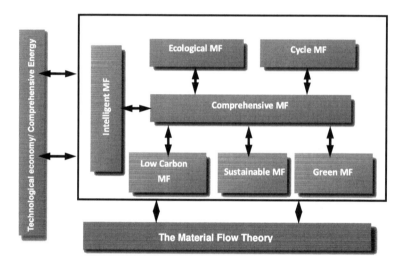

Author Biography

Prof. Shoubo Xu was born in Shaoxing (a city which located at Zhengjiang Province of China). He obtained a Bachelor's Degree in Power Engineering from Nanjing Institute of Technology in 1955. Then he graduated from the Energy Institute of the Academy of Science of USSR in 1960, with an Associate Doctorate Degree of Technological Science. Now he is honored as a professor, consultant and PHD supervisor of Economics and Management school at Beijing Jiao Tong University. And he also works as the Director of China Center of Technological Economics Research, the President of Comprehensive Energy Institute, Honorary Dean of the Material Flow School at Beijing Jiaotong University, and named as Chinese Director of the Sino-Austria Innovation Research Center. At the same time, Dr Xu is also regarded as the Chairman of Professors Association in the Economics and Management department. Besides that, he was the core initiator and co-founder of the Chinese Technological Economics and Comprehensive Energy Engineering, the pioneer of our nation's Comprehensive Material Flow Engineering and the science of Managing According to Reason MR.

For more than 50 years, academician Xu has made 422 achievements in theoretical and application aspects in the three new scientific fields of TE, ETE/CEE and MFTE/CMFE. More than 50 of his achievements have received awards, including the National Science Congress Award, the National Science & Technology Progress Award and various awards from the Chinese Academy of Science, Chinese Academy of Social Sciences, National Development and Reform Commission and City of Beijing, etc., he has received National Science & Technology Progress Award nine times (the first prize provincial, one time National Science & Technology Progress first prize and one time third prize, four times provincial second prize and three times third prize).

Chapter 2
Green Supply Chain Design and Management

Zuo-Jun Max Shen

Increasing environmental awareness has resulted in great interest in supply chain sustainability. Studies have shown that a large proportion of the carbon emission actually comes from the upstream and downstream members of the supply chain, so there is great need to access the greenness of a supply chain as a whole. For example, Wal-Mart found out that 90 % of their carbon emission comes from their suppliers. However, many activities in the supply chain are dependent and thus identifying the carbon emissions from each activity is difficult. To overcome those difficulties, we need to understand the important issues in green supply chains and the state-of-art research. In this talk, I will review various issues in green supply chain design and management and point out some research opportunities.

Author Biography

Prof. Dr. Max Shen obtained his Ph.D. in Industrial Engineering and Management Sciences from Northwestern University. He joined the department in July 2004. Before that he taught at the Industrial and Systems Engineering Department at the University of Florida. His primary research interests are in the general area of integrated supply chain design and management, and practical mechanism design. He has published more than 70 papers, and he is also an associate editor for Operations Research, Naval Research Logistics and Journal Omega, area editor for IIE Transactions, senior editor for Production and Operations Management.

Z.-J. M. Shen (✉)
University of California, Berkeley, USA
e-mail: shen@ieor.berkeley.edu

Chapter 3
Integrating the Environment, Urban Planning, and Transport: Where Does Economics Fit in?

Kenneth Button

This paper considers the role that economics can play in integrating the environment, urban planning and transportation. The main challenge in moving towards sustainable development is in developing the holistic framework that integrates the environment, economy, and social stability highlighted in the Brundtland Report of 1987. There is a frequent misunderstanding that economics is purely concerned with the market and with prices in markets; e.g. the setting of transit prices and the financial viability of airlines. This has never been the case, and indeed, much of the current debate about the environment can be found, albeit in the context of the knowledge of the time in economic writings from at least the 1920s or earlier. The paper highlights some of the contributions that economics can make in the particular case of handing environmental matters when concerned with urban planning and transport. Within this framework we are concerned with matters of social equity as well as narrower, and more traditional notions of efficiency. It will make use of a number of case studies to provide illustrations of where economics has proved useful.

Author Biography

Prof. Kenneth Button is a Professor of Public Policy at the George Mason School of Public Policy and a world-renowned expert on transportation policy. He has published, or has in press, some 80 books and over 400 academic papers in the field of transport economics, transport planning, environmental analysis and industrial organization. Some of his recent books include: Airline Deregulation: An International Perspective (David Fulton Publishing), Flying into the Future: Air Transport Policy in the European Union, Edward Elgar Publishing), Handbook of Transport Modelling, (Pergamon Press); Transport, the Environment and Sustainable Development (E & FN Spon Publishing); Meta-analysis in

K. Button (✉)
School of Public Policy, George Mason University, Fairfax, USA
e-mail: kbutton@gmu.edu

Environmental Economics (Kluwer); Air Transport Networks (Edward Elgar Publishing). He is editor of the leading international academic journals Transportation Research D: Transport and the Environment and of the Journal of Air Transport Management and is on the editorial boards of nine other journals. He is on the scientific committee of the World Conference on Transport Research and the Advisory Board of the Air Transport Research Group.

Chapter 4
The History and Challenges of Japan's Low-Carbon Transportation Systems

Takayuki Morikawa

Japan has been building the rail-based transportation system since the modernization of the nineteenth century. Yet it experienced the motorization starting the early 1970s and has been struggling for the low-carbon transportation system mainly by introducing transportation demand management schemes. The recent car technologies help lower CO_2 emission by applying efficient engines, new generation vehicles and ITS.

The main research of Prof. Takayuki Morikawa concerns the areas of Travel Behavior, Transportation Demand Analysis, Transportation Policies, ITS, etc. And he has published over 150 academic papers in various international journals and conferences.

Author Biography

Prof. Takayuki Morikawa got his Master of Engineering from Kyoto University in 1983 and Master of Science from Massachusetts Institute of Technology (MIT) in 1987. Thereafter, he obtained his Ph.D. under the supervision of Prof. Moshe Ben-Akiva in 1989. In 1991, he began to work as Associate Professor at Nagoya University. From 1996 to 1997, he made his research as Visiting Associate Professor at the Department of Civil and Environmental Engineering, MIT. Since 2000, he has been a Professor at Nagoya University. Now Prof. Takayuki Morikawa is affiliated with Graduate School of Environmental Studies and Green Mobility Cooperative Research Center at Nagoya University.

T. Morikawa (✉)
Graduate School of Environmental Studies, Nagoya University, Nagoya, Japan
e-mail: morikawa@nagoya-u.jp

Chapter 5
Low Carbon Urban Design

Peter Boelsterli

Abstract Municipalities are administrative entities who manage a full set of key elements that are highly relevant to influence the future carbon-carbon footprint in regional contexts. In various approaches all over the world, cities are taking action and share their knowledge systematically. They combine a systematic change and quality management approach with the development of sector-specific tools as well as policy recommendations. A complete new approach in planning technologies will be needed to fully cope with the challenges, rising by the given growth of complexity in future urban developments. Based on cloud computing solutions, integrative planning can be reached by combining the different key layers into one overall approach that allows rapid prototyping and the parametric development of complex environments.

Author Biography

Prof. Peter Boelsterli was born in 1961 in Winterthur, Switzerland. He studied architecture at ETH Zurich and founded his own practice in 1993; today the company is called X6. X6 started applied research on sustainability in architecture in a time when most people weren't familiar with this term. X6 has created projects from single buildings to urban planning. Since 2008, Peter Boelsterli is the appointed China Delegate of the Swiss Rector's Conference of Universities of Applied Sciences & Art and supports the Development of Sino-Swiss Science and Technology Cooperation's. Today he combines higher education activities with teaching and research. In addition he continues to develop his practice as an architect in themes such as smart urban low carbon planning, future buildings, convenience urbanism, etc. via different networks of collaboration and partners all over the world. Peter Boelsterli is a member of SIA, BSA and RIBA; he is professor of architecture at Bern University of Applied Science & Art. With his wide field of activities he is one of the key persons of the Swiss Higher Education, Architecture and Urban Planning Landscape.

P. Boelsterli (✉)
Berne University of Applied Sciences, Burgdorf, Switzerland
e-mail: boe@x6.com

Chapter 6
Interdisciplinary Behavior Studies for Cross-Sector Energy Policies

Junyi Zhang

It is very difficult to reduce the energy consumption in domestic and passenger transport sectors, where households make use of various in-home and out-of-home energy-powered appliances (in-home: e.g., refrigerator, air-conditioner, and washing machine; out-of-home: e.g., passenger car and motorcycle) to meet their daily life needs. Decisions on the ownership of energy-powered appliances and the resulting amount of energy consumption are behaviorally interrelated with each other from appliance to appliance, from time to time, from household to household, and from context to context. To explain such complicated behavior phenomenon, various decision-making models have been proposed in different research disciplines, which are usually based on different behavioral assumptions. These existing studies will be first reviewed and then it will be argued how important to integrate various scientific insights from different disciplines to represent the household energy consumption behavior. Furthermore, it will be illustrated how different behavioral aspects can be jointly modeled in an interdisciplinary way. For the sake of the understanding, some case studies linked with residential choice, time use, and travel behavior will be introduced. Finally, the issues of behavior studies contributing to the design of cross-sector energy policies will be discussed.

Author Biography

Prof. Junyi Zhang's main research topics include travel behavior survey and modeling, household behavior modeling in transportation, theory of citizen's life decisions and behavior, tourist behavior modeling, household car ownership and usage, integrated urban and transportation modeling,

---

J. Zhang (✉)
Graduate School for International Development and Cooperation,
Hiroshima University, Higashi-Hiroshima, Japan
e-mail: zjy@hiroshima-u.ac.jp

sustainable urban development and transportation, intelligent transport systems, traffic safety, public transportation policies, urban and transportation issues in developing countries, city center development and pedestrian behavior, environmental and energy policies in transport sector, and low-carbon urban system design and so on. Prof. Zhang published 266 refereed academic papers and 211 non-refereed papers, and was awarded as Best Paper/Research Awards for 7 times and Outstanding Paper Awards for 3 times by international/domestic associations/conferences.

Chapter 7
International Journal of Shipping and Transport Logistics: An Insider's Perspective

Y. H. Venus Lun

7.1 Impact Factor

- The latest release of the ISI Journal Citation Reports (JCR) by Thomson Reuters reported that *IJSTL* secured an impact factor of 1.844:
    - 46/166 in "Management" category
    - 7/24 in "Transportation" category

7.2 Research Methodologies

Key research methodologies:

- Analytical
- Empirical
- Optimization

  Others: Case Study, Review, Simulation…..

7.3 Hints to Authors

- Originality
    - phenomenon, theory, model, method, review
    - Contribution

Y. H. V. Lun (✉)
The Hong Kong Polytechnic University, Hongkong, China
e-mail: venus.lun@polyu.edu.hk

- academic, practical, managerial
- Significance
  - meaningful, useful, extendable
- Rigour
  - data, analysis, conclusion
- Lucidity
  - clear, logical, smooth, comprehensible

Author Biography

Dr Y.H. Venus Lun is an assistant professor in department of Logistics and Maritime Studies, The Hong Kong Polytechnic University. She is the Director of the Shipping Research Centre, The Hong Kong Polytechnic University and the Deputy Programme Director of MSc in International Shipping and Transport Logistics, The Hong Kong Polytechnic University.

Dr Y.H. Venus Lun is the Editor-in-Chief, International Journal of Shipping and Transport Logistics, and the Editor of Springer Series in Shipping and Transport Logistics.

Chapter 8
The Roles of Railway Freight Transport in Developing the Low-Carbon Society and Relevant Issues

Guoquan Li

With the socio-economic changes, the transport system in developed countries has been formed adequately. Users can freely choose their expected transport means according to relevant needs. Railway transport as one of the most available means with environmental-friendliness has been required to play more important roles in developing eco & low-carbon society. In this study, the situations of surface freight and potential demands to be suitable for railway container are analyzed, and the possible predominance ranges of railway in transport cost are estimated by the comparative analysis between railway and truck freight rates in transport distance. Moreover, by a case study, the effects of railway freight transport in the reductions of logistics costs and CO_2 emissions, and the savings of energy are derived. Finally, the relevant issues concerning the actual conditions are discussed.

Author Biography

Dr. Li Guoquan Dr. Li Guoquan is a senior researcher in the transport planning, at Railway Technical Research Institute, mainly working for the followings.
(1) Infrastructure and railway planning
(2) Freight transport and Logistics system management
(3) Regional public transport system based on the socio-economic changes
(4) Transportation policy and developing strategy
(5) Laws, acts, regulations concerning transportation.
Dr. Li obtained the Ph.D. in Engineering, laboratory of urban system planning from Kyushu University, and joined the laboratory of transport system engineering as an assistant professor in April 1996. He had/has been a senior researcher in the Institute for Transport Policy Studies from 1999, Railway Technical Research Institute from 2005. Also he is a part time teacher in logistics system, at Tokyo City University.

G. Li (✉)
Transport Planning and Marketing Laboratory,
Railway Technical Research Institute, Tokyo, Japan
e-mail: ligq@rtri.or.jp

Chapter 9
Chinese Condition Must be Considered on Developing Green Building in China

Youguo Qin

China is a big developing country in the world. There are huge local differences between different areas in China on the conditions of geography, climate, population and on economical and social development. How to consider the Chinese conditions on developing green building in China is discussed in this paper from seven aspects: pay more attention on ordinary houses; energy and CO_2; "green" not be replace by "low carbon"; solar energy; "footprint" and "ecological value of land"; rethinking of Chinese condition "large population and little land"; insist the policy of architecture design: "applicability, economy, and attention to beauty with possible conditions".

Author Biography

Prof. Youguo Qin is a full professor in school of Architecture, Tsinghua University. He is the Chairman of National Board of Architectural Education Accreditation of China, the Chairman of Architectural Physics Committee, Architectural Society of China, and the Chairman of Green Architecture Committee, Architectural Society of China.

Y. Qin (✉)
Tsinghua University, Beijing, China
e-mail: saqyg@mail.tsinghua.edu.cn

Part II
Low Carbon Transportation

Chapter 10
Study on Traffic and Infrastructure Construction Performance Assessment Based on Sustainable Development

Jie Zhang, Huibing Xie, Minghui Liu and Kai Liu

Abstract The road transport, as an important part of the whole contemporary transport system, has many advantages such as the flexibility, promptness, good accessibility and so on. However, the contradictions between the traditional development of road transport and natural environment and resource have been apparent. Specifically speaking, there are three aspects: (1) high consumption of resource; (2) land resource wasting; (3) severe pollution. Therefore, the sustainability assessment on transport infrastructures is highly necessary. This paper analyzes four problems of transport infrastructures including natural resource, environment, society and economy, and then a theory of sustainability assessment on transport infrastructures based on the quality and load of transport infrastructures is proposed. The assessment system contains 4 level-one indicators, 14 level-two indicators and 61 assessing criteria. The questionnaire method and AHP (Analytic Hierarchy Process) method are used to define the weights of every indicator. Finally, SDI (Sustainable Development Index) is introduced to express the sustainability of the infrastructure evaluated.

Keywords Traffic and infrastructure construction · Sustainable development · Performance assessment · Index system · Analytic hierarchy process

---

J. Zhang · H. Xie (✉) · M. Liu · K. Liu
Beijing Jiaotong University, Beijing, China
e-mail: dragen1987@163.com

10.1 Introduction

The road transport, as an important part of the whole contemporary transport system, has many advantages such as the flexibility, promptness, good accessibility and so on. It is the symbol which can show the level of social economic development of a country or a region (http://www.jttj.gov.cn/shownews.asp?id=1756).

With the improvement of people's living standard, the road transport in China has changed greatly. The contradictions between the traditional development of road transport and natural environment and resource, which can be classified as the following three parts: (1) high consumption of resource; (2) land resource wasting; (3) severe pollution, has become apparent (World Commission on Environment and Development 1987). Hence, it is necessary to assess the sustainability of transport infrastructure projects.

Extensive research is being carried out in this area. Xiong et al. (1999) discussed the relationship between the transport and sustainable economic development from a perspective of sustainable economic development and acknowledges that "the traffic development is closely related to the implementation of the strategy of sustainable development". Zhu (1999) pointed out that an important criterion to measure the sustainable transport is whether the transport system could maintain long-term dynamic social net income or welfare maximum. Qian and Minghuai (2001) analyzed the negative impact of transportation infrastructure on the land and resources and suggested to eliminate the impact of transportation from three aspects of the economic policy, legal management and publicity and education. Zhou Jun (Jun and Yisheng 2005) analyzed the structure of the urban transport infrastructure and its collaborative relationship with the city and internal rules, proposed the integration mode of the sustainable development of urban transport infrastructure.

In this study, a useful assessment framework is established and the importance of the indicators in the framework is analyzed. A good expression of the assessment result is proposed to reflect the sustainability of the infrastructure.

10.2 Indicators for the Assessment System

The Aim of the sustainability assessment of transport infrastructures is to appraise the quality and performance provided by the road transport and the environment loading caused by it, finally propose some suggestions for improvement (Liu 2006). For the sake of assessing from various perspectives, a multi-level indicator system is established, which can reflect the characteristics of road transport and integrate with its sustainability performance.

In the assessment system, the indicators can be divided into two groups: (1) "L"- Load, namely consumption of energy and resource and pollution, which is further divided into LR1 Resource Consumption and LR2 Environment Influence; (2) "Q"- Quality, namely the functions and service provided by transport infrastructures which is further divided into Q1 Social Economy, Q2 Service Quality (Steele and Cole 2003). All assessment criterias, which should be scored as 0–5, are listed hereinafter, by which the meaning of every indicator is reflected.

Table 10.1 Assessment indicators related to society and economy

Secondary indicator	Assessment criterias
Promotion of society (Xiang 2006)	Promotion of job opportunity in local area
	Contribution to infrastructures, road net and transport
	Contribution to the culture, education and sanitation
	Contribution to effectiveness of local transport
Promotion of political stability	Projects complying with the national and regional planning
	Contribution to shorten the gap between urban area and countryside
	Contribution to the resistance to natural disasters
	Contribution to the national defense
Promotion of regional economy	Importance of the infrastructure for the local economic development
	Foundation of regional economy: the level of industrial production, the level of agricultural production, investment on basic infrastructures, the situation of labor productivity, the situation of local finance and etc.
	Resource and technologies: the indicator reflects the potential incidents of economy development, labor, source of construction materials and technologies. It contains water resource, mine resource, tourist resource, labor resource, integrated natural conditions and technologies
	Urban condition: the level of industrial production in urban area, the scale of urban area and the level of infrastructures in urban area
	Comparison among beneficial areas: investment amount, construction duration, period for recovery of investment and indirect beneficial areas

Table 10.2 Assessment indicators related to function of transport

Secondary indicator	Marking scheme
Traffic capacity	Lane capacity
	Safety
	Applicability
Integrate into network	Construction quality of road
	Construction quality of landscape
	Construction quality of safety installation
	Construction quality of monitoring system
	Assess to the road network
Operation	Maintenance planning
	Risk identification
	Risk management

Table 10.3 Assessment indicators related to resource

Secondary indicator	Marking scheme
Land resources (Jianshe et al. 2008)	Utilization of land resource is controlled
	The land selected has not been used
	Measures of land improvement is used
Resource recovery	Recoverability of structure <15 %
	15 % $\leq$ Recoverability of structure <20 %
	20 % $\leq$ Recoverability of structure <25 %
	25 % $\leq$ Recoverability of structure <30 %
	Recoverability of structure $\geq$30 %
Sustainable wood	0 < Utilization ratio of sustainable wood $\leq$20 %
	20 % $\leq$ Utilization ratio of sustainable wood <50 %
	50 % $\leq$ Utilization ratio of sustainable wood <70 %
	70 % $\leq$ Utilization ratio of sustainable wood <90 %
	Utilization ratio of sustainable wood $\geq$90 %
Environmental friendly materials	4 (or more) appointed building materials is used
	3 appointed building materials is used
	2 appointed building materials is used
	1 appointed building material is used
	No appointed building material is used
Material saving	Material saving has not been considered in the structure design
	Material saving has been considered in the structure design
	Material saving has been considered adequately in the structure design

Table 10.1 shows the assessment criterias of the "Social Economy", it mainly contains promoting social development, promoting political stability and promoting regional economic.

Table 10.2 shows the assessment criteria of the "Service Quality", it mainly contains Transport function, Road net and Maintenance.

Table 10.3 shows the assessment criteria of the "Resource Consumption", it mainly contains Land resource, Resource reuse ratio, Environment-friendly timber, environment-friendly construction materials and Consumption of materials.

Table 10.4 shows the assessment criteria of the "Environment Influence", it mainly contains Ecology, Landform and topography and Culture heritage.

10.3 Indicator Weight

Different indicators have different contribution to the assessment system; actually they are not equally important to each other. So they should have different weights in an assessment system. Weights mainly depend on two aspects: one is reliability of indicators; the other is how much attention decision makers pay. In this study, the weight system is established by the questionnaire survey and AHP method.

10 Study on Traffic and Infrastructure Construction

Table 10.4 Assessment indicators related to environmental

Secondary indicator	Marking scheme
Ecological investigation (Tan et al. 2002)	Taking actions to recover the ecological condition
	Improving the biodiversity
	Establishing the path for animals migrating and breeding
	Not taking any measures to protect the mature woods
	Taking action to protect the existing old trees
	Taking action to protect the other mature trees
	Taking action to improve the condition of existing plants
Landform	Existing plants and topography are suitable for the transport infrastructure construction
	Keeping the original landform as much as possible
	Considering the protection of topsoil
	The surface runoff does not damage land surface
	Taking actions to recover the unavoidable damage to the land surface
Cultural heritage	The protection of heritages
	Making plan for protection heritages
	The selection of road line avoiding unmovable heritages; if the requirement cannot be met, making protection plan
	The permit from the government for moving or tearing down heritages

A total of 60 questionnaires, which aimed at investigate the efficiency and significance of the indicators, had been issued and 49 effective questionnaires were got. The respondents are stakeholders or experts in this field. It is assumed that their standpoints are more effective to reflect the objective condition.

The Analytic Hierarchy Process (AHP) is a structured technique for organizing and analyzing complex decisions. Based on mathematics and psychology, it was developed by Thomas L. Saaty in the 1970s and has been extensively studied and refined since then. It has particular application in group decision making, and is used around the world in a wide variety of decision situations. In this study the AHP method is applied to analysis the statistic data achieved from the questionnaires to ensure the consistency of the data.

Table 10.5 listed the weight of level-1 and level-2 indicators.

10.4 Assessment Result

All the qualitative indicators can be assessed quantitatively and integrated based on the weight of every indicator. In order to make the assessment result accessible, the SDI (Sustainable Development Index) is introduced as the result.

$$\text{SDI} = Q/L \qquad (10.1)$$

Table 10.5 Weight for assessment indicators

First Grade indicators	Weight	Secondary indictors	Weight
Q1 society and economy	0.6	Promotion of society	0.3
		Promotion of political stability	0.2
		Promotion of regional economy	0.5
Q2 service	0.4	Traffic capacity	0.45
		Integrate into network	0.32
		Operation	0.23
LR1 resource	0.4	Land resources	0.6
		Resource recovery	0.15
		Sustainable wood	0.05
		Environmental friendly materials	0.1
		Material saving	0.1
LR2 environment	0.6	Ecological investigation	0.3
		Landform	0.3
		Cultural heritage	0.4

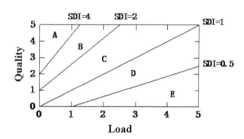

Fig. 10.1 Criteria for classifying of SDI

where Q and L are final score of the Quality and the Load respectively.

According to the Q, L and SDI, the sustainability of transport infrastructures can be divided into 5 categories (A, B, C, D, and E) as can be seen in Fig. 10.1.

Grade A means the best performance of transport infrastructures, while Grade E means the worst one.

The bigger Q is and the smaller L is, the bigger the gradient would be, in other words, which means the better level of sustainability of road transport infrastructures.

10.5 Concluding Remarks

The sustainability of traffic and infrastruction was studied and an assessment framework was established in this paper. Four critical issues about sustainability of traffic and infrastructure construction, named nature resource, environment, society and economy, are analyzed in the assessment framework. A theory of the sustainability of transport infrastructure based on performance and load was proposed and an indicator system was established. Questionnaire was designed and handed out to stakeholders such as managers, designers et al. Analytic hierarchy process was used to determine the weight of assessing criterias valid. Notwithstanding, the weight is not verified by the social investigation, It is meaningful for the late-comer to consult. Finally, SDI is introduced to express the sustainability of the infrastructure evaluated.

References

Chen Q, Wang M (2001) Transportation infrastructure construction and land resource protection. Yunnan Commun Sci Technol 17:62–64
Information on http://www.jttj.gov.cn/shownews.asp?id=1756
Jun Z, Liu Y (2005) Network analysis of urban transport infrastructure and sustainable development. City Traffic 3:21–24
Liu F (2006) Research on post evaluation of expressway construction project. PhD Thesis, Hohai University
Long X (2006) Feasibility study and establishment of engineering project. Urban Roads Bridges Flood Control 6:159–163
Steele K, Cole G (2003) Highway bridges and environment sustainable perspectives. Civil engineering, pp 176–182
Tan R, Culaba A (2002) Environmental life-cycle assessment: a tool for public and corporate policy development, American Center for Life Cycle Assessment, Washington. Available. http://www.lcacenter.org/library/pdf/PSME2002a.pdf
Tian J et al (2008) Study on the situation and trend of safety assessment technology for transportation construction project. China Saf Sci J 18(6):171–176
World Commission on Environment and Development (1987) Our common future, chapter 2: towards sustainable development. Oxford University Press, Oxford
Xiong Y, Xiong A (1999) Transportation and economic development. China Railway Publishing House, China
Zhu Z (1999) Externality theory and its application analysis in transportation economics. Railway Publishing House, China

Chapter 11
Sustainable Development of China's Road Transportation Infrastructure: Situation and Prospect

Jie Zhang, Kai Liu and Yurong Zhang

Abstract Along with the rapid development of the national economy, as well as the increasing of transportation on road, China's road traffic infrastructure faces major challenges. It has been already widely accepted that it is imperative to protect the environment in the process of road transportation infrastructure construction and further realize the sustainable development of the transportation infrastructure. In this paper, the current situation of China's road transportation infrastructure is analyzed; and the philosophy and connotation of the sustainable development of road transportation infrastructure is elaborated; at last, the corresponding countermeasures concerning the legal, fiscal and demonstration policies are provided.

Keywords Road traffic infrastructure · Sustainable development · Current situation · Policy countermeasure

11.1 Introduction

Since the twentieth century, the human beings have faced serious environment challenges, including that the resources are increasingly exhausted, ecological destruction and frequent accidents of all kinds of pollution. Therefore, the environmental problems have become one of the global problems which the human beings faced today, countries all over the world have set sustainable development

J. Zhang (✉) · K. Liu
School of Traffic and Transport, Beijing Jiaotong University, Beijing 100044, China
e-mail: zhangyr1988@126.com

Y. Zhang
School of Civil Engineering, Beijing Jiaotong University, Beijing 100044, China

as a strategic target; claiming that in order to achieve the harmonious development between human beings and the environment, it is necessary to persist on the sustainable development.

Transportation infrastructure construction is the basic industry of national economy, and it is the support and guarantee for the development of economy and society. The level of transportation infrastructure construction development is directly related to the rate of national economic development. Non-sustainable of transport infrastructure will be a direct result of non-sustainable of urban economic development, environmental and social equity. At present, the Chinese government has been gradually increasing the intensity of investment in the road transportation infrastructure. In 1978, the China's highway mileage was only 890,000 km. However, by 2006, the China's highway mileage has reached 3.457 million km, which is over three times than that before the opening up. In 2008, the newly increased highway mileage of China reached 100,000 km, including 6,433 km expressway which reached 60,300 km, ranking the second longest expressway in the world. And the construction of road infrastructure investment increased by 2.4 % to 664.5 billion Yuan. However, there are still a lot of problems in the road transportation infrastructure of China. Although the road transportation infrastructure can promote the economic and social development, it still had a huge impact on the ecological environment. For example, in 2008, the annual carbon emission of China was about 5 billion tons. And the greenhouse gas emission accounted for about 10 % of the total national greenhouse gas emission. Besides, the construction of road transportation infrastructure will destroy the natural environment, landscape and water resources and cause the natural environment pollution around, especially in some fragile environmental areas. In addition, in recent years, the road accidents happened frequently in China, both in the construction phase and in the operation phase. Among these impacts, the road bridge accidents caused the largest damage, the most serious loss and the most severe social influence.

Therefore, we should establish the concept of circular economy, rely on technological progress and develop high efficiency and low energy consumption transport equipment and new technologies for traffic-related environmental protection. We should conserve resource, reduce energy consumption and protect the environment to build economical transport industry and get to clean transport and green transport, thus achieving sustainable development of road transportation infrastructure in our country.

11.2 The Status of China's Transportation Infrastructure

For a long time, the development of road traffic infrastructure in China shows a model that: Economic scale expansion→Traffic demand to expand→Passive increase of road infrastructure→Re-expansion of the economies→Stimulating traffic demand to expand. In China, with the sustained growth of urban industry and the size of the population, it made the road transport needs growing, which led to the road transportation infrastructure in China relatively weak. It mainly reflected in the following points.

11.2.1 Road Transport Planning is Seriously Constrained by Land

The development of road transport infrastructure has been severely hampered, due to the land constraints. According to the Beijing's land use planning, in the year of 2010, the land for traffic is 434.7 km^2, only accounts for 13 % of the total construction land. In addition, the high cost of land use increased the cost of the transport infrastructure significantly. According to the related statistics, the average price of the land in Tokyo's main block is about 24,000 Yuan/m^2, while it is 34,330 Yuan/m^2 in Guangzhou.

11.2.2 The Lack of Comprehensive Consideration of Transport Infrastructure Planning

Currently, the irrationality of the road network structure is widespread in China, such as the proportion of low-grade highway is large, the road grade is low and the road function is not well defined, all of these making the traffic efficiency greatly reduced. In addition, in China, the rural highway construction task is heavy, and the rural highway maintenance is backward. Therefore, it's urgent to consider the integrated planning of rural transport infrastructure and the construction of custody mechanism.

11.2.3 The Lack of Comprehensive Management Mechanism of Transport Infrastructure

There are still serious traces of the planned economy in our transportation infrastructure management. Owing to the lack of full consideration of the repair and maintenance in the operation period of infrastructure, a lot of infrastructure deterioration problems and greatly reduced the structure life.

In the face of the current status of our urban transport infrastructure, some scholars carried out some related researches for sustainable development in China's transportation infrastructure. In 1998, Xiong Yongjun discussed the relationship between the transport and sustainable economic development from a perspective of sustainable economic development and held that "the traffic development is closely related to the implementation of the strategy of sustainable development" (Xiong and Xiong 1999). In 1999, Zhu Zhongbin pointed out that an important criterion to measure the sustainable transport is whether the transport system could maintain long-term dynamic social net income or welfare maximum (Zhu 1999). In 2001, Chen Qian analyzed the negative impact of transportation infrastructure on the land and resources and suggested to eliminate the impact of transportation from three aspects of the economic policy, legal management and publicity and education (Chen and Wang 2001). In 2006, Zhou Jun, who analyzed the structure of the urban transport infrastructure and its collaborative relationship with the city and internal rules, proposed the integration mode of the sustainable development of urban transport infrastructure (Zhou and Liu 2005).

In addition, some academic activities related to the traffic sustainable development have been carried out, such as Asia-Pacific conference of the sustainable development on traffic and environmental technology, Sustainable Development Strategy and Construction Forum on urban rail transit of China, etc., various types of traffic engineering journals and newspapers such as Chinese Journal of Highway, Chinese Journal of Engineering and Management of Road Traffic have been published. These academic activities have greatly attracted industry experts and scholars, they communicate and discuss on the environment and sustainable development issues in the transport industry, and they promote new theories and new results published.

11.3 The Concept and Connotation of the Sustainable Development of Transportation Infrastructure

11.3.1 The Concept of Sustainable Development of Transportation Infrastructure

In 1987, a report named submitted "Our Common Future" by the World Environment and Development Commission (WCED) put forward a widely accepted definition of sustainable development, i.e. 'Sustainable development is development that meets the needs of the present without compromising the ability of future generations to meet their own needs (World Commission on Environment and Development 1987).

Sustainable development of the transport is put forward under the sustainable development theory based on the fact that there are unsustainable factors in the existing transportation system, therefore, the focus of traffic sustainable development is: how to realize the harmonious development of the economy, society and ecology as well as the transportation department's own development according to the basic requirements of sustainability. As a result, we should understand the basic connotation of sustainable transport from the following three aspects.

(1) **Economic sustainable development of transport**

Transport system as a subsystem of the socio-economic system, whose sustainable development is an important part of sustainable social and economic development. The traditional mode of transport development is unsustainable, mainly displays in the discordance among the transportation system supply capacity, resource consumption, environmental and ecological protection with the requirement of economic sustainable development.

Therefore, the economic sustainable development of traffic should include two meanings. First, in the view of the relationship between transport and national economic, the transport system should meet the demand of economic and social development, i.e. the traffic system should coordinate with the national economic and social development; Second, in the view of the internal transport system, it is necessary to realize the transportation efficiency, i.e. the sustainability of the economy of the transport system. The transport system should be ensured that it can improve the people's material condition, provide an economic traffic which can meet the continuous changing needs, pursuit the traffic economic benefits, and achieve the benign circulation of the traffic assets.

(2) **Social sustainable development of transportation**

Traffic social sustainable, i.e., make full use of transportation functions to eliminate poverty, adjust and improve the social justice, and at the same time, all the society members can share the benefits of transport development equally.

To achieve sustainable social development of the transport, we must change our values. i.e. we must shift from the traditional concept of the simple pursuit of the quantity expansion to the new concept of sustainable development which lays emphasis on the comprehensive benefits and the long-term impact; we must shift from the traffic consumption concept of the individual's desire to the public interest; we must shift from the traffic management of a single decentralization to the bilateral control which equally emphasis on the source and flow.

(3) **Ecologically sustainable development of transport**

Ecologically sustainable development of the transportation, which called for promoting the construction and development of transportation systems, meanwhile emphasizing on the protection of the ecological environment and development and utilization of resources reasonably (mainly refer to the non-renewable resources); emphasizing on the expansion of the road network and channel, and at the same time paying attention to the supervision of the transport system, ensuring that the traffic, environment and ecology keep coordination and compatible relationship. Besides, it is necessary to minimize the negative impact of transport development on the environment and ecology, especially the negative impact on the human life and health.

11.3.2 Connotation of Sustainable Development of Transport Infrastructure

Transportation infrastructure system is accompanied by the development of economic systems, social systems and environmental systems, therefore, the development of transport infrastructure must also meet the need for economic and social development at present and in the future, improve resource utilization efficiency, improve the environmental quality, realize the coordination development among transport and economic, society and environment. In the meantime, it is urgent to enhance the security and stability of the transportation system. The basic pattern of the sustainable development of transport infrastructure systems is shown in Fig. 11.1.

In this integrated model, firstly, transport infrastructure system acts as the core part, and then gradually spread to form a stable system including economic, social, and environmental system elements which associated with the transport infrastructure system. Secondly, the structural elements associated with the transport infrastructure of economic, social and environmental systems act as carriers, gathered into a coherent system, and then through the continuous exchange of the elements, a coordinated whole is formed. This integrated structure mentioned above in essence is the basic mode of sustainable development of transportation infrastructure system. This development model will directly promote the economic progress healthily and steadily, and then forma sustainable development situation, finally promote the coordinated development of economy, society and environment.

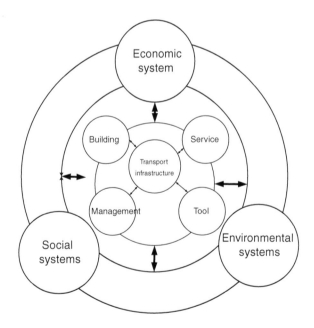

Fig. 11.1 The integrated model of transportation infrastructure sustainable development

11.4 Some Measures of the Transport Infrastructure Sustainable Development

The focus of transport development in future is to expand the capacity, optimize the structure, improve the quality, improve the service, guarantee the safety and protect the environment, which is a very difficult task. To achieve the sustainable development of transport infrastructure, on the one hand, the development of transport infrastructure must meet China's economic and social development needs, and lay the material foundation for sustained, healthy and rapid economic society development; on the other hand, the development of transport infrastructure must adapt to environmental capacity and resources reserves. Therefore, this article puts forward several countermeasures of the sustainable development of transport infrastructure in China.

(1) **Update the laws and regulations of transportation infrastructure, then further strengthen the management of environment protection**
In order to promote the sustainable development of the transport infrastructure, the Ministry of Transportation (MOT) has issued a series of laws and regulations, for example "Traffic construction project environmental protection management regulations" (Ministry of communications of the people's Republic of China 2003)," Specifications for Environmental Impact Assessment of Highways"(JTGB03 2006), and " Design Specifications of Highway Environmental Protection" (JTGB04 2010), etc. However, the related standards with the sustainable traffic infrastructure are very limited, there is no description of the design, construction and acceptance specifications about road greening, landscape protection, noise control. Therefore, we should further improve the transportation infrastructure sustainable construction laws and regulations to provide the basis for promoting the sustainable development of the transport infrastructure.

(2) **Improve operational marketization of the transport infrastructure**
Improving the traffic infrastructure marketization helps to create and enhance the value of the transportation infrastructure, and guarantee the normal repair maintenance costs of the transportation infrastructure, so as to achieve the purpose of capital circulation, which is conducive to perfect the regulation of further attract more capital into.

(3) **Increase the propaganda of sustainable development of the transport infrastructure**
Propaganda work is an important means of improving public realization to sustainable development of transport infrastructure, through the use of the media publicity, technical training, the seminar will be held, can effectively improve the social awareness and supervision.

11.5 Conclusion

In view of the present situation and existing problems of China's highway transportation infrastructure development, we should be aware that if we want to achieve highway traffic infrastructures' sustainable development, we must change the traditional traffic models, carry out an integrated transportation infrastructure planning to ease the pressure of resource and environment. The saving transportation industry must be established to improve the quality of transport services, and make the integration of transport to come true, improve the people's living standards and the competitiveness of the national economy. Related industry management must be strengthened to ensure that the transportation infrastructure quality in the design and construction.

References

Chen Q, Wang M (2001) Transportation infrastructure construction and land resource protection. Yunnan Commun Sci Technol 17:62–64
JTGB03-2006 (2006) Specifications for environmental impact assessment of highways
JTGB04-2010 (2010) Design specifications of highway environmental protection
Ministry of communications of the people's Republic of China (2003) Environmental protection measures for the administration of transportation construction projects
World Commission on Environment and Development (1987) Our common future, chapter 2: towards sustainable development. Oxford University Press, Oxford
Xiong Y, Xiong A (1999) Transportation and economic development. China Railway Publishing House, Beijing
Zhou J, Liu Y (2005) Network analysis of urban transport infrastructure and sustainable development. City Traffic 3:21–24
Zhu Z (1999) Externality theory and its application analysis in transportation economics. China Railway Publishing House, Beijing

Chapter 12
Energy Demand and Emission from Transport Sector in China

Yin Huang and Mengjun Wang

Abstract This paper aims to present a comprehensive overview of the current status and future trends of energy demand and emissions from transportation sector in China. Firstly, a brief review of the national profile of energy demand and the CO_2 emission is presented to serve as background for this study. Secondly, the current status of Chinese transportation sector, including the energy demand and emission for transportation sector and sub-sector, is analyzed. At last, a conclusion summarizes and analyses the findings from the study, with the aim to raise further discussion or research interests in this area. This study serves as a guideline for further investigation and research to implement and improve the transportation sector.

Keywords Transportation · Energy demand · CO_2 emission · Energy policy · China

12.1 Introduction

Rapid growth in more than two decades has made China the second largest energy-consuming nation after the US. Rapidly growing energy demand and emission from China's transportation sector in the last two decades have raised concerns over energy security, urban air pollution and global warning (Yan and Crookes 2007, 2010). This paper aims to present a comprehensive overview of the current

---

Y. Huang (✉) · M. Wang
School of Civil Engineering, Central South University, Changsha, Hunan Province, China
e-mail: share0122@126.com

Y. Huang
School of Traffic and Transportation Engineering, Central South University, Changsha, Hunan Province, China

status of energy demand and emission from transportation sector in China. It is organized as follows: A brief review of the national profile of energy demand and the CO_2 emission is presented to serve as background for this study; the current status of Chinese transportation sector, including the energy demand and emission for transportation sector and sub-sector, is analyzed; the policies to control transport energy and CO_2 emission are then summarized; after that a discussion section summarizes and analyses the findings from the study, finally conclusions are drawn.

12.2 Current Status of Chinese Transport Sector and Sub-Sector

With the fast-growing economy, transportation has become a crucial component of modern life in China. Urbanization, transport congestion and unbalanced distribution of energy resources and demand have led to the ever-greater passenger and freight mobility, and the transport sector is gaining a rising share in the total (Yan 2008). The energy intensity for transportation sector is equal to the transition intensity multiplied by energy efficiency for transportation sector. The transition intensity, energy efficiency and energy intensity for transportation sector are calculated as follows (See Table 12.1).

The energy consumption in sub-sector for transportation is increasing rapidly. From the trend of energy consumption in sub-sector for transportation from 1990 to 2008, the energy consumption of highways increased with the annual growth of 9.8 %, which is higher than the annual growth rate of the energy consumption for transportation sector—8.3 %. The energy consumption in railways is declined from 1990 to 2008 with the annual growth rate of −0.1 %. The reduction of the energy consumption of the railways is mainly attributed to the phase-out of the steam locomotives. With the development of economics and technology in China, airplane becomes the main transport tool for long journey. The energy consumption of airplane increased rapidly with the annual growth rate of 14.7 % in the period of 1990–2008.

The CO_2 emission has increased steadily over the past 20 years and is still moving upwards. It is estimated there were about 4.93 hundred million tons of CO_2 emission in 2009 which is more than four times from that in 1990 with annual growth rate of 7.8 % (Oh and Chua 2010). It is inevitable that pollutant emission will continue to climb as long as fossil fuels remain as the main contributor in transport sector. Figure 12.1 shows the intensity of carbon and energy in China from 1990 to 2009.

The CO_2 emission of each sub-sector in 2009 is presented in Fig. 12.2. The CO_2 emission of transportation mainly focuses on highways. The CO_2 emission of

Table 12.1 The transition intensity, energy efficiency and energy intensity for transportation sector in China

Year	Transition intensity (km/person)	Energy efficiency (Standard oil (kg)/10,000 ton-km)	Energy intensity (standard oil(ton)/person)
1990	2087.3	134.2	0.028
1991	2187.6	132.2	0.029
1992	2329.1	130.5	0.030
1993	2492.4	133.2	0.033
1994	2664.7	124.0	0.033
1995	2736.7	124.5	0.034
1996	2833.6	203.1	0.058
1997	2728.7	210.3	0.057
1998	2722.0	209.9	0.057
1999	2782.2	215.5	0.060
2000	3129.0	199.2	0.062
2001	3144.4	203.3	0.064
2002	3364.5	200.1	0.067
2003	3521.0	217.8	0.077
2004	4127.8	217.8	0.090
2005	4538.8	217.6	0.099
2006	4993.0	216.9	0.108
2007	5639.8	206.8	0.117
2008	7597.9	159.4	0.121
2009	8069.7	154.4	0.125

The table shows that the transition intensity increased with the annual growth rate of 6.46 % from 1990 to 2009 and the energy intensity increased with the annual growth rate of 6.63 %

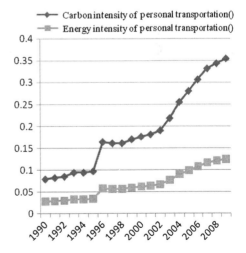

Fig. 12.1 The intensity of carbon and energy in China from 1990 to 2009. (Calculated by author)

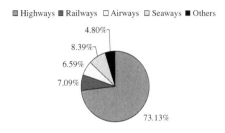

Fig. 12.2 The CO_2 emission of each sub-sector in 2009. *Source* Annual Review of Low-carbon Development in China: 2010

transportation is 456940 thousand ton. Highways transport is the biggest CO_2 emission contributor with a share of 73.13 %, seaways, railways, airways and others, ranked second, third, fourth and fifth, making up 8.39, 7.09, 6.59 and 4.80 % of the total respectively. Hence, if the Chinese government wants to control the energy consumption and emission, the first step is to control the energy use of highways.

12.3 Policy Recommendation

The rapid-growing use of transportation energy in China during the 1990–2009 periods raised the question of China's oil security in the early 90s and led to a high priority of oil conservation in China's energy policy. Trucks and private cars accounted for the majority of transport oil consumption in China at that time. Energy conversation strategies for trucks and private cars were thus explored. These included technical improvement of engines, improvement of size composition of vehicles, promotion of diesel vehicles, improvement of highways conditions and transportation management (He et al. 1995).

12.3.1 Fuel Economy Regulation and Emission Standard

Fuel economy improvement is thus considered as an important option to reduce vehicular energy use and emission in 2005. The Chinese government therefore implemented its first vehicle fuel economy standard, targeting light-duty passenger vehicles. Moreover, vehicle emission standards are necessary to promote advanced technologies of emission control technologies for new vehicles (Gallagher 2006). In the 1990s, China implemented 17 standards for vehicular emission based on international control standards in the mid-1970s (He et al. 2002).

12.3.2 Public and Non-motorized Transport

Chinese government adopts a lot of policies to control the number of private cars and promote the public transport for wider usage to meet the goal of reducing energy consumption.

The government has realized the importance of public transport and started to take actions in recent years. At present, urban metro systems exist in 12 Chinese cities and there are plans calling for expanding and upgrading existing systems and building new ones in 15 other cities (Gan 2003; Cervero and Day 2008).

12.4 Conclusion

The problem of the transportation energy and environment are the major challenges faced globally in the twenty first century and are especially serious pollution for China. In this study, energy demand and emission from transport sector in China have been analyzed by considering the energy consumption and emission form sector and sub-sector for transportation. As a result of this study, the following conclusions are reached:

(1) In China, the energy consumption of highways is of the highest increasing mode with the annual growth of 9.8 %.
(2) In China, highways transport was the biggest CO_2 emission contributor, accounting for 73.13 % of total transport CO_2 emission.
(3) Highways transport has dominated Chinese oil consumption and is one of the fastest growing energy users in China. Consequently, CO_2 emission from this sector has also risen in an alarming rate.
(4) There are some economic policies in China implemented for transportation sector to reduce energy consumption.

From the conclusions above, there is an urgent need to adopt suitable energy policy to balance the energy consumption and transport service and reduce the CO_2 emission in transportation sector.

Acknowledgments I am very grateful to the Postdoctoral Science Foundation of Central South University (NO. 7604130003) and The Fundamental Research Funds for the Central Universities (NO. 721500041) whose support is essential for the completion of this project.

References

Cervero R, Day J (2008) Suburbanization and transit-oriented development in China. Transp Policy 35:315–323

Gallagher KS (2006) Limits to leapfrogging in energy technologies? Evidence from the Chinese automobile industry. Energy Policy 34:383–394

Gan L (2003) Globalization of the automobile industry in China: dynamics and barriers in greening of the road transportation. Energy Policy 31:537–551

He J, Zhang A, Xu Q (1995) Strategies for energy conservation for China's trucks and buses. Energy Sustain Dev 2:52–55

He KB, Huo H, Zhang Q (2002) Urban air pollution in China: current status, characteristics, and progress. Ann Rev Energy Environ 27:397–431

Oh TH, Chua SC (2010) Energy efficiency and carbon trading potential in Malaysia. Renew Sustain Energy Rev 14:2095–2103

Yan X (2008) Life cycle energy demand and greenhouse gas emissions in China's road transport sector: future trends and policy implications. PhD thesis. University of London, London

Yan X, Crookes RJ (2007) Study on energy use in China. J Energy Inst 80:110–115

Yan X, Crookes RJ (2010) Energy demand and emissions from road transportation vehicles in China. Prog Energy Combust Sci 36:651–676

Chapter 13
Exploring the Effect of Inter-Stop Transport Distances on Traction Energy Cost Intensities of Freight Trains

Xuesong Feng, Haidong Liu and Keqi Wu

Abstract With a computer-aided simulation approach, this research analyzes the change of the traction energy cost intensity of a typically formed Chinese freight train hauled by representative locomotives for different target speeds. It is found that the inter-stop transport distance of a freight train should be longer than 20.00 km to decrease the traction energy cost per unit transport. Moreover, the railway freight transport work should organize heavy load trains hauled by respectively various locomotives with different traction performances which are suitable for different target speeds in view of different inter-stop transport distances.

Keywords Freight train · Target speed · Stop-spacing · Computer-aided simulation · Traction energy cost intensity

13.1 Introduction

Railway is one of the major freight transport modes in China. About 19.50 % of the total 14,183.70 billion freight ton-kilometers (t-km) in 2010 (National Bureau of Statistics of China 2011) are completed by railway freight trains in China and

---

X. Feng (✉) · K. Wu
Integrated Transport Research Center of China, Beijing Jiaotong University,
No.3 Shangyuancun, Haidian District, Beijing 100044, People's Republic of China
e-mail: Xuesong.Feng@bjtu.edu.cn; xsfeng@bjtu.edu.cn

X. Feng · H. Liu
MOE Key Laboratory for Urban Transportation Complex Systems Theory and Technology,
Beijing Jiaotong University, No.3 Shangyuancun, Haidian District, Beijing 100044,
People's Republic of China

these freight trains consume much energy (Li and Mao 2001; Xi and Chen 2006; He et al. 2010). It is well known that the traction energy cost intensity of a train is much concerned with its target speed (Uher et al. 1984; Hoyt and Levary 1990; Liu and Golovitcher 2003; Huang and Qian 2010). Many researchers have been continuously making effort to interpret the relationship between them (Chui et al. 1993; Lukaszewicz 2001; Miller et al. 2006; Bocharnikov et al. 2007; López et al. 2009). However, valuable existing research findings which are in practice incompletely examined by previous studies still cannot clarify the detailed influence of the target speed of especially a (conventional locomotive hauled) freight train upon its traction energy cost intensity in view of the important impacts of multi-factors such as the traction performance of the locomotive, the transport distance between neighboring stops, etc. from a comprehensive perspective. Based on the computer-aided simulations of the freight transports by a typically formed Chinese freight train hauled by respectively various types of the representative locomotives in China on a hypothetically straight and smooth railway line, this study attempts to explore the accurate effect of the length of stop-spacing (i.e. the inter-stop transport distance) on the intensity of the Traction Energy Cost (TEC) (i.e. energy consumed by traction and braking) of the train for different target speeds in a quantificational manner.

The contents of this paper are organized as follows. The formation of the studied freight train in this research and the computer-aided simulation approach utilized in this work to compute the TEC of the train are explained in Sect. 13.2. Next, Sect. 13.3 studies the detailed changes of the TECs per 10,000 t-km of the train with the increase of its target speed for different inter-stop transport distances by utilizing the simulation approach introduced in Sect. 13.2. Finally, Sect. 13.4 draws conclusions, makes some suggestions for the transport work of the freight trains in China and proposes some future research issues.

13.2 Train Formation and Simulation Approach

With referring to the work of Feng (Feng 2011), the general framework of the computer-aided simulation approach shown in Fig. 13.1 is applied in this research to calculate the TEC of the transport of a train.

A typical Chinese freight train today is usually formed by 60 coupled wagons hauled by 1 locomotive. One of the major types of the wagons is the C70. 60 coupled and fully loaded C70 wagons hauled by one locomotive compose the freight train studied in this research. The transport processes of the studied freight train hauled by respectively two major types of the locomotives for the railway freight transport work in China, i.e. the (electric) SS1 and the (electric) SS4, are simulated in this research. The whole trip of the train from one stop to the next is simulated for one calculation interval after another. The traction force and operating condition (i.e. motoring, coasting or braking) of the train are considered to be unchanged in one calculation interval in this research. Various types of

Fig. 13.1 Framework of the simulation approach to calculate the TEC

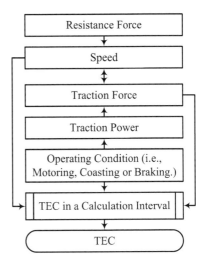

locomotives have different traction performances to overcome the resistance force in their traction processes on the same rail line for the same transport work and the same target speed at the cost of different traction energy consumption. The TECs of all the calculation intervals of the simulation work from the startup of the train at a station to its stop at another station are summed into the TEC of the trip between these two stops.

13.3 Analysis of Traction Energy Cost Intensity

The TEC per 10,000 t-km is defined by Eq. (13.1) to evaluate its traction energy cost intensity.

$$e_{ij}^v = E_{ij}^v / PKM_{ij}^v \tag{13.1}$$

where,

e_{ij}^v TEC per 10,000 t-km of the train with the target speed of v from station i to station j, Unit: kWh/10,000 t-km,

E_{ij}^v TEC of the train with the target speed of v from station i to station j, Unit: kWh, and

PKM_{ij}^v Passenger-kilometers completed by the train with the target speed of v from station i to station j, Unit: %

The transport distance (Unit: 10,000 km) from the nth stop ($S(n)$) to the ($n + 1$)th stop ($S(n + 1)$) ($n = 1, 2, \ldots, 20$) of the hypothetically straight and smooth railway line in this research is interpreted by Eq. (13.2).

$$D_{S(n),S(n+1)} = 5.00 \times 10^{-4} \times n \tag{13.2}$$

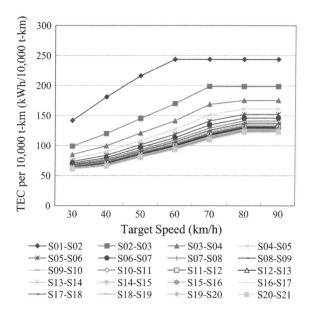

Fig. 13.2 TECs of the freight train hauled by the SS1

The train studied in this work is fully loaded. The changes of the TECs per 10,000 t-km of the freight train hauled by respectively the SS1 and the SS4 with the increases of the target speeds between different neighboring stops are revealed in Figs. 13.2 and 13.3 correspondingly. It is found in these two figures that the TEC per 10,000 t-km is obviously increased with the decrease of the length of stop-spacing from about 20.00 km for the same target speed. Moreover, the increase of the TEC per 10,000 t-km with the improvement of the target speed is accelerated by the decrease of the transport distance between neighboring stops until the inter-stop transport distance or the traction performance of the locomotive starts to prevent the train from achieving a relatively high target speed especially for a comparatively short stop-spacing. It is evidently shown that making stop-spacing shorter than 20.00 km ceases more early the achievement of the target speed and meanwhile the increase of the TEC per 10,000 t-km with the improvement of the target speed. In contrast, the inter-stop transport distances longer than 20.00 km have little effect on the TEC per 10,000 t-km.

It is revealed in Fig. 13.2 that if the target speed of the train hauled by the SS1 is e.g. 70.00 km/h, the decrease of the length of stop-spacing from 20.00 to 10.00 km additionally consumes approximately 50.00 kWh per 10,000 t-km. Because of the restriction of the traction performance of the SS1, the target speed of the train hauled by the SS1 cannot exceed 80.00 km/h. As a result, the TECs per 10,000 t-km become stable for all the transport distances between neighboring stops after the target speed reaches 80.00 km/h. In comparison, the relatively superior traction performance of the SS4 has no impact on the increase of the TECs per 10,000 t-km of the train with the improvement of the target speed for different inter-stop transport distances, as shown in Fig. 13.3. The decrease of the

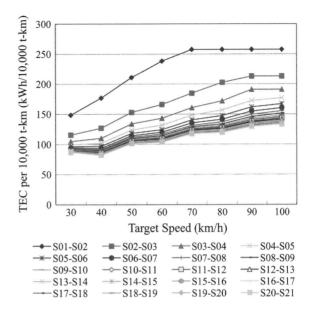

Fig. 13.3 TECs of the freight train hauled by the SS4

inter-stop transport distance from 20.00 to 10.00 km additionally consumes around 40.00 kWh per 10,000 t-km when the target speed of the train hauled by the SS4 is 70.00 km/h.

13.4 Conclusions

It is empirically confirmed that the TEC per unit transport of a freight train with an unchanged utilization ratio of its loading capacity increases apparently in an enlarging scale with the decrease of the length of stop-spacing from 20.00 km for the same target speed. On the contrary, the increase of the TEC per unit transport with the improvement of the target speed of the train is slightly decelerated by the increase of the transport distance between neighboring stops. Therefore, the inter-stop transport distance of a freight train had better be no shorter than 20.00 km. In addition, the railway freight transport work should organize heavy load trains hauled by respectively various locomotives with different traction performances suitable for different target speeds in view of different inter-stop transport distances.

Only the transports of a typically formed freight train hauled by respectively two major types of the locomotives in China on a hypothetically straight and smooth rail line are analyzed in this research to explore the effect of the length of the stop-spacing on the TEC per 10,000 t-km for different target speeds of the train. The impact of the track alignments on the traction energy cost intensities of freight trains formed by different types of railway freight cars should be studied together with the effect of inter-stop transport distances, traction performances of locomotives, etc. for different target speeds from a more comprehensive viewpoint to further validate the conclusions of this study in the future.

Acknowledgments This research is financially supported by the National Basic Research Program of China (2012CB725406) and the National Natural Science Foundation of China (71131001; 71201006).

References

Bocharnikov YV, Hillmansen S, Tobias AM, Goodman CJ, Roberts C (2007) Optimal driving strategy for traction energy saving on DC suburban railways. IET Electr Power Appl 1(5):675–682

Chui A, Li KK, Lau PK (1993) Traction energy management in KCR. IEE Conf Publ 1(388):202–208

Feng X (2011) Optimization of target speeds of high-speed railway trains for traction energy saving and transport efficiency improvement. Energy Policy 39(12):7658–7665

He J, Wu W, Xu Y (2010) Energy consumption of locomotives in China railways during 1975–2007. J Transp Syst Eng Info Technol 10(5):22–27

Hoyt EV, Levary RR (1990) Assessing the effects of several variables on freight train fuel consumption and performance using a train performance simulator. Transp Res Part A Gen 24(2):99–112

Huang Y, Qian Q (2010) Research on improving quality of electricity energy in train's traction. Control Decis 25(10):1575–1579

Li Q, Mao Y (2001) China's transportation and its energy use. Energy Sustain Dev 5(4):92–99

Liu R, Golovitcher IM (2003) Energy-efficient operation of rail vehicles. Transp Res Part A Pol Pract 37(10):917–932

López I, Rodríguez J, Burón JM, García A (2009) A methodology for evaluating environmental impacts of railway freight transportation policies. Energy Policy 37(12):5393–5398

Lukaszewicz P (2001) Energy consumption and running time for trains: modelling of running resistance and driver behavior based on full scale testing (Doctoral Dissertation). Royal Institute of Technology, Stockholm

Miller AR, Peters J, Smith BE, Velev OA (2006) Analysis of fuel cell hybrid locomotives. J Power Source 157(2):855–861

National Bureau of Statistics of China (2011) China statistical yearbook 2011. China Statistics Press, Beijing

Uher RA, Sathi N, Sathi A (1984) Traction energy cost reduction of the WMATA metrorail system. IEEE Trans Ind Appl 20(3):472–483

Xi J, Chen GQ (2006) Energy analysis of energy utilization in the transportation sector in China. Energy Policy 34(14):1709–1719

Chapter 14
A Freeway/Expressway Shockwave Elimination Method Based on IoT

Ling Huang and Jianping Wu

Abstract Shockwave is one of the most complex recurrent traffic flow phenomena on freeway/expressway, whose characteristics are not fully understood. With the field data, we compared the driving behaviors (headways and reaction times) before and during the propagation of shockwaves. The drivers seemed to change their driving strategies when they "recognized" a shockwave, thus a *Fuzzy Logic based Shockwave Recognition Algorithm* was proposed, and last we proposed a shockwave elimination method applying the ideas of the Shockwave Recognition Algorithm and Internet of Things.

Keywords Shockwave · Fuzzy logic · Internet of things

14.1 Introduction

Shockwave is commonly recognized as a sudden compression of traffic flow, which acts as an active or moving bottleneck (Homburger and Kell 1984). Shockwave has a great influence on highway capacity, reducing speeds of traffic flow upstream and downstream. Shockwave also contributes 15 % road traffic accidents (rear-end collisions) in China (Xuan 2000), and cause more fuel consumption, increasing environment damage.

L. Huang (✉)
South China University of Technology, Guangzhou 510640, China
e-mail: Hling@scut.edu.cn

J. Wu
Tsinghua University, Beijing, China

So far, the studies on shockwave are mainly focused on the following aspects: (1) highway shockwave characteristics based on Improved L-W fluid theory (Kuhne et al. 2000); (2) the highway bottleneck queue length or delay time based on shockwave theory (Smith et al. 2003; Munoz and Daganzo 2002); (3) traffic flow stability with small disturbance by macroscopic continuum model and micro car-following models (Zhang 1999, Jingang et al. 2003); (4) Highway shockwave phenomena and traffic accidents (Golob et al. 2004); (5) shockwave prevention and control measures (Breton et al. 2002, Hegyi et al. 2005). Besides, the recent developments of Internet of Things (IoT) provide new ideas for shockwave solutions. Researchers have evaluated the performance of inter-vehicle communication in a unidirectional dynamic traffic flow with shockwave (Bao et al. 2009), and studied the impact of traffic-light-to-vehicle communication on fuel consumption and emissions (Tielert et al. 2010). Yet, the data used in the above studies mostly were from dual loop detectors, or generated by microscopic traffic flow simulation models. As a major road traffic characteristic, shockwave received relatively less studies by traffic engineers and researchers, because of the lack of effective data.

This paper reports a recent research project and preliminary results on shockwaves by microscopic dynamic vehicle data obtained from video tapes. First introduced the microcosmic dynamic analysis on shockwave; then presented a fuzzy logic based shockwave anticipation algorithm, and based on this algorithm proposed shockwave elimination/reduction method applying Internet of Things technology. Last were the conclusions.

14.2 Microscopic Analysis on Shockwaves

A video image process system—Vspeed—has been developed for video data detection and collection (Bai et al. 2006). In this way, we got about 80 cases of shockwaves and collected moving data of over 1000 vehicles. Then, we made comparative analysis on the driving behavior (time headways and the reaction time) before and during shockwave situations.

14.2.1 Comparative Analysis on Time Headways

Time headway refers to the time that elapses between the arrival of the leading vehicle and the following vehicle at the test point. In normal situations without a shockwave, the average and standard deviation of time headways were both significantly smaller than those during the shockwave propagations (Table 14.1).

Table 14.1 Time headways statistics

Time Headway	Mean (s)	Standard dev. (s)	Max (s)	85 % value (s)	15 % value (s)	Min (s)	Sample size
Normal situations	2.07	0.87	5.89	2.88	1.28	0.62	1479
Shockwave situation	3.40	6.76	22.40	4.48	1.98	0.76	1380

14.2.2 Comparative Analysis on Drivers' Reaction Time

In this case we assumed that during shockwave propagation the rear light of the leading vehicle is the stimuli of the following vehicle, then the reaction time of the driver *i* would be:

$$T_i = (t_s)_i - (t_s)_{i-1} \tag{1}$$

where: T_i is the reaction time of driver i $(t_s)_i$ and $(t_s)_{i-1}$ are the times when the rear lights of vehicle *i* and *i + 1* become on respectively.

The mean value divers' reaction time in shockwave situations seemed larger than that of the normal situations in other research findings, showing that in shockwave situations the drivers' behaviors changed. The further analysis of the data revealed that in the early stage of shockwave propagation, the average drivers' reaction times were as long as 2.26 s. As shockwaves propagated through about 10–15 vehicles, the average drivers' reaction times dropped dramatically to 1.16 s. It seems that during the propagation, downstream drivers recognize that the upstream vehicles are in shockwave situations and *adjust their driving behaviors according*. Yet our results seem still larger than those Perception-Reaction Time (PRT) in early studies (Koppa 2000) (Table 14.2). The possible explanation is that our measurement of reaction time contains the *movement time* (0.26 s averagely), and in shockwave situations with lower speed and larger headways, drivers are not so "urgent" as those in experiment situations of former studies.

Microscopic data analysis showed that some of the driving characteristics such as the reaction time, time headways were changing during shockwave propagations. As Ranney points out, humans are very adaptable and will develop new strategies for new situations (Ranney 1999). We are reasonable to consider that perhaps

Table 14.2 Comparisons of the reaction time statistics

Reaction Time	Mean (s)	Standard dev. (s)	85 % value (s)	15 % value (s)	Sample size (s)
Reaction time (early stage of shockwave propagation)	2.26	1.88	4.34	6.00	254
PRT in early studies (surprise)*	1.31	0.61	1.87	2.45	NA
Reaction time (later stage of shockwave propagation)	1.16	1.09	1.27	1.35	262
PRT in early studies (expected)*	0.54	0.1	0.64	0.72	NA

drivers are adjusting their driving strategies during shockwave propagation to eliminate/reduce the negative effects. That means the drivers could recognize the propagation of a shockwave by their observations on the front one or two vehicles, which gives inspirations to the elimination of shockwave.

14.3 Fuzzy Logic based Shockwave Recognition Algorithm

For better understanding of the drivers' behaviors when they meet a shockwave, we interviewed more than 100 drivers of different age groups, driving experience and genders in Beijing, March 2006. Most of the interviewees thought that they could recognize a shockwave simply by motions (speed, headway) of the leading car. Early recognition of shockwave propagation is the key to shockwave elimination/reduction (Yuan 2006). Researchers have successfully used dual loops detecting the traffic-flow occupancy (density) to recognize shockwaves. With the development of IoT and intelligent vehicle, more and more cars will equipped with smart devices which could sense situations around, assistant driving and communicate with other intelligent infrastructures and other intelligent vehicles. Thus it is quite reasonable to develop a shockwave recognition algorithm applying fuzzy logic learned from drivers' experience as main method; the flowchart is shown in Fig. 14.1.

The Shockwave recognition fuzzy Inference system has three inputs: v_A—speed of the following car, v_R—relative speed between the leading and the following car and $T_c^{-1} = v_R/R$—the anticipation variable, where R is the space headway; and one output: a—the level of driver's anticipation of the shockwave propagation, $0 \leq a \leq 1$.

The membership functions of the input and output variables are common triangular function distributions, and the fuzzy sets of all variables see Table 14.3.

Fuzzy rules can be summed up from the drivers' practical experience, for example:

IF v_A is Very Fast (V5), v_R is Opening (RV2), T_c^{-1} is Much Too Far, THEN the anticipation of a shockwave (a) is Very Strong (a1).

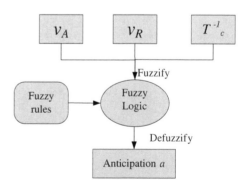

Fig. 14.1 Fuzzy logic flowchart for shockwave recognition

Table 14.3 Fuzzy set terms used in the model

v_A	v_R	$T_c^{-1} = v_R/R$	a
Very slow (V1)	Opening fast (RV1)	Much too far (T1)	Very strong (a1)
Slow (V2)	Opening (RV2)	Too far (T2)	Strong (a2)
Normal (V3)	About zero (RV3)	Satisfied (T3)	Normal (a3)
Fast (V4)	Closing (RV4)	Too close (T4)	Light (a4)
Very fast (V5)	Closing fast (RV5)	Much too close (T5)	Very light (a5)

Theoretically there are 125 fuzzy rules, as some of fuzzy rules are not suitable for actual situations; we finally established 85 fuzzy rules, which could be referred to (Yuan 2006).

14.4 Shockwave Elimination/Reduction Method

The Internet of Things (IoT) is a novel paradigm that is rapidly gaining ground in the scenario of modern wireless telecommunications. From the shockwave recognition algorithm of intelligent vehicle device, a shockwave elimination/reduction method based on the IoT technology is proposed in Fig. 14.2.

First, the Intelligent Vehicle (IV) detects dynamic data, using the proposed fuzzy logic based shockwave recognition algorithm, when the level of anticipation (a) reaches to a certain degree, IV will give an alert of shockwave situation, and start the corresponding shockwave driver assistance program (if there any). At the same time to start the communication program, contacting the near downstream IVs and/or Intelligent Infrastructures, sending the information on shockwave recognition, time, location and so on. Hence other IVs would change their driving

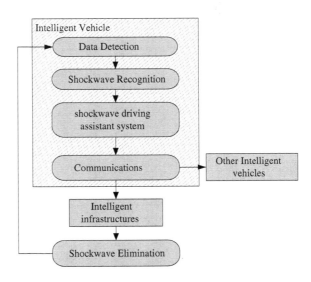

Fig. 14.2 Shockwave elimination method based on the IoT

strategy accordingly, and the Intelligent Infrastructures would start the elimination program, for example coordination of variable speed limits (Breton et al. 2002; Hegyi et al. 2005), ramp metering, information releasing by VMS, etc. to reduce the downstream traffic flow density, till the shockwave is eliminated.

14.5 Conclusions

For better understanding the phenomenon of shockwave, we develop specialized VIPS software to obtain the microscopic shockwave data, by comparative analysis on vehicles headways and the drivers reaction time before and during shockwave situations, we found that in the shockwave propagation process the drivers are often able to identify the shockwave and adjust their driving strategy accordingly. Then a fuzzy logic based on shockwave recognition algorithm is developed by interviewing over 100 different drivers, and a shockwave elimination method based on IoT technology is proposed, provide a thought to the future applications of IoT in Intelligent Transportation System.

Acknowledgments This work was supported in part by the Fundamental Research Funds for The Central Universities and The National Natural Science Foundation of China (No. 51108192).

References

Bai H, Wu J, Liu C (2006) Motion and haar-like features based vehicle detection. 12th institute of electrical and electronics engineers multimedia modelling conference, Beijing, China, Jan (2006)
Bao Y, Sun X, Wang Z (2009) Performance evaluation of inter-vehicle communication in a unidirectional dynamic traffic flow with shockwave. In: Proceeding of ultra modern telecommunications and workshops, 2009. ICUMT, pp 1–6
Breton P, Hegyi A, De Schutter B, Hellendoorn H (2002) Shockwave elimination/reduction by optimal coordination of variable speed limits. The IEEE 5th international conference on intelligent transportation systems. pp 225–230
Golob TF, Recker WW, Alvarez VM (2004) Freeway safety as a function of traffic flow. Accident Anal Prevention 36:933–946
Hegyi A, Schutter BD, Hellendoorn J (2005) Optimal coordination of variable speed limits to suppress shock waves. IEEE Trans Intell Transp Syst 6(1):102–112
Homburger WS, Kell JH (1984) Fundamentals of traffic engineering, Traffic stream characteristics 12th edn. p 4
Koppa RJ (2000) Monograph of traffic flow theory, Human factors, Chapter 3. FHWA (2000) pp 4–6
Kuhne R, Michalopoulos P (2000) Continuum flow models. Traffic flow theory, chapter 5 FHWA(2000)
Munoz JC, Daganzo CF (2002) The bottleneck mechanism of a freeway diverge. Transp Res Part A 36:483–505
Ranney TA (1999) Psychological factors that influence car-following and car-following model development. Transp Res Part F Traffic Psychol Behav 2(4):213–219

Smith BL, Ling Q, Venkatanarayana R (2003) Characterization of freeway capacity reduction resulting from traffic accidents. J Transp Eng 123:362–368

Tielert T et al (2010) The impact of traffic-light-to-vehicle communication on fuel consumption and emissions. Proceedings of the conference of internet of things (IoT), Nov 29–Dec 1 2010, Tokyo

Xuan J (2000) Usual traffic accident types. Safety and Health at Work 5:17

Yi J, Lin H, Alvarez L, Horowitz R (2003) Stability of macroscopic traffic flow modelling through wave front expansion. Transp Res B 37:661–679

Yuan Y (2006) Study on the expressway shockwave based on video capture technology. MSc Dissertation, Beijing Jiaotong University

Zhang HM (1999) Analyses of the stability and wave properties of a new continuum traffic theory. Transp Res B 33:399–415

Chapter 15
Allocating the Subsidy Among Urban Public Transport Enterprises for Good Performance and Low Carbon Transportation: An Application of DEA

Qianzhi Dai, Yongjun Li, Qiwei Xie and Liang Liang

Abstract This paper proposes a stimulating mechanism for allocating subsidies to urban public transport enterprises. The allocation method is based on data envelopment analysis and the satisfaction degrees of urban public transport enterprises. It first finds the set of subsidy allocation that can keep the Pareto efficient for both the whole urban public transit industry and each urban public transport enterprise to reflect the efficiency principle, and then yields a unique subsidy allocation scheme from the set of subsidy allocations with considering the equity of satisfaction degrees. The allocation mechanism can reflect the market competition regulation on some level and benefit to achieve the goal of Green Transport in urban public transit industry. An example of allocating the subsidy among urban public transport enterprises is illustrated.

Keywords Data envelopment analysis (DEA) · Subsidy allocation · Efficiency · Satisfaction degree · Equity · Low carbon transportation

---

Q. Dai (✉) · Y. Li · L. Liang
School of Management University of Science and Technology of China,
Hefei 230026, Anhui Province, People's Republic of China
e-mail: qianzhi@mail.ustc.edu.cn

Y. Li
e-mail: lionli@mail.ustc.edu.cn

L. Liang
e-mail: lliang@ustc.edu.cn

Q. Xie
Institute of Automation Chinese Academy of Sciences, Beijing 100190,
People's Republic of China
e-mail: qwxie2012@gmail.com

15.1 Introduction

The common-weal is one of features in urban public transit industry, which causes the generally losses of urban public transport enterprises. Therefore, most governments provide the financial support, namely subsidy, to these enterprises. However, it is difficult to supervise the totals of actual operating losses and besides, it is also hard to distinguish the actual operating losses caused by the common-weal, thus some problems might be generated, such as higher operating losses, more fuel consumption and lower satisfaction degrees of passengers, etc. In other words, the subsidy may be lead to the many problems, such as resource wastes, high pollution emissions and low operating efficiencies, if it can't be reasonable allocated. Therefore, it is significant to design a scientifically subsidy stimulating mechanism to cultivate the competitiveness of public transport enterprises, which is advantageous to achieve low carbon transportation and Green Transport.

Data envelopment analysis (DEA) (Charnes et al. 1978) is a well-established non-parametric methodology for measuring the performance of peer decision making units (DMUs) with multiple inputs and outputs. There are some studies to measure the performance of transportation system in recent years, such as (Karlaftis 2004; Zhao et al. 2011; Lao and Liu 2009), etc. Furthermore, the equivalence relationship between the DEA efficient and Pareto efficient has been demonstrated (Yan and Wei 2011). However, to our knowledge, there is no previous study to allocate the subsidies among urban public transport enterprises based on DEA methodology. According to DEA methodology, the efficiency is the ratio of outputs and inputs. If the subsidy is regarded as an independent input to impact on the performance of urban public transport enterprises, then the problem of allocating the subsidy can be transferred as the problem of allocating the fixed input (resources or costs), and the latter problem is an important application of DEA. Many previous DEA literatures have studied the problem, such as (Cook and Kress 1999; Cook and Zhu 2005; Beasley 2003; Lins and Gomes 2003; Gomes and Lins 2008; Lozano and Villa 2005; Lozano et al. 2004; Lozano and Villa 2004; Asmild et al. 2009; Avellar et al. 2007; Milioni et al. 2011; Li et al. 2009; Li et al. 2012), etc.

We consider that the subsidy allocation criterion should balance the efficiency and fairness among urban public transport enterprises (i.e., DMUs). Kolm (1971) first demonstrated that there is a fundamental conflict between the two axiomatically criterions. Based on Li et al. (2012), the developed approach in this paper first finds the set of subsidy allocations that can ensure the Pareto efficient for both each DMU and the whole industry to reflect the efficiency principle. Then an efficient allocation interval can be obtained. The interval is directly related to the performance of DMUs. Furthermore, the satisfaction degree is defined based on the efficient allocation interval to evaluate the psychology perception of DMUs. At the end, the allocation approach is proposed with considering the fairness of satisfaction degrees, and a unique subsidy allocation scheme can be generated from the set of efficient subsidy allocations. The subsidy allocation mechanism can reflect the market competition regulation on some level and benefit to achieve the goal of Green Transport in urban public transit industry.

The rest of this study is organized as follows. In Sect. 15.2, we propose the allocation methodology based on DEA and the satisfaction degrees. An example of allocating the subsidy among urban public transport enterprises is presented in Sect. 15.3. Conclusions are given in Sect. 15.4.

15.2 Methodology

Suppose there are n independent rational DMUs, and each DMU_j ($j = 1, 2, \ldots, n$) consumes m inputs x_{ij} ($i = 1, 2, \ldots, m$), to generate s outputs y_{rj} ($r = 1, 2, \ldots, s$). The total of allocated subsidy and the subsidy allocated to DMU_j are denoted as R and R_j ($j = 1, 2, \ldots, n$) respectively, then the efficiency E_d^{before} for any given DMU_d $d \in \{1, 2, \ldots, n\}$ under evaluation before allocating the subsidy can be calculated by the following model (15.1). It is an input-oriented CCR DEA model, where u_r, v_i are unknown multipliers attached to rth output and ith input respectively. It allows each DMU under evaluation to select a set of input and output weights under the relatively constraints to maximize its efficiency. If $E_d^{before} = 1$, then DMU_d is called efficient.

$$\begin{aligned} Max \sum_{r=1}^{s} u_r y_{rd} / \sum_{i=1}^{m} v_i x_{id} &= E_d^{before} \\ s.t. \sum_{r=1}^{s} u_r y_{rj} / \sum_{i=1}^{m} v_i x_{ij} &\leq 1, \forall j \\ u_r, v_i &\geq 0, \forall r, i \end{aligned} \quad (15.1)$$

$$\begin{aligned} Max \sum_{r=1}^{s} u_r y_{rd} / \left(\sum_{i=1}^{m} v_i x_{id} + R_d \right) &= E_d^{after} \\ s.t. \sum_{r=1}^{s} u_r y_{rj} / \left(\sum_{i=1}^{m} v_i x_{ij} + R_j \right) &\leq 1, \forall j \\ \sum_{j=1}^{n} R_j &= R \\ u_r, v_i, R_j &\geq 0, \forall r, i, j \end{aligned} \quad (15.2)$$

Considering the allocated subsidy R, we can easily transfer model (15.1) to model (15.2). E_d^{after} is the optimal efficiency value of DMU_d. Based on the constrains of model (15.2), we can easily obtain the subsidy allocation set ensuring both each DMU and the whole organization CCR efficient under a common set of weights as giving in the following Eq. (15.3) (Li et al. 2012):

$$R_j = \sum_{r=1}^{s} u_r y_{rj} - \sum_{i=1}^{m} v_i x_{ij}, \forall j$$

$$\sum_{j=1}^{n} R_j = R \quad (15.3)$$

$$u_r, v_i, R_j \geq 0, \forall r, i, j$$

$$\rho_d = \frac{R_d - \underline{R_d}}{\overline{R_d} - \underline{R_d}}, \forall d \quad (15.4)$$

We can obtain the interval of efficient allocation by setting "*Max R_d*" and "*Min R_d*", respectively, as the objective function and Eq. (15.3) as the constraints. Denote the interval as $[\underline{R_d}, \overline{R_d}]$, then we want to obtain a unique subsidy allocation from the interval, i.e., $R_d \in [\underline{R_d}, \overline{R_d}]$. Rational DMU_d is selfish to receive the maximum $\overline{R_d}$, but unwilling to accept the minimal one $\underline{R_d}$, thus we can define the satisfaction degree of DMU_d to the subsidy allocation as given in Eq. (15.4), where $\rho_d \in [0, 1]$. DMU_d satisfies with the allocation completely when $\rho_d = 1$, which means that it affords the maximum $\overline{R_d}$. On the contrary, it affords the minimal one when $\rho_d = 0$. Thus the psychology perception of DMUs can be quantificational evaluated.

The equity of satisfaction degrees among DMUs is advantageous to implement the allocation. Several interpretations about the fairness are existed in different research areas, but none of them is dominated accepted. The max–min fairness can reflect the Rawlsian justice and be widely accepted in practice (Bertsimas et al. 2011) (Rawls 1971), thus we use it in this paper. Accordingly, by setting "$\underset{u,v}{Max} \underset{1 \leq d \leq n}{min} \rho_d = (R_d - \underline{R_d})/(\overline{R_d} - \underline{R_d})$" as the objective function and the Eq. (15.3) as the constraints, we can get the allocation model. The allocation criteria are maximizing the environment efficiency and the fairness of satisfaction degrees, which is the efficiency-equity tradeoff. Apparently, the final allocation model is a multi-objective programming. To solve it, we can let $\underset{1 \leq d \leq n}{min} \rho_d = \beta$, and the detailed algorithm can refer to (Li et al. 2012).

15.3 Allocating the Subsidy to Urban Public Transport Enterprises

The original data of seven urban public transport enterprises (DMU A-G) in 2009 is from a Provincial Communications Department. Considering the goal of this paper and the suggestions from several experts in urban public transport area, we select the input and output variables as follows:

Input 1: the number of standardized operating buses. The input 1 of DMU A-G is followed by 3453, 1355, 572, 945, 346, 289 and 261 respectively.

Input 2: the costs per 1,000 km. The input 2 of DMU A-G is followed by 5642.28, 5196.7, 5137.98, 4135.83, 4171.54, 3620.49 and 3534.53 respectively.

Output 1: the satisfaction degrees of passengers. The output 1 of DMU A-G is followed by 60.87, 63.77, 55.56, 50.48, 56.68, 61.83 and 61.93 respectively.

Output 2: the number of transporting passengers per kilometer. The output 2 of DMU A-G is followed by 2.86, 2.92, 2.80, 2.12, 2.47, 1.73 and 1.00 respectively.

Besides, we assume the total of allocated subsidy $R = 10,000$ Yuan.

The relevant results of the proposed approach are shown in Table 15.1. As shown in the last column, there are two groups of efficiencies. The left column is the performance of public transport enterprises before the subsidy allocation, which is obtained by model (15.1). The right one is the efficiencies based on the subsidy allocation, in which we can find that all enterprises are Pareto efficient. The third column is the allocation interval, which shows that it can't allocate the subsidy with the upper bound 17660.04 since it is larger than the total allocated subsidy 10,000. The subsidy allocation is shown in the second column and shows that better performance can generally lead to larger allocated subsidy for public transport enterprises, such as DMU E, F and G. The worst performance, DMU A, causes the lowest allocated subsidy 687.52. Although the performance of DMU B is less than G, however, it is still an enterprise with good performance and besides, the enterprise scale of DMU B is far larger than G. Therefore, it is allocated the third highest subsidy 1739.17. As shown in the fourth column, the satisfaction degrees well reflect the psychology perception of each enterprise and the fairness principle (the interval of satisfaction degree is [0.3811, 0.5595], and the difference is very small). The relatively large difference of the allocated subsidy provides motivation to the enterprises with poor performance to improve their performance by decreasing the inputs and increasing the outputs. Some measures can be taken to achieve the goal, such as purchasing the buses with low carbon technology since this type buses would be obtained a larger weight in the process of calculating the standardized operating buses by China transportation policy; decreasing the administrative cost and other costs to avoid wastes and paying more attention on

Table 15.1 The relevant results based on the proposed approach

DMU	Allocation	Allocation interval Minimum	Allocation interval Maximum	Satisfaction degree	Efficiency Before	Efficiency After
A	687.52	0	1803.93	0.3811	0.8561	1
B	1739.17	1071.22	2300.53	0.5434	0.9490	1
C	982.98	18.14	2549.70	0.3811	0.9204	1
D	1058.87	814.76	1376.73	0.4344	0.8797	1
E	2056.14	1378.67	3156.23	0.3811	1.0000	1
F	2214.62	1007.38	3165.07	0.5595	1.0000	1
G	1260.70	0	3307.86	0.3811	1.0000	1
Totals	10000.00	4290.17	17660.04			

high quality service to improve the satisfaction degrees of passengers and increase the attraction of passengers to select their buses. Thus, the proposed subsidy allocation methodology is in favor of stimulating urban public transport enterprises to initiatively decrease the operating losses and benefitting to achieve the goal of low carbon transportation by a way of economic incentives.

15.4 Conclusions

How to improve the performance of public transport enterprises and achieve the goal of low carbon transportation are two difficult problems. This paper first finds the interval of efficient subsidy allocation, which reflects the efficiency principle. Then we define the satisfaction degree based on the interval. To yield a unique subsidy allocation, we consider the fairness of satisfaction degrees, which reflects the equity principle. The relevant results of the illustrated example demonstrates that the subsidy allocation mechanism can provide incentives for urban public transport enterprises to initiatively improve their performance and can provide a guidance to achieve the goal of Green Transport.

Acknowledgments This research is supported by National Natural Science Foundation of China under Grants (No. 70901070 and 61101219), Science Fund for Creative Research Groups of the National Natural Science Foundation of China (No. 70821001) and the Funds for International Cooperation and Exchange of the National Natural Science Foundation of China (No. 71110107024).

References

Asmild M, Paradi JC, Pastor JT (2009) Centralized resource allocation BCC models. Omega 37:40–49

Avellar JVG, Milioni AZ, Rabello TN (2007) Spherical frontier DEA model based on a constant sum of inputs. JOper Res Soc 58:1246–1251

Beasley JE (2003) Allocating fixed costs and resources via data envelopment analysis. Eur J Oper Res 147:198–216

Bertsimas D, Farias VF, Trichakis N (2011) The price of fairness. Oper Res 59(1):17–31

Charnes A, Cooper WW, Rhodes E (1978) Measuring the efficiency of decision making units. Eur J Oper Res 2:429–444

Cook WD, Kress M (1999) Characterizing an equitable allocation of shared costs: a DEA approach. Eur J Oper Res 119:652–661

Cook WD, Zhu J (2005) Allocation of shared costs among decision making units: a DEA approach. Comput Oper Res 32:2171–2178

Gomes EG, Lins MPE (2008) Modelling undesirable outputs with zero sum gains data envelopment analysis models. J Oper Res Soc 59:616–623

Karlaftis MG (2004) A DEA approach for evaluating the efficiency and effectiveness of urban transit systems. Eur J Oper Res 152:354–364

Kolm S (1971) Justice et Equite. CEPREMAP, Paris

Lao Y, Liu L (2009) Performance evaluation of bus lines with data envelopment analysis and geographic information systems. Comput Environ Urban Syst 33:247–255

Li YJ, Yang F, Liang L, Hua ZS (2009) Allocating the fixed cost as a complement of other cost inputs: a DEA approach. Eur J Oper Res 197:389–401

Li YJ, Yang M, Chen Y, Dai QZ, Liang L (2012) Allocating a fixed cost based on data envelopment analysis and satisfaction degree. Omega (in press)

Lins MPE, Gomes EG, Mello JCCBS, Mello AJRS (2003) Olympic ranking based on a zero sum gains DEA model. Eur J Oper Res 148:312–322

Lozano SA, Villa G (2004) Centralized resource allocation using data envelopment analysis. J Prod Anal 22:143–161

Lozano SA, Villa G (2005) Centralized DEA models with the possibility of downsizing source. J Oper Res Soc 56(4):357–364

Lozano SA, Villa G, Adenso-Diaz B (2004) Centralized target setting for regional recycling operations using DEA. Omega. 32:101–110

Milioni AZ, Avellar JVG, Gomes EG, Mello JCCBS (2011) An ellipsoidal frontier model: Allocating input via parametric DEA. Eur J Oper Res 209:113–121

Rawls J (1971) A theory of justice. Harvard University Press, Cambridge

Yan W, Wei Q (2011) Data envelopment analysis classification machine. Inf Sci 18:5029–5041

Zhao Y, Triantis K, Murray-Tuite P, Edara P (2011) Performance measurement of a transportation network with a downtown space reservation system: A network-DEA approach. Transp Res Part E 47:1140–1159

Chapter 16
What Counts in the Bus Use for Commuting? A Probe Survey Based on Extended Theory of Planned Behavior

Wen Wu, Dong Ding and Ping Wu

Abstract In order to lead people to pro-environmental transportation modes, it is necessary to learn about the psychology and behavior of users. Based on the extended theory of planned behavior—incorporating habit, descriptive norm and personal norm, this paper analyzes the relationship between these elements and intention to use bus, and the influence of habit on intention-behavior. The result of survey conducted in Chengdu City reveals different outcomes from previous study, especially for the attitude and perceived behavior control. In addition, the intention to use bus and habit can explain 41.7 % variance of the bus use ratio to the total travel. However, the interaction of habit and intention is insignificant.

Keywords Extended theory of planned behavior · Habit · Bus use

16.1 Introduction

The global warming and emissions have been the global issue. Public transportation becomes one of the strategies to deal with this problem. However, only when more people are willing to take bus, will the strategy play its role. The demand is closely

---

W. Wu (✉) · D. Ding
School of Transportation and Logistics, Southwest Jiaotong University,
Chengdu 610031, China
e-mail: wuwen88hope@gmail.com

P. Wu
Key Laboratory of Traffic Engineering, Beijing University of Technology,
Beijing 100124, China

P. Wu
Transport Management Institute Ministry of Transport
of The People's Republic of China, Hebei 101601, China

related to people's psychology and behavioral habit. Much research has been done in this field. Theory of planned behavior is one of the most widely used models to tackle the cognitive determinants of mode use. According to this theory, intentions are based on combination of attitude toward the behavior, subjective norm and perceived behavior control; intention has effect on behavior; beliefs about the consequences of behaviors frequently influence attitude (Ajzen 1991). Many researchers have extended this theory by incorporating descriptive norm, personal norm (moral norm), awareness of responsibility for problems, such as (Heath and Gifford 2002; Eriksson and Forward 2011; Gardner 2009; Abrahamse et al. 2009). In addition, empirical study shows that habit has moderating effect on these elements (Eriksson et al. 2008; Klockner and Matthies 2004).

Some research conclusions have been reached. For instance, the influence of attitude, perceived behavior control and descriptive norm on intention to use bus is significant (Heath and Gifford 2002). These conclusions are useful for policymakers to reduce emissions and lead to more use of public transportation. However, such research is conducted in countries where car is the necessity of family. In China, despite the rapid increase of car ownership, most people have to rely on bus, which may result in different conclusions.

In this paper, we want to explore two questions, (1) the relationship of cognitive determinants with intention to use bus in China; (2) whether bus use habit influences intention-behavior. The motivation to deal with the second question is to explore whether the long time use will have impact on people's mode choice. For example, if the bus use habit would moderate intention-behavior, the authority could encourage people by some policies to use bus for a period to form the habit; then the bus use habit can help them keep on taking bus.

16.2 Method

The study was conducted in Chengdu City, the capital of Sichuan Province, China. As people all need to go shopping despite their ages, employment and travel modes, and people in leisure are more willing to fill the questionnaire, the survey was done in and around Wal-Mart, Parkson and Ren Ren Le supermarkets. Before the final survey, the pilot survey was also conducted. It showed that participants could not clearly distinguish some statements of self-report habit indexes (SRHI) among 13 questions. So we revised some questions and deleted some quite similar statements. In the end, 8 statements are left (see Appendix). In the subjective norm, two items are integrated because many think only one of them have the influence.

In the survey, 200 questionnaires were sent randomly to the people in and around the markets. There were 171 valid response questionnaires. More women than men responded (104 female). 91 participants' age are among 18–25. No participant is older than 65. This is primarily due to the poor vision and limited understanding of senior citizens. People with higher education are more willing to respond (121 with B.S or higher education degree). These phenomena correspond with other studies reporting that female and people with higher education are more willing to answer questionnaires (Eriksson and Forward 2011).

In the questionnaire, participants had to consider the total commute trip number and the bus use number for commuting in the last 2 weeks. Thus, the bus use ratio to the total trip could be obtained. In the measurement of attitude, the combination of behavioral beliefs and outcome evaluations is used (Eriksson and Forward 2011). Participants rated consequences of using bus on a 5-point scale (1 = strongly disagree, 2 = disagree, 3 = uncertain, 4 = agree, 5 = strongly agree). There are 11 behavioral belief items, including "Taking bus is environment-friendly" "Taking bus can reach the destination quickly" "The risk of traffic accident in taking bus is low" "Taking bus costs low" "Walking distance is short before and after taking bus" "It takes a long time to wait for the bus" "It is crowded in the bus" Taking bus risks being stolen" "It is noisy in the bus" "It is inconvenient to transfer the bus" "Taking bus risks bus self-burning danger". Then participants had to rate the outcome evaluations on a 5-point scale (1 = not important at all, 2 = not important, 3 = uncertain, 4 = important, 5 = very important). There are 8 outcome evaluation items, including environment-friendliness, travel time, safety, travel cost, walking distance, waiting time, comfort, transfer convenience. The attitude can be obtained by multiplying behavioral belief value and outcome evaluation value then dividing 2. So the attitude range is 0.5–12.5. The alpha value of attitude is 0.70. After recoding, the higher value indicates the more positive attitude toward bus.

As for perceived behavior control, participants were asked directly "Taking bus as the main commute mode for me is difficult" and gave their opinion on a 5-point scale (1 = strongly disagree, 5 = strongly agree). The lower the value, the easier it is for participant to use bus as the main mode. In subjective norm, participants were asked that to what extent they agreed with the statement "My friends and family suggest me to take bus for commuting" (1 = strongly disagree, 5 = strongly agree). Descriptive norm was measured in the similar way except that family member and friends are distinguished. In personal norm, participants rated on a 5-point scale for the statement "I think I'm obliged to take bus in order to make my own contributions to the low-carbon society". Intention to use bus was evaluated by asking "I intend to use bus as the main mode for commuting in the following 2 weeks". The higher value means more willingness to take bus.

There is only 1 item in perceived behavior control, subjective norm, personal norm and intention, and 2 descriptive norm items, which may reduce the reliability(the descriptive norm alpha = 0.67). However, the pilot survey showed that participants were not able to distinguish the similar items. In addition, one item is commonly used by other research, such as (Klockner et al. 2003; Nordiund and Garvill 2003). The habit measurement includes 8 items, where SRHI is used. The higher value reflects stronger habit of bus use. The alpha value is 0.86.

Table 16.1 Means and standard deviations for attitude, perceived behavior control, subjective norm, descriptive norm, personal norm, intention to use bus and habit

	Mean	SD
Attitude	6.16[a]	3.43
Perceived behavior control	2.78	1.24
Subjective norm	3.46	1.19
Descriptive norm	3.47	1.20
Personal norm	3.96	1.03
Intention	3.82	1.28
Habit	3.48	1.29

[a] Scale: 0.5–12.5, higher value indicates more positive attitude towards bus

16.3 Results

In Table 16.1, citizens' attitude toward bus is almost neutral (M = 6.16, SD = 3.43). Personal norm value (M = 3.96, SD = 1.03) is higher compared to other elements, which means that people tend to think they have obligation to take bus for reducing emissions. The value of habit of bus use is similar to that of intention.

In Table 16.2, the bivariate correlation is conducted. Interestingly, the correlation between bus use ratio and attitude is not significant as other research reveals (Heath and Gifford 2002; Eriksson and Forward 2011). So is the intention. In these studies, attitude is strongly related to bus use or intention to use bus. This difference may result from the different life styles. In western countries, car is the family necessity, and people can make choices according to their attitudes. While in Chengdu, according to the data from Chengdu Vehicle Management Institute and Chengdu Statistics Department, there are about 2.579 million cars and 11.25 million people up to March, 2012. Most people have to depend on bus. So the higher bus use does not mean more positive attitude toward bus. The smaller the value of PBC is, the easier for participant to use bus, the higher bus use ratio will be, which leads to the negative relationship between PBC and bus use ratio.

Table 16.2 Bivariate correlations of bus ratio, attitude, perceived behavior control (PBC), subjective norm (SN), descriptive norm (DN), personal norm (PN), intention to use bus and habit

	Bus ratio	Attitude	PBC	SN	DN	PN	Intention
Attitude	0.045	–	–	–	–	–	–
PBC	−0.186*	−0.066	–	–	–	–	–
SN	0.298**	0.165*	−0.048	–	–	–	–
DN	0.385**	0.078	−0.044	0.492**	–	–	–
PN	0.078	0.124	−0.057	0.385**	0.303**	–	–
Intention	0.589**	0.085	−0.105	0.533**	0.520**	0.387**	–
Habit	0.604**	0.020	−0.156*	0.428**	0.395**	0.246**	0.707**

*Correlation is significant at the 0.05 level (2-tailed)
**Correlation is significant at the 0.01 level (2-tailed)

16 What Counts in the Bus Use for Commuting?

Personal norm has insignificant correlation with bus use ratio though it has strong correlation with intention, which means even though people show strong obligation, they not necessarily choose bus in practice. High environmental obligation may not guarantee the environmental action. Subjective norm and descriptive norm both have strong correlation with intention. People who are important to the traveler are closely related with traveler's mode. Of course, this may result from the alternative mode availability. In addition, habit and intention are strongly related. But correlation cannot tell us the casualty. It's hard to say whether people intend to use bus because of habit, or they have no choice but to use bus so the habit forms and they intend to use bus before other modes being available.

In order to predict the intention to use bus, hierarchical multiple regression was conducted. The 3 elements in the conventional TPB entered first, then the habit, finally other independent variables entered.

Before regression analysis, the assumptions were tested. The sample size is 171, the independent variable number is 6. So the ratio of cases to independent variables (28.5:1) is higher than 5:1, and it is also higher than the preferred value 15:1. Kolmogorov–Smirnov values are all between $-1.0 \sim +1.0$; Durbin-Watson value is 2.143, which falls in the interval of 1.50–2.50; tolerance values are larger than 0.10. The results are shown in Table 16.3.

In Table 16.3, the 3 elements in conventional PTB can explain 29.1 % variance in intention to use bus. After incorporating the habit, 56.5 % variance could be

Table 16.3 Attitude (A), subjective norm (SN), perceived behavior control (PBC), habit, descriptive norm (DN), personal norm (PN) as predictors of intention to use bus

	Intention to use bus			
	R^2	ΔR^2	β	t
Step 1	0.291	0.291		6.044
Attitude			−0.008	−0.126
PBC			−0.080	−1.220
SN			0.531	8.029***
Step 2	0.565	0.274		1.848
Attitude			0.006	0.122
PBC			0.001	0.014
SN			0.281	4.901***
Habit			0.586	10.224***
Step 3	0.611	0.047		−0.496
Attitude			0.005	0.104
PBC			0.002	0.043
SN			0.160	2.615***
Habit			0.529	9.419***
DN			0.190	3.272***
PN			0.137	2.561*

***$p < 0.001$
**$p < 0.01$
*$P < 0.05$

Table 16.4 Intention, habit, intention × habit as predictors of bus use ratio

	Bus ratio			
	R^2	ΔR^2	β	t
Step 1	0.347	0.347		
Intention			0.589	9.479***
Step 2	0.418	0.070		
Intention			0.324	3.893***
Habit			0.375	4.505***
Step 3	0.426	0.008		
Intention	–	–	0.545	3.322***
Habit	–	–	0.604	3.580***
Intention × habit	–	–	−0.425	−1.559

***$p < 0.001$

explained, which shows the habit is a significant predictor of the intention. After the inclusion of descriptive norm and personal norm, R^2 only increases slightly compared to the second step. Independent variables all have significant influence on intention to use bus except attitude and perceived behavior control. The insignificant influence of perceived behavior control may come from the "floor effect" or "roof effect", as the individual perceived behavior control on bus differs slightly when almost everyone could easily have access to bus.

Table 16.3 shows that habit is a significant predictor of the intention. Next, the influence of habit on intention-behavior is analyzed. Taking bus use ratio as the dependent variable, the hierarchical regression is executed, where intention entered first, then the habit, intention × habit entered finally.

In Table 16.4, intention explains 34.7 % variance in bus use ratio; intention and habit together can explain 41.7 % variance. But the influence of intention × habit on bus use ratio is not significant. This means that in bus mode, habit has no significant moderation on intention-behavior and the long time bus use may not influence people's intention-behavior. The policy to help citizens form the bus use habit so they can automatically use bus later on will not work.

16.4 Discussion

It is necessary to understand people's behavior of bus use for executing public transportation policy. This paper analyzed the influence of common elements on intention to use bus in Chengdu City based on extended theory of planned behavior. The result shows that people's attitude toward bus is neutral instead of positive. On the other hand, intention to use bus is insignificantly related to attitude. Even people show high obligation to reduce emissions, they may not practice in real. This can be seen from the insignificant correlation between personal norm and intention. The regression result shows that all independent variables significantly influence the intention to use bus except attitude and perceived behavior control. The influence of habit on intention-behavior is insignificant. So the long-time bus use cannot guarantee the insistence on bus use.

This is the preliminary research. Further study has to be done to explain the reason for the inconsistence with previous research conclusions that attitude and perceived behavior control has significant influence on intention. The different results show that travel pattern in China is different from that of other countries and different policies are needed for the specific situation in China. Other factors, such as alternative mode availability and intention to buy a car, have to be closely examined as well.

References

Abrahamse W, Steg L, Gifford R, Vlek C (2009) Factors influencing car use for commuting and the intention to reduce it: a question of self-interest or morality? Transp Res Part F: Traffic Psychol Behav 12:317–324

Ajzen I (1991) The theory of planned behavior. Organ Behav Hum Decis Process 50:179–211

Eriksson L, Forward SE (2011) Is the intention to travel in a pro-environmental manner and the intention to use the car determined by different factors? Transp Res Part D: Transp Environ 16:372–376

Eriksson L, Garvill J, Annika M (2008) Interrupting habitual car use: the importance of car habit strength and moral motivation for personal car use reduction. Transp Res Part F: Traffic Psychol Behav 11:10–23

Gardner B (2009) Modeling motivation and habit in stable travel mode contexts. Transp Res Part F Traffic Psychol Behav 12:68–76

Heath Y, Gifford R (2002) Extending the theory of planned behavior predicting the use of public transportation. J Appl Soc Psychol 32:2154–2189

Klockner CA, Matthies E (2004) How Habits interfere with norm-directed behavior: a normative decision-making model for travel mode choice. J Environ Psychol 24:319–327

Klockner CA, Matthies E, Hunecke M (2003) Problems of operationalizing habits and integrating habits in normative decision-making models. J Appl Soc Psychol 33:396–417

Nordiund AM, Garvill J (2003) Effects of values, problem awareness, and personal norm on willingness to reduce personal car use. J Environ Psychol 23:339–347

Chapter 17
Evaluation Study on the City Bicycle Rental System

Jianyou Zhao, Yunjiao Zhang and Cheng Zhang

Abstract Evaluation study on city bicycle rental system plays an extremely important role in the work of the whole city's transportation. The quality and the service level of the rental system's construction will influence the effects of citizens' using it and the improving of the city's transportation problems. However, there still has not a very independent and perfect evaluation system of the public bicycle rental studies so far in China. Based on the principle of selecting index, this thesis is going to divide the evaluation studies on city bicycle rental system into three index systems, namely the rental network design, infrastructure system and management system, by which it can build a scientific and rational evaluation system. By using the Analysis Hierarchy Process (AHP), transacting qualitative index into quantitative index, confirming every index weight, finally there could be a bicycle rental system evaluation value after a comprehensive analysis. Also proceeding an empirical study from the applicative angle to verify the practicality and the feasibility of this article's evaluation theory and method.

Keywords Public bicycle rental system evaluation · Analysis hierarchy process (AHP) · Weight verification · Verify

---

J. Zhao (✉) · Y. Zhang · C. Zhang
School of Automobile, Chang'an University, Xi'an 710064, China
e-mail: jyzhao@chd.edu.cn

Y. Zhang
e-mail: 178560167@qq.com

C. Zhang
e-mail: 303413252@qq.com

17.1 Introduction

With the rapid development of the social economy, citifying process, continuous enlargement of the scale of the city, the rapid growth of population, our city welcome in the unprecedented high speed development time of the motorization. Take Shenzhen for example, motor vehicle population of this city breakthrough 1,700,000 and the annual growth number of vehicle amounts is over 150,000 (Hou 2012).The fact above led the city's transportation problem became more and more important and the problem mainly includes: road resources' saturation, serious traffic jam, the speed of the vehicle is getting slower, the time of the residents trip is getting longer and so on and all of these is becoming the major problem that restrict the development of the city. Meanwhile the huge amounts of the cars brought up the environment pollution and the resources consumption that had a seriously affect the city's ecological environment, social fairness, life quality and sustainable development.

Since 1922, United Nations conference on environment and development had pass 'agenda of twenty-first century', the agenda had promoted 'sustainable development' to a practical objective that we all human beings should pursuit together. Under its theoretical direction, Chris Bradshaw (Su and Luo 2011) brought up the green traffic idea in the 1994 that we should take the green classification to our transportation means and the priority of the classification from high to low followed as walking, bicycle, public transportation and the last is individual automobile driving. The government or the relative company can set couples of the public bicycle rental stations around the city, the citizens or the tourist can rent the bike at any station and give it back at any other stations after finishing using it (Li 2010). It mainly solves the problem like the citizens 'short haul trip', the 'last mile' after the bus and the tourists visiting problem, etc. Moreover it is still an energy-saving, environment protecting and healthy product that can improve the resource utilization ratio remarkably, relief the pressure of the city's transportation jam and reducing the air pollution, lead to a better place for people to live in the city.

17.2 Establishment of City Bicycle Rental System Evaluation

17.2.1 The Principle of Selecting Evaluation Indexes

An accurate evaluation of the bicycle rental system must be based on the index that can fully reflect the usage rate of the public bicycle. The settings of the index system ought to from the actual conditions and should obey the following rules: (a) Scientific and effectiveness character principle. (b) Comparability and flexibility character principle. (c) The combination of dynamic and static principle. (d) The combination of the qualitative index and the quantitative index principle.

According to the contents of this article, through analysis and combine the character, application scope and the evaluation requirement of the index itself to establish the relevant index system, finally use the confirming of the index weight and Analysis Hierarchy Process (AHP) to calculate all the index value weight so as to confirm the specific level of the evaluation of the bicycle rental system.

17.2.2 Construction of the Evaluation Index System

We should establish a comprehensive and omni-directional evaluation system by cluster analysis that based on all the effective factors of the public bicycle before evaluate the city's bicycle rental system. Based on the functions of the bicycle rental system, it's going to divide the system into three parts, that is, the rental network design, infrastructure system and management system.

The specific evaluation index system of city bicycle rental system shown in Fig. 17.1.

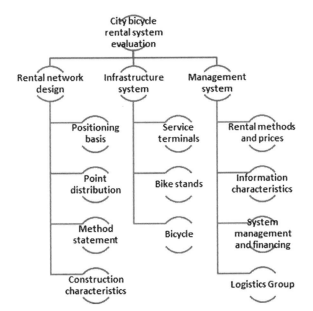

Fig. 17.1 City bicycle rental system evaluation indexes

17.3 The Establishment of the Indexes Weights Based on the AHP

17.3.1 The Overview of Analytic Hierarchy Process

The basic steps to calculate the weight of each index are as follows:

(1) *Building the hierarchical structure*

Based on the investigation and analysis, the range, goals, contained factors and their internal relations of the issue would be figured out. The hierarchical structure could be established.

(2) *Constructing judgment matrix*

By judging a series of twinning factors in each hierarchy, based on certain ratio scale, the judgments were quantified and the comparative judgment matrix was built.

(3) *Determination of hierarchy weights and consistency check*

To measure the reliability and consistency of the evaluation, an index $CR = CI/RI$ was introduced to measure the judgments' deviation. When $CR < 0.10$, the consistency is good.

Each hierarchy index weight could then be $P = \sum_{i,j,k} P_{ijk} \cdot W_{ijk}$ obtained by calculating the weight based on the method shown in the Fig. 17.1.

17.3.2 Comprehensive Evaluation

Corresponding to specific index evaluation criteria, a concrete evaluation conclusion could then be given.

$$\sum_i W_i = 1, \quad \sum_i \sum_j W_{ij} = 1, \quad \sum_i \sum_j \sum_k W_{ijk} = 1$$

where P_{ijk} is the evaluation value of the simple index.

Based on the Table 17.1, by analyzing the evaluation value of the simple index, the detail evaluation of the system in different parts could be obtained. According to the evaluate result, the corresponding countermeasures could be set to improve and enhance the quality of the system.

Table 17.1 City bike rental system evaluation standard

Excellent	Fine	Middling	Qualified	Disqualified
90–100	80–90	70–80	60–70	<60

17 Evaluation Study on the City Bicycle Rental System

Table 17.2 Excel algorithm for computing process based on the analytic hierarchy process

City bicycle rental system	Rental network design	Infrastructure system	Management system	According to multiply the line	Open n power	Wi	Awi	Awi/Wi		
Rental network design	1	2	3	6	1.8171	**0.5278**	1.6118	3.0536		
Infrastructure system	1/2	1	3	1.5	1.1447	**0.3325**	1.0153	3.0536	CI=(λ-n)/(n-1)	CR=CI/RI
Management system	1/3	1/3	1	0.1111	0.4807	**0.1396**	0.4264	3.0536	0.0268	0.0517
					3.4425			3.0536	Awi/Wi	
Rental network design	Positioning basis	Point distribution	Method statement	Construction characteristics	Open n power	Wi	Awi	Awi/Wi		
Positioning basis	1	3	5	7	105	3.2011	**0.5756**	**2 1/3**	4	
Point distribution	1/3	1	3	5	5	1.4954	**0.2689**	**1.1032**	4.1028	CI=(λ-n)/(n-1)	CR=CI/RI
Method statement	1/5	1/5	1	1/2	0.02	0.3761	**0.0676**	**0.2805**	4.1478	0.0328	0.0368
Construction characteristics	1/7	1/5	2	1	0.0571	0.4889	**0.0879**	**0.3592**	4.0853		
						5.5614			4.0985		
Infrastructure system	Service terminals	Bike stands	Bicycle	According to multiply the line	Open n power	**Wi**	Awi	Awi/Wi		
Service terminals	1	2	2	4	1.5874	**0.4934**	1.5066	3.0536	CI=(λ-n)/(n-1)	CR=CI/RI
Bike stands	1/2	1	2	1	1	**0.3108**	0.9491	3.0536	0.0268	0.0517
Bicycle	1/2	1/2	1	0.25	0.6299	**0.1958**	0.5979	3.0536	Awi/Wi	
					3.2174			3.0536		
Management system	Rental methods and prices	Information characteristics	System management and financing	Logistics Group	Open n power	**Wi**	Awi	Awi/Wi		
Rental methods and prices	1	1/3	1/4	1/5	0.0167	0.3593	**0.0741**	0.2992	4.0368	CI=(λ-n)/(n-1)	CR=CI/RI
Information characteristics	3	1	3	1/4	2.25	1.2247	**0.2527**	1.2421	4.9163	0.0359	0.0402
System management and financing	2	1/3	1	1/5	0.1333	0.6043	**0.1247**	0.4668	3.7449		
Logistics Group	5	2	5	1	50	2.6591	**0.55**	2.0478	3.733		
						4.8475			4.1078		

Table 17.3 The weights of city bike rental system evaluation index system

Project	Level 1 of evaluation content		Level 2 evaluation content	
	Evaluation index	Weight	Evaluation index	Weight
City bicycle rental system evaluation	Rental network design	0.528	Positioning basis	0.575
			Point distribution	0.269
			Method statement	0.068
			Construction characteristics	0.088
	Infrastructure system	0.332	Service terminals	0.493
			Bike stands	0.311
			Bicycle	0.196
	Management system	0.140	Rental methods and prices	0.074
			Information characteristics	0.253
			System management and financing	0.125
			Logistics Group	0.548

17.3.3 Determination of the System Index Weight

At this thesis, the software Excel was used to calculate the evaluation value of all steps for its powerful operating function (Xian 2012). The simplification of the consistency check and the adjustment of the judgment matrix were realized. The method was called as the Excel arithmetic of the analytic hierarchy process. The calculation results and data of the method were all keep the accuracy of the highest digit and the error was minimized. All steps of the calculation were concise as shown in the Table 17.2.

Table 17.4 the operation process of Excel algorithm based on the AHP.

Then the weight of each evaluation index could be determined as shown in the Table 17.3.

17.4 Case Analysis

From the 1980s to 1990s, most people of Wuhan city use bicycles (3.8 million), other than motors as their daily vehicles. With the rapid devolvement of the economy of the 1990s, more motors ware used to replace bicycles. The number of bicycles in Wuhan city has decreased to 1 million in 2008 (Li et al. 2009). In order to solve the increasingly serious problem of traffic jam, in May 2009, the municipal government of Wuhan City carried out of policy to support the public bicycle rental system, which is very effective. In the thesis, the author gives a comprehensive evaluation of the public bicycle renting system, based on the combination of different index and the usage of leveled analysis, as in Table 17.4.

17 Evaluation Study on the City Bicycle Rental System

Table 17.4 Evaluation of PBRS in Wuhan

Item	Level 1 Index	Value	Level 2 Index	Conclusion	Value
City bicycle rental system evaluation 82.931	Rental network design	91.050 (0.528)	Positioning basis	The positioning mentality is total quantity control, classified processing, scale balancing, flexible adjusting. The positioning basis is connected to public transportation, giving service to the short distance and large scale travel	92
			Point distribution	Point distribution mainly for transit point, public point, settlements, recreation point, campus point. Set transit point, public point, campus point as center, the average distance to settlements is less than 300 m	90
			Method statement	Construction manual is relatively perfect, reached the construction design standards	95
			Construction characteristics	Bicycle equipped with diversification, service system point is excessive with big range	85
	Infrastructure System	80.483 (0.332)	Service terminals	Service terminal facilities are simple. Search function need to be completed	75
			Bike stands	Reasonable designed, the quantity of equipments reach the standard	92
			Bicycle	The bicycle is of convenience, comfort and safety, but lack of management control system device	76
	Management System	58.118 (0.140)	Rental methods and prices	Residents could open the "Integrity Card" by their ID card for free to rent the bicycle without the rental	95
			Information characteristics	Bicycle information system need to be completed	50
			System management and financing	Operating the company as an enterprise with marketization which benefited by the bicycle body float advertising and carport advertising with indeterminacy for rent. Sources of funds need to be expanded	62
			Logistics group	For less the service hotline, issues of the residents are not resolved in time; Maintenance is not in time, bikes are cumulative loss; Plenty System fault with incomplete stability; have not introduced the wireless network mobile service points yet	56

The value of city bicycle rental system gets 82.931 points in Wuhan. Based on the criterion of Table 17.3, we can say that this is a relatively high value-fine. Specifically, this system has a good designation of network and infrastructure. But there is potential problem in its management system, which need further improvement. The result of the evaluation of city bicycle rental system is in accords with our investment of the real situation, which means that our methods are sound and effective.

17.5 Conclusion

It has been a general tendency and consensus to develop the environmentally friendly public transportation system, such as the bicycle and bus system, and to build their evaluation system. In this thesis, after the consideration the public bicycle system of the main cities in the world, the author chooses the rental grids design, the infrastructure system, the management system as the mean evaluation index. Using analysis hierarchy process, we are able to value the performance of a system in different index precisely, and get a final conclusion by assigning the indexes with different weight. This method not only enables us to analysis the system thoroughly, but also enables us to find the potential problem and come up with an effective solution. And it is a fundamental solution to the city trips "last mile", provide protection for the residents travel, and achieve the fundamental needs of residents of efficient, comfortable, fast, affordable.

References

Hou Z (2012) Public bicycle operation mode [J]. Transp Technol Econ (70)
Li L, Chen H, Sun X (2009) The public bicycle rental points layout plan of Wuhan city [J]. City Transp 7(4)
Su F, Luo S (2011) China's urban low-carbon transport method of propulsion research [J]. Integr Transp
Xian F (2004) Simple calculation of the analytic hierarchy process (AHP) in a spreadsheet (EXCEL) [DB/OL]. Chinese scientific and technological papers online
Li Z (2010) Scale analysis of long-term development of urban public bike rental station [J]. Transp Energy Environ Prot (2)

Chapter 18
The Green Traffic Strategy in Low Carbon Community

Zesong Wei, Xia Wang and Xiaolong Pang

Abstract This paper starts from green traffic design perspective of the low carbon community; studies the different layers as followed: planning low carbon traffic, constructing the community mode guided by no motor vehicle, encouraging low carbon transport, using environmental protection fuel and road materials; advocating low carbon traffic etc. The intention of this paper is to develop low carbon life style for city community, to reduce carbon emissions of overall city, and create a healthy, comfortable life environment for the residents.

Keywords Low carbon · Community · Low carbon traffic · Green transportation

18.1 Introduction

Community is the place for people's life and dwelling, it is not only the basic function unit of the city, but also the basic space of realizing environmental change and social low carbon development. Low Carbon Community refers to develop low carbon economy in urban living community, innovate low carbon technology,

---

Z. Wei
Tianjin University, Tianjin, China

Z. Wei (✉)
School of Architecture and Design, Beijing Jiaotong University, Beijing, China
e-mail: wzs1003@263.net

X. Wang
School of Architecture, Zheng Zhou University, Henan, China

X. Pang
Zhengzhou University Multi-functional Design and Research Academy, Henan, China

change the way of life, through the land, building, energy, resources, transportation, and other comprehensive method, maximum limit to reduce the greenhouse gas emissions of residential planning construction, and process of use and management, realize clean, efficient and sustainable development of the city residential community.

18.2 Planning Low Carbon Traffic

18.2.1 Realizing Convenient Connection of Public Transportation with Community

It is essential to ensure convenient contact of public traffic system of community and nearby for the low carbon community traffic planning. Community should have at least one or two public transit lines around it, and the distance between the nearest transit site and residential gateway is about 400–500 m (time should be less than 5 min on foot), to promote the resident's will of transportation mode choice with public transport system as the priority.

For example, according to the urban strategy of "living first", Vancouver proposed to develop the complete residential community in foot path scale. Masdar's low carbon community connects traffic with community outside by the planning traffic system of three levels: underground transportation, ground traffic and personal rapid transit system. In the ground walk system, the longest distance that people arrive at the bus site is controlled within 200 m. Personal rapid transit system organized by 1,700 motorized personal rail MRT coaches and automatic control system meets the demand of private travel.

18.2.2 Organizing Mixed Walk Network in the Community

The traditional squares nets road system used for residential house, although convenient to car traffic, but it will be too wasteful, and also against the security and the construction of community's sense. Although ended road style is a single path, and contact shortage, it is quiet, safe and benefit for building the sense of place. The combination of square nets road and ended grid road will be complementary advantages, through the "permeability of filter" principle to form a new network form— Fused Grid. The network structure is discontinuous for the car,but it is linked for pedestrians and bicycle, so can limit motor vehicle traffic but be beneficial to the mobile pedestrians and bicycle. This means that the network loves in positive transportation means, selectively "filter" off the car (Fig. 18.1).

Located in Alberta province, Canada, Calgary Saddlestone Community uses mixed road network. On one hand, optimize the land use streets, encourage

Fig. 18.1 (**a**) Traditional squares grids (**b**) traditional ended type (**c**) fused grid structure

walking, and at the same time actively discourage the short distance driving, to achieve peace and security of the community, increasing the potential of the social interaction. Covers an area of 38 hectares, with a population of 5,500, Vauban District in Germany Freiburg is the world's first "car free community". To a certain extent, resident's walking or cycling is attributed to path planning of mixed road network.

18.2.3 Reducing Parking Volume within Community

The increase of the car parking land must be at the expense of reducing the other land, from the practice of low carbon community, reducing the internal parking of community (or do not set up the parking lot), setting the centralized parking lot in the peripheral is the commonly used methods. This way has significant advantage in the development of public transport, increasing the comfort and security and residential neighborhood communication. For example, there are only two parking spots at residential edge in Germany Vauban District. Walk or cycling is the main means of transportation. There are tramcars leading to the city centre on the Residential traffic main roads; it will take 15 min to reach the city centre. All the houses are layout along the bus route; it takes a few steps to arrive parking lot. "Commuter Formula" covering the range of 60 km radius provides a convenient travel for the community residents.

18.3 Constructing the Community Mode Guided by Non-motorized Vehicle

18.3.1 Constructing Multi-Function Mixed Community

Multifunctional mixed community sets business, entertainment, food and many kinds of function as one, perfects supporting the community living, can provide the mature supporting facilities and reasonable life distance, meets varied life need of residents; it makes the slow transit as the main transportation of the community interior, and reduces the long distanced travel from the source.

Sonoma settlements, located in the north of San Francisco, plan for living range of 5 min from the overall planning of new urbanism, ensure the residents can arrive stores and other supporting facility within five minutes walking distance, and enjoy the local, sustainable, fair trade products and services provided by the community. 60 % of the daily traffic can be solved without private cars. Parking need is also decrease.

18.3.2 Setting Appropriate Residential Scale

Large, adopting the close residential property management pattern, it is not convenient for the residents from home to the bus station, the narrowing of the residential scale, is benefit for people to abandon their travel by private cars. According to research, the closed community, scale in the 200 × 200 m, the area is in 4 ha or so is suitable for residents to choose to travel by walk and bike.

So that appropriate scale community can be seen as one or more live "cells", public transport station can be seen as "nuclear", "cells" layout forms has high adaptability on public transportation service, if we combine many kinds of travel targets for the residents such as going to work, going to school, entertainment, shopping and so on, design them in the same route, the residents can finish a variety of activities at the same time by one trip, to be able to reduce travel times and the distance of the residents.

18.3.3 Advocating the Pedestrians and Bike Oriented Development Mode

By "the public transport oriented development (TOD)" has many advantages, but it is more suitable for big cities with many population; for small and medium-sized community, pedestrian-oriented or bicycle-oriented development (POD, pedestrian-oriented development) (BOD, Bicycle oriented development) have more potential than public transportation as the direction of development. People oriented and the most suitable community for walk and cycling often has the following features:

A center: the community suitable for walk generally has a very strong recognized center, or a shopping area, or a main street, or a public space.

Density: settlements are compact and not the scattered, this makes people more close to the shop and working place, make public traffic be more cost-effective.

Mixed function: work place and home is not far apart, the school and workplace is also very near, and most residents can walk to go.

Public space: having enough public space for residents' assembly or recreation.

The network on pedestrian: network structure provides the most easy environment to walk, it mixes the natural form of residence, commerce, public buildings of the community, mixed walking district can greatly reduce travel times and length, especially car trip.

18.4 Encouraging Low Carbon Transport

18.4.1 Promoting the Use of Low Carbon Transport Tool

Using Low cost zero carbon personal transportation system, such as the using of zero energy consumption electric bicycles, The electric bicycle appearance is different from ordinary electric car sold on the market, it is more similar to bicycle, because it installs two pieces of lithium battery, which can run up to 120 km.

18.4.2 Improving the Green Transportation Supporting Measures of Transport

According to the residential density of road network, reserve bus land, construct land of clean energy vehicle facilities, such as LPG stations, transformer substation, charging field, charging pile, and so on. Planning wide distribution, public bicycles of convenient service, hire point electric car, and combined with public transportation site plan and construct bicycle site. These stop point also has the function to charge and change electricity for electric car. It will install a solar energy photovoltaic panel (Photovoltaic Solar Panels) of tree shape, transformed electrical energy will charge for electric bicycle in time. At the same time, it can also charge for spare lithium batteries stored in the dock point.

18.4.3 Building Friendly Traffic System of Walking and Cycling

Setting walking and bicycle travel system in the community, safe, continuous string priority to the premise should be the first step, second to consider the demand of pedestrian or riding and vegetation gardeners and landscape. Bicycle lanes of Shenzhen Overseas Chinese Town (oct) community can be divided into daily traffic routes and leisure fitness line. The total length of bicycle lanes on traffic routes is 16 km, divided into bilateral respectively driving and unilateral two-way driving, using the ground and mark of different color to distinguish.

18.4.4 Encouraging the Using and Restoring Public Transportation

Encouraging the using or restoring public transportation can effectively reduce the travel of the private cars. For example, an Diego California Housing Works community in the United States, contacts transportation departments and schools, restores the school bus service, there are 120 households benefit from the recovery of public school bus, effectively reduce traffic and corresponding expenses for parents ferrying their children.

18.5 Using Environmental Protection Fuel and Road Materials

18.5.1 Using the Technology of Warm Mix Asphalt

The so-called Warm Mix Asphalt technology (referred to as: WMA), is through the certain technical measures, make the asphalt can be mixed and constructed at relatively low temperatures, while maintaining its asphalt mixture construction technology not under FIMA using performance. With the warm mix asphalt technology, the road can save 20–30 % heating fuel when construction, can make the carbon dioxide emissions be reduced by 46 %, carbon monoxide reduces about two-thirds, sulfur dioxide reduces 40 %, nitrogen oxide gas reduces nearly 60 %, while toxic "asphalt smoke" when paving to produce, can reduce 80 %, largely protecting environment and construction technical personnel's physical health.

18.5.2 Use Environmental Protection Construction Materials

The increasing use of materials causes a lot of resource waste; we should use environmental protection green building materials in the construction of the road, reduce material energy consumption in the process of constructing, and enhance the recycling use and the use of recycled material ratio. At present, it basically has the following kinds of material saving technology, such as asphalt pavement recycling technology; Cement road surface reconstruction technique that increases laid; waste materials recovery for road using technology; the road consolidation technology with construction waste.

18.5.3 Using Environmental Protection Fuel

In order to alleviate the tension escalated of oil resource and pollution problems, it is an important topic to develop of cheap, clean alternative sources of energy for environmental protection. Samso Island in Denmark, on the one hand, introduces biofuel to gradually eliminate gasoline; on the one hand, set up hydrogen factories to manufacture hydrogen as substitute fuel for car.

18.6 Advocating Low Carbon Traffic

Residents participate in the planning design and developing process of the community is an important link of building low carbon community. The implement of "Public participation", can let people feel "belonging" and "tenderness" by the process of communication, conferring with planning design. In the case of absence of low carbon concept, there may not be conscious low carbon travel behavior, the cultivation of the residents low carbon transport style needs not only increase residents' knowledge of low carbon traffic, but also need to guide them set up low carbon travel view as soon as possible, through policy, education, publicity and many kinds of ways, to improve self-consciousness of the low carbon travel.

18.7 Conclusion

Reducing carbon emissions of city requires long-term efforts, the carbon emissions of city traffic is more than 30 % of the total emissions, resident land of city is more than 30 % of the total land. Advocating residential low carbon traffic is beneficiary for carbon abatement of the whole city, at the same time, it can provide quiet, health, and comfortable life environment for the residents. In the residential planning process, creating conditions for low carbon travel way is the future direction of the residential community planning.

Chapter 19
Discussion on Countermeasures of China's Low-Carbon Tourism Development

Xuefeng Wang and Hui Zhang

Abstract Low-carbon tourism is gradually emerging currently, which caused Extensive attention by the government, academia and society. However, study of low-carbon tourism is still in initial stage. On the basis of reviewing of research progress at home and abroad, this paper analyzes the necessity and realistic foundation of China's development of low-carbon tourism, and finally countermeasures are proposed to promote the development of low-carbon tourism in our country.

Keywords Low-carbon tourism · Connotation · Development of countermeasures

19.1 Introduction

The concept of Low-carbon is proposed in the context of responding to global climate change and advocating reducing the greenhouse gas emissions from living activities of human. As the proposed new concept of "low-carbon tourism", there is a growing concern about energy-saving and emission reduction, which related to the sustainable development of human society. As a kind of low power consumption and low pollution tourism, low-carbon tourism has become the focus of attention of the world tourism industry, and the future direction of tourism recognized.

X. Wang (✉) · H. Zhang
School of Economics and Management, Beijing Jiaotong University,
Beijing 100044, People's Republic of China
e-mail: Tour6@163.com

Low-carbon tourism, which has brought new opportunities for today's tourism style transformation, is becoming the trend of the future development of tourism. However, a related area of research has lagged far behind in the development of the practice, low-carbon tourism call for an urgent need for theoretical guidance. In this context, through comparison and analysis on domestic and international low-carbon tourism, and summarizing existing research experience, the author investigate the main research directions of low-carbon tourism in the coming period.

In order to seek beneficial inspiration for the domestic low-carbon tourism research and promote the progress of the industry as a whole, ultimately to achieve the strategic goals of sustainable tourism development.

19.2 Research of Low-Carbon Tourism in Domestic and Abroad

Low-carbon tourism-related research is carried out earlier and the literature of which is relatively abundant abroad. Moreover, the foreign study of low-carbon tourism is mainly concentrated in the determination of carbon emissions in the tourism sector and process of the tourism. For example, Becken and Simmons (2002) published a series of articles for depth study of the energy use of tourism from various angles, such as use patterns, energy use and modes of transportation (Becken and Simmons 2002; Susanne Becken 2002).

With the rapid development of transportation technology, there are more and more modes of transportation Available for tourists, and different way to travel has different carbon emissions. According to the World Tourism Organization statistics of carbon dioxide emissions in the tourist sector, the aircraft's carbon emissions by 40 % of the total, accounted for 32 % of car and other transport accounted for 3 %, carbon emissions from the transport sector accounted for 75 % of the total carbon emissions in the tourism sector, so the control of carbon emissions from the transport sector is the key to energy saving.

In the study of New Zealand Tourism traffic, Becken et al. (2003) found that the amount of energy consumed by the cross-border travel is four times the domestic tourism, so change the way of tourist travel can greatly reduce their energy demand and carbon emissions (Becken et al. 2003). Peeters et al. (2007) analyzed the environmental impacts of European tourist traffic—we should focus on reducing the environmental impact of tourism aviation and intercontinental travel in order to reduce the external costs of travel in Europe (Peeters et al. 2007). Measured the carbon emissions of the five national parks' tourist transport in Taiwan, Lin (2010) believes that government departments should take active management measures to reduce the carbon emissions of tourist transport effectively (Lin 2010).

Compared with the tourist traffic's carbon emissions, the tourist destination' structure is more complex, and its determination of carbon emissions is more difficult, therefore, fewer numbers of studies related to tourism destination in carbon emissions. The existing research focused on the determination of travel carbon emissions and the study of tourism destination's carbon footprint. Kelly and Williams (2007) constructed a conceptual framework and energy use model to determine energy consumption and emission levels of the tourism destination, which provides an effective method for the management and decision-making (Kelly and Williams 2007). Dick Sisman and his colleagues carried out system and in-depth research on the carbon footprint of tourism; they think that air travel is clearly the most important part of the tourism greenhouse gas emissions, as for the tourist destination, It is necessary to conduct a detailed review and give sufficient attention on hotel energy consumption and waste disposal (Dick Sisman & Associates 2007).

Domestic research literature on low-carbon travel is less than in foreign. Xiao (2009) first proposed the concept of low-carbon tourism, which should be the target of continuous development of tourism because it has a clear vision and objectives (Xiao 2009).Thereafter, from September 2009, due to the low-carbon concept continues to heat up, sharp increase in reports and research papers on low-carbon tourism, the contents of which are introducing visitors the concept of low-carbon tourism and related knowledge, as well as propagandizing the construction of scenic in low-carbon, in order to guide the tourists to choose a low-carbon tourism. Mature cases are rare because of the low-carbon tourism is still being explored. Huang (2009a, b) described the case of Pinglin in Taipei's and Bama in Sichuan's low-carbon tourism development, and lessons are drawn from the successful experience and proposals for low-carbon tourism development (Huang 2009a, b). Cai and Wang (2010) discussed the low-carbon tourism connotation and proposed the main path of achieving low-carbon tourism on analyzing the elements in the process of tourism development, such as tourist attraction, tourist facilities, tourist experience environment and tourism consumption etc. (Cai and Wang 2010). For the actual situation in the suburb of Beijing, Liu (2010) proposed a corresponding low-carbon tourism development ideal model.

Overall, it can be seen that our stage of low-carbon tourism study is still in its infancy, remain in the level of qualitative analysis through the comparison of domestic and international research. The total amount of relevant literature is rare, and news reports accounted for the majority. In the course of the study, many of them carried out a comprehensive analysis of low-carbon travel in a macro perspective and their research focused mainly on the low carbon way of tourism. Thus, the perspectives of these studies are relatively homogeneous and lack of quantitative research. Although many scholars have carried out positive and useful explorations, and achieved some achievements, but there are many fields remains to be in-depth study and discussion.

19.3 China's Low-Carbon Tourism Development Status and Issues

19.3.1 The Meaning and Characteristics of Low-Carbon Tourism

As the name suggests, Low-carbon tourism is a kind of tourism to reduce the "carbon", that is, tourists try to reduce carbon dioxide emissions in their travel activities. Based on low energy consumption and low-pollution, green travel advocated to minimize the carbon footprint and carbon dioxide emissions in the travel, which is the deep-level performance of eco-tourism. The contents of low-carbon tourism Include environmentally friendly low-carbon policy and low carbon tourist routes launched by the government and travel agencies, environmental baggage carried in personal mobility, live in environmental hotel, choose lower carbon dioxide emissions of transport or a bicycle and on foot.

For travelers, low-carbon tourism is emerging forms of tourism, the tourism industry, is a new management philosophy. The essence of low-carbon tourism is a new type of tourist consumption patterns of low-carbon emission intensity, which performance for low power, low emissions, low pollution, high-grade, high experience and high responsibility. Low-carbon travel applies to all classes of travel activities, including the three key points: First, to change existing travel patterns, and promote public transport and hybrid vehicles, electric cars, bicycles and other low-carbon or carbon-free way, but also rich tourist life, increase tourism projects. Second, to reverse the culture of extravagance and wastefulness, and strengthen the clean, convenient, comfortable and functional, enhance the culture of the brand. The third is to strengthen the tourism development of intelligent, Improve operational efficiency, timely and comprehensive introduction of energy saving technology to reduce carbon consumption, and ultimately form the whole industry chain of circular economy mode.

19.3.2 China's Policy on Development of Low-Carbon Tourism

In 2009, China has committed that the emissions of carbon dioxide per unit of GDP in 2020 decreased by 40–45 % of 2005, so there is huge pressure to reduce emissions in the future. Low-carbon in tourism industry will be an important component of China's emission reduction targets. In conformity with the responsibility of the tourism industry on global climate change, Chinese tourism industry must implement the requirements of "Accelerate the changes of economic growth pattern" that put forward by General Secretary Hu jintao. We should Clear the low carbon responsibility of tourism industry, catch the good development opportunities of low carbon tourism, reduce carbon emissions of tourism and energy consumption, So as to promote the realization of the emissions reduction commitment in our country. We will realize the goals of transformation of the tourism industry's development way and the world's leading tourist destinations in the construction, eventually promote the whole society of common prosperity and sustainable development.

At the government level, in December 2009, in the "Views of the State Council on Accelerating the development of tourism" It clearly pointed out the need to "promote energy conservation and environmental protection, the implementation of the tourism saving energy and water reduction projects", which is the first official energy saving requirements for tourist hotels and scenic spots, and provided policy support for the development of China's low-carbon tourism (State Council 2009). Compared to other industries, tourism is known as a "low-pollution industry" reputation, it belongs to the service sector, characteristics of small footprint and selling environment and culture, which is precisely consistent with the emissions-reduction targets.

19.3.3 China's Practical Basis of the Low-Carbon Tourism Development

Although the low-carbon tourism is a new thing in the world and its concept is just understood by the public, but on a practical level, Chinese folk of low-carbon tourism have long been. Development of green tourism resources, construction of the green tourism products, green tourism operators, the implementation of the Green Tourism management and cultivate a green tourism consumption have become the consensus of the industry and market. As early as in 1997, Ping lin scenic spots in Taiwan, which was created as the first low-carbon tourism demonstration area, tried a lot of new efforts in transportation, consulting, tourism, tourist behavior, etc. Mount Emei Scenic Area, as the old "low-carbon scenic spot", is the pioneer of low-carbon tourism practices. By strengthening the digitization project of the Mount Emei scenic monitoring the area of air and water quality, vegetation, to achieve a co-coordinated development of scenic spots, transportation, hotels, dining, entertainment, travel agencies. Moreover, Years ago motor vehicles were prohibited from entering a lot of tourist attractions like Jiuzhaigou, Yuntaishan, Zhangjiajie scenic spots, Hybrid buses and battery cars were used for scenic transportation to reduce carbon emissions. At the beginning of the year 2010, Yanzi Gou scenic area in western part of Si Chuan first introduced the new concept of the "low-carbon travel", in order to reduce greenhouse gas emissions, reduce the burden of virgin forests to absorb carbon emissions, and strive to build the first low-carbon tourism area of western China. All these successful low-carbon tourism practice provided people with the further development of low-carbon tourism experience (Liu 2011).

19.3.4 Analysis of the Problems of Low-Carbon Tourism Development in China

There are still faced with many problems of low-carbon economy's further development. First of all, many tourists do not understand the low-carbon tourism, once mention "low-carbon tourism", the majority of people in the industry are at a loss. Secondly, in the tourism production enterprises, the higher cost of low-carbon technologies will inevitably lead to many companies reluctant to adopt low-carbon development model; third, low-carbon tourism products will also increase tourists' travel expenses, so the travel agency designed "low-carbon tourism" products are difficult to promote. For example, it is undoubtedly more low-carbon travel that if the tour group to give up taking the bus transfer to the battery car in the scenic, However, the cost of electric vehicles will be apportioned to the tourist, which is bound to be resisted by tourists; In addition, many travelers' luxurious habits are difficult to change, and for economic benefits, some tourism enterprises will cater to the preferences of these visitors.

19.4 Countermeasures of China's Low-Carbon Tourism Development

19.4.1 Governments: To Strengthen the Propaganda of the Low-Carbon Tourism

Government departments are the initiation power to promote the low-carbon economy and low carbon travel. By formulating policies and regulations, government plays an exemplary role in guiding and promoting tourism enterprises to actively introduce and use low-carbon technologies. For example, combined with the region's own resources characteristics, tourism enterprises provide high-quality low-carbon tourism experience for visitors; at the same time, the government departments' propagandizing tourism enterprises' low-carbon products can also promote understanding of the tourists on the concept of low-carbon, thereby increasing the possibility of receiving low-carbon tourism products and the subjective will of them. In turn, the change of the tourists' chooses can also stimulate and promote enthusiasm for tourism enterprises to launch low-carbon products, thus forming a decisive force (Chen and Mo 2011). If low-carbon tourism to be really implemented, It must be timely introduction of low-carbon tourism industry policies and standards by government, which play constraints for all tourism participants, regulate, guide, adjust, and the role of incentives to develop low-carbon travel incentives and a new industry standard.

19.4.2 Enterprise: To Encourage the Development of Low-Carbon Tourism Products

Tourists decided to travel supply, tourists' tends to low-carbon tourism consumption will ultimately lead to a comprehensive design and development of low-carbon travel experience products for tourism enterprises, which contributing to the full realization of the tourism low-carbon.

First of all, with the low-carbon awareness in-depth public mind, the travel agencies should develop and design new low-carbon travel routes, and introduce the concept of carbon consumption in tourists' food, housing, transportation, tours and other elements; to increase low-carbon tourism project design; for example, you can organize volunteer trips, tree planting and travel. Secondly, to construct low-carbon hotel and carbon offsets to these hotels. Construction of the Low-carbon hotel can be divided into the hotel itself, building materials used in low-carbon and low-carbon services (green hotel is one) in the management process, that is the so-called low-carbon of hardware and low-carbon of software. Last but not least, the tourist attractions should adhere to a low-carbon tourism development, as far as possible construct scenic spots with original ecological materials and introduces intelligent devices to improve operational efficiency, reduce energy consumption, in order to form the circular economy model of the whole industry chain.

19.4.3 Tourists: To Promote Low-Carbon Tourism Consumption

Tourists are one of the most important tourist participants, and whose travel wishes, consumer preferences, travel behaviors play important roles in low-carbon tourism. Low-carbon tourism consumption style refers to the tourists to reduce personal travel carbon footprint through a variety of ways and means in the process of tourism consumption. In the same course of traveling, different tourism consumption style will lead to significant differences of personal travel carbon footprint. Low-carbon tourism has attracted the participation of a number of travel enthusiasts began to personally choose it, such as the use of public transport; when they go out as much as possible to take a carpool; take more walking and cycling in tourist destination. There are a variety of methods to realize the low-carbon development in air traffic, such as cuts in hand luggage, reducing the allowance on the plane; Encouraging airlines to provide free bus or rail rewards mileage, instead of free frequent flyer miles, In order to facilitate passenger choose more environmentally friendly public transport. So, through the public efforts to achieve traffic carbon compensation.

19.4.4 Tourist Destinations: Construction of Low-Carbon Tourism Demonstration Area

Relies on low-carbon tourism city and consists of low-carbon tourism area, Low-carbon tourism destination is a kind of tourist destination which fully implement the concept of low-carbon tourism. The development of a tourist destination can not be separated from the tourist city construction, low-carbon tourism destination should be fully implemented the concept of low-carbon from all parts of the planning and construction, the development strategy, market operations and external marketing.

Build low-carbon tourism destination must strictly follow the circular economy and sustainable consumption principles from planning to production. Implementation of "clean production" for tourism products, and "green marketing" for Tourist destination will contribute to create a "low-carbon tourism image" in the entire tourist destination. In particular, tourism community, tourist attractions and tourist city can adopt different measures to promote low-carbon travel. Promoting low-carbon tourism and priority constructing low-carbon tourism development demonstration areas in a number of conditions are suitable areas are strategic choices to achieve tourism development mode transformation (Dong and Yang 2011).

19.5 Conclusions

With the low-carbon technologies become more sophisticated in the field of socio-economic penetration, And low-carbon lifestyle is widely advocated in the whole society, Low-carbon tourism as a new tourism development, will demonstrate its powerful demonstration effect and guiding values. Low-carbon tourism and low-carbon economy is of the same strain, which emphasis minimizing carbon emissions and reducing their carbon footprint in the tourism process. The core of the low-carbon tourism is to reduce carbon dioxide emissions, which leads a harmonious relation between tourism development and natural environment. On the basis of the protection of the natural environment, low-carbon tourism is intended to ensure the sustainable development of health tourism. Therefore, developing low-carbon travel and building a low-carbon tourism destination is a transformation strategy for tourism development mode.

References

Becken S (2002) Analysing international tourist flows to estimate energy use associated with air travel[J].J Sustain Tour 10(2):114–131

Becken S, Simmons DG (2002) Understanding energy consumption patterns of tourist attractions and activities in New Zealand[J]. Tour Manag 23(4):343–354

Becken S, Simmons DG, Frampton C (2003) Energy use associated with different travel choices[J]. Tour Manag 24(3):267–277

Chen H, Mo L 2011 a study on low carbon tourism' concept, features and force mechanism[J]. J Beijing Int Stud Univ 7:34–36

Dick Sisman & Associates (2007) Tourism destinations carbon footprints[R], vol 3. Cambridge University Press, Cambridge

Dong Y, Yang X (2011) Domestic and international low-carbon tourism research[J]. J Southwest Agric Univ Soc Sci Ed 12:5–11

Huang W (2009a) Study on implementation of low-carbon tourism patterns for Bama tourism[J]. Fortune Today 9(10):104–105

Huang W (2009b) On the low-carbon tourism and the creation of low carbon tourist attractions[J]. Ecol Econ 24 (11):100–102

Kelly J, Williams PW (2007) Modelling tourism destination energy consumption and greenhouse Gas emissions: Whistler, British Columbia, Canada[J]. J Sustain Tour 15(1):67–90

Lin T-P (2010) Carbon dioxide emissions from transport in Taiwans national parks[J]. Tour Manag 31(2):285–290

Liu X (2009) Discuss on low-carbon economy and low carbon tourism[J]. China Collect Econ 16(13):81–82

Liu X (2010) Low carbon tour—a future rural tourism model of Beijing[J]. Social Science of Beijing 101(1):42–46

Liu X (2011) Low-carbon tourism and its development[J]. Commer Res 53(2):175–179

Meng C, Wang Y-M (2010) Low-carbon tourism: a new mode of tourism development[J]. Tour Tribune 25(1):13–17

Peeters P, Szimba E, Duijnisveld M (2007) Major environmental impacts of European tourist transport[J]. J Transp Geogr 15(2):83–93

State Council (2009) The views of the State Council on Accelerating the development of tourism[Z]. Natl Dev 41:1–4

Chapter 20
A Study of Vehicle Tax Policy Adjustment Based on System Dynamics in the Background of Low-Carbon Transport

Feifei Xie and Xuemei Li

Abstract Vehicle exhaust emission is one of the main factors which cause global warming. Low-carbon transport should be developed without delay to mitigate global warming. This paper discusses the effect of tax policy adjustment on private vehicle ownership, that is, how tax policy adjustment guide private car consumption and uses, as a result to reduce CO_2 emission and promote the development of low-carbon transport. In the paper with system dynamics model, the extent of tax policy adjustment's effect on private vehicle ownership will be simulated. Finally the important role which vehicle tax policy adjustment plays in the achieving low-carbon transport will be found.

Keywords Low-carbon transport · Tax reform · System dynamic

20.1 Introduction

In order to promote the development of low-carbon transport and reduce the CO_2 emission in China, controlling private travel size and reducing car emissions intensity should play their respective roles. From international experience, both in controlling private travel size as well as on reducing private vehicle emissions intensity, tax-fee pricing policies have a significant impact. Therefore this paper focuses on tax policy reforms' effects on private car ownership, that is, study the effect of adjustment of tax and fee on controlling private travel size and reducing private vehicle emissions intensity. Thus tax and fee adjustment of vehicle's impact on achieving low-carbon transport can be explained.

F. Xie (✉) · X. Li
School of Economics and Management, Beijing Jiaotong University, Beijing 100044, China
e-mail: skyingxx@163.com

Due to vehicle tax adjustment is placed in the low-carbon transport system in this paper, and the relationships between variables of urban low-carbon transport are complex and dynamics, general quantitative research methods are not suitable. System Dynamics can not only solve the dynamic, multiple circuits and nonlinear problem, but also find and study the impact factors of relevant extent in the system proceeding from the overall, and solve non-stability system problem under the incomplete information condition (Liu 2010). This paper will apply system dynamics theory and method to study on the private vehicle consumption system in Beijing and simulate the future development trend of private vehicle ownership based on historical data and planning data. On this basis, we will explore the inside influence and effect mechanism of the reform of taxes and fees on the development of private cars in Beijing, provide a decision-making basis for guiding private car consumption to achieve low carbon transport.

20.2 Vehicle Tax Adjustment Model Based on System Dynamics

Here we study the tax adjustment's influence on private vehicle consumption in Beijing. Private vehicle consumption is not only affected by tax adjustments, but by the combined effect of demographic, economic, environmental policy the consumer environment and many other factors as well. In this article after analyzing system structure and all kinds of factors' cause-feedback relationship, we develop a system dynamics model on a private car consumption system in Beijing, and feedback relationships between factors shows in Fig. 20.1. Arrows in the figure represent causal relationships, and positive and negative signs refer to the positive effects or negative effects.

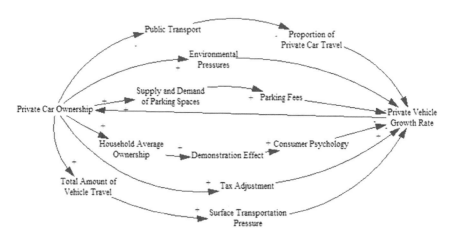

Fig. 20.1 System feedback diagram

20.2.1 Analysis of Causal Feedback

System Causal feedback loops is an effective means of system dynamics approach for constructing system model, and it uses feedback loops to describe dependency relationships among the various factors (Xing et al. 2000).

Taking into account the effect of the adjustment of tax on private car ownership we get loop I: private car ownership → (+) tax adjustment → (−) proportion of private car travel → (−) private vehicle growth rate → (+) private car ownership. The loop is negative feedback. Considering that the government use adjusting the parking cost as an economy means we can get loops II: private car ownership → (−) supply and demand of parking spaces → (+) parking fees → (−) private vehicle growth rate → (+) private car ownership. The loop is negative feedback. Considering ground transport pressure we can set up loops III: private car ownership → (+) total amount of vehicle travel → (+) surface transportation pressure → (−) private vehicle growth rate → (+) private car ownership. The cumulative effect of the loop is negative. Considering the impact of consumer psychological factors on private car ownership, loop IV is established: private car ownership → (+) household average ownership → (+) demonstration effect → (+) consumer psychology → (+) private vehicle growth rate → (+) private car ownership. The loop feedback effect is positive. Considering the effect of environmental factors on private consumption, loop V is established: private car ownership → (+) total amount of vehicle travel → (+) environmental pressures → (−) private vehicle growth rate → (+) private car ownership. The loop feedback is negative. Considering the impact of urban public transport on the development of private cars, we can get loops VI: urban public transport → (−) proportion of private car travel → (+) private vehicle growth rate → (+) private car ownership. The loop feedback also is negative.

20.2.2 System Flow Chart

System flow chart is the organic combination of basic variables and symbols in system dynamics. According to above qualitative analysis, we will establish the SD equation to present the system structure and its feedback mechanism with the DYNAMO language. DYNAMO language uses L, R, A as a description of the type, respectively identifying the state variables, rate variables and auxiliary variables (Zhang and Yu 2010). DYNAMO is split time through system simulation, so each variable has a time subscript and J, K, L represents the past, present and future point in time, JK, KL, said in the past and future sessions.

The system flow diagram is involving many variables and complex relationships with each other, including six loops in Fig. 20.2. Taking account of the per capita disposable income, fuel fees, vehicle purchase price, tax rate and other external economic factors' impact on the demand for private cars, we select the

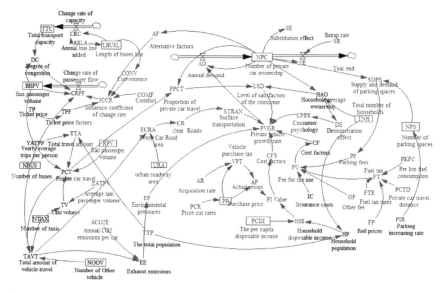

Fig. 20.2 Private vehicle consumption system flow diagram

value of PI(motor vehicle sales price/annual income of households) as Measurable criteria. Feedback loop I contains the main equation as follow one.

$$A\ FU.K = IC.K + PF.K + OF.K + FT.K. \tag{20.1}$$

$$A\ FT.K = FTR.K * PCTD.K * PKFC.K * FP.K. \tag{20.2}$$

$$L\ PP.K = PP.J * (1 - PCR). \tag{20.3}$$

$$A\ AP.K = PP.K + PP.K * VPT. \tag{20.4}$$

$$A\ PI.K = AP.K / HSI.K. \tag{20.5}$$

Control the number of parking spaces and parking fees is one of the important regulation of economic instruments for the consumption of private cars in Beijing, main equation which consists loop II is given below.

$$L\ NPS.K = NPS.J + PIR * NPS.J. \tag{20.6}$$

$$A\ SDPS.K = NPC.K / NPS.K. \tag{20.7}$$

$$A\ PF.K = 5500 + 2000/SDPS.K. \tag{20.8}$$

The private vehicle growth rate is mainly decided by economic factors, policy factors, road traffic, rail traffic, use factors, consumer psychology and consumer satisfaction. DT calculates interval time from J to K moment.

Feedback loop III contains main equation as follows.

$$L\ NPC.K = NPC.J + DT * (AD.JK - YE.JK). \qquad (20.9)$$

$$R\ AD.KL = NPC.K * PVGR.K. \qquad (20.10)$$

$$R\ YE.KL = NPC.K * SR. \qquad (20.11)$$

$$A\ PVGR.K = CF.K * CFS.K * CPSY.K * EEP.K * LSD.K * STRAN.K. \qquad (20.12)$$

$$A\ ECRA.K = URA.K / TAVT. \qquad (20.13)$$

Among above equation, the demand growth rate of private cars is decided by fee factors, cost factors, consumer psychology, environmental pressures, consumer satisfaction and ground transportation.

Loop VI displays psychological factors' effect on the consumption of private cars, involving following main equations.

$$A\ HAO.K = NPC.K / TNH. \qquad (20.14)$$

In addition, environmental factors in the environmental pressures module of the private car consumption SD model is reflected by the motor vehicle exhaust emissions. Because of investigation of low carbon transportation system, emissions of pollutants hypothesis CO_2, its column has an effect on environmental protection policy, and the main equation is given below.

$$TTA.K = YATPP.K * TTP.K. \qquad (20.15)$$

$$A\ PCT.K = TTA.K - TV.K - BPV.K - RPV.K. \qquad (20.16)$$

$$A\ TAVT.K = PCT.K + NBUS.K + NTEXI.K + NOOV.K. \qquad (20.17)$$

$$A\ EE.K = TAVT.K * ACEPC. \qquad (20.18)$$

Among above, the buses and taxis are operational nature, travel rate of simple computing is set to 1, bus and taxi travel amount equal to the ownership. With increasing number of bus passenger, public transportation is more and more crowded. As a result, some passengers will chose other modes of transport such as by car. Here bus transport capacity can be partially determined by the length of bus lines which positively correlated with the bus coverage which has a direct impact on the convenience of the public to take the bus. We establish the dynamic equations for loop VI based on historical data as follows.

$$L\ BPV.K = BPV.J + CRPF.JK * DT. \qquad (20.19)$$

$$L\ TTC.K = TTC.J + CRC.JK * DT. \qquad (20.20)$$

$$L\ LBUSL.K = LBUSL.J + LBUSL.J * ABLAR.K. \qquad (20.21)$$

$$\text{R CRPF.KL} = \text{BPV.K} * \text{ICCR.K} * (1 - \text{DC.K}). \tag{20.22}$$

$$\text{R CRC.K} = \text{ABLAR.K}. \tag{20.23}$$

$$\text{A DC.K} = \text{BPV.K} / \text{TTC.K}. \tag{20.24}$$

$$\text{A ICCR.K} = \text{TPF} * \text{COMF} * \text{AF} * \text{CONV.K}. \tag{20.25}$$

$$\text{A TPF.K} = \ln(\text{TP.K})/\ln(\text{TP0.K}). \tag{20.26}$$

$$\text{A PPCT.K} = \text{TAPCT.K}/\text{NPC.K}. \tag{20.27}$$

20.3 System Simulation

In this paper, system dynamics software Vensim PLE is used to simulate the system and study the effect of the taxes adjustment on Beijing's future consumption of private cars. We set system simulation time from 2003 to 2023 and simulation step for 1 year. The initial values of state variables in the model are determined in accordance with the relevant data of the Beijing Statistical Yearbook and the Beijing traffic Yearbook published. Here, it is assumed that the external environment is not much shock. The value which is on behalf of the rate of change is determined through parameter fitting and regression analysis based on the obtained historical and statistical data. With the help of Eviews, one of the most popular statistical software, we will calculate the growth rate growth rate parameters which show in Table 20.1.

In order to test whether the model can better reflect the true situation, five indicators including private car ownership, number of bus, rail passenger and urban roadway area are selected. Compared between simulation data in 2006 and real data to test the goodness of fit, the results show in Table 20.2. All these errors are less than 5 %, indicating that the analog system and the actual behavior of the system closer, so the model is effective.

Table 20.1 Model parameters

Parameters (growth rate)	Value (%)	Parameters	Value
Urban roadway area	3.7	Fuel tax	1 Yuan/liter
Population	4.7	Average trips per person	2.76 Time/person
The per capita disposable income	6.2	Insurance costs	950 Yuan/vehicle
Total number of households	2.1	Acquisition rate	10 %
Number of parking spaces	9.1	Scrap rate	6.7 %
Bus	3.6		
Taxi	0.9		
Other vehicle	1.2		
Length of bus line	2.2		

Table 20.2 Test of model results

Item	Actual value	Analog value	Absolute error (%)
Private car ownership	3002748	303.718	1.15
Number of bus	21716	21178.7	2.47
Rail passenger	142268	138284	2.80
Urban roadway area	9179	9133.32	4.98

(1) The effect of acquisition tax on private car ownership

Vehicle acquisition tax is an important economic lever to guide private car payment. Figure 20.3 shows the sensitivity analysis of the effect of the acquisition tax on the ownership of private cars. Due to the tax burden is quite heavy in the stage of purchase in China, acquisition tax is not suitable for blindly increasing (Zhang and Liu 2009). Here, only a theoretical analysis of the impact of the acquisition tax adjustment on private car ownership. In fact, the space in increasing acquisition rate is very small, the AR1 here on behalf of the purchase tax rate of 10 %, AR2 on behalf of the purchase tax rate of 12 % and AR3 on behalf of the purchase tax rate of 14 %. It can be seen from the figure, with the increase of the acquisition tax rates, the growth of private car ownership will be suppressed.

(2) The effect of fuel tax on private car ownership

Fuel tax was introduced in early 2009, it adopted quantitative collection. At the same time car tolls and other five charges have been cancelled (Liu 2007). But there is still a lot of room to improve the adjustment function of fuel tax. Figure 20.4 shows the sensitivity analysis of the effect of the fuel tax on the ownership of private cars. FTP1, FTP2 and FTP3 respectively represent the initial state of charge 1, 2 and 3 Yuan fuel tax per liter of petrol. We can see from the figure that with the fuel tax increasing, the growth of private car ownership has been inhibited.

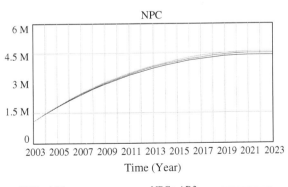

Fig. 20.3 Private vehicle consumption system

Fig. 20.4 The effect of fuel tax on private car flow diagram ownership

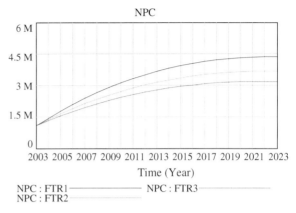

20.4 Conclusions and Recommendations

In summary, the implementation of tax policy adjustments can promote to effectively control the ownership of private cars, make private car consumption achieve a healthy and orderly development. Reasonable tax policy reform, through the reduction of private car travel behavior and control the consumption of private cars, can fully play a function of guiding. Tax and fee reform will guide rational decision-making on the private car consumption. As a result, it reaches its purpose to control the number of road passengers, alleviate traffic congestion, reduce greenhouse gas pollution, and realize the ultimate goal to achieve low-carbon transport.

In this paper, with the help of system dynamics modeling, model of private cars' consumers in transport system is built, and purchase tax and fuel tax adjustment is simulated, then the simulation results strongly suggest that the reform of the tax and fee can promote rationalization of the consumption of private car. However, due to the limit study level of the author, the equation involved in the simulation system is not scientific enough, and further research needs to be rationalization.

In short, considering from the perspective of regulating automobile consumption and energy savings, the relevant departments can try to further improve the ratio of the fuel tax and reduce the toll fees, both to ensure that the local fiscal revenue, and can more effectively reflect the guiding role of the tax on automobile consumption. Automobile tax system should be introduced to more of the criteria in China. For example, it is irrational to decide the consumption tax rate only by displacement, because there are many factors that affect vehicle fuel consumption and emissions. Although there are quite a lot of difficulties in exactly considering these factors, at present it is unfair to develop the related tax policy only by displacement. For the travel tax, in order to play its regulation of automobile consumption's role, the introduction of the existing international tax system may have reference value. So the taxes and fees can be charged by the engine power, cylinder capacity and fuel consumption to greenhouse gas emissions as representative to calculate. In Austria, the travel tax is based on engine power and press the annual payment (Kloessn and Müller 2011). Furthermore, the frequency of use of the vehicle can also be considered, that is, considering the mileage to determine the payable amount. However, these adjustments will greatly increase the cost of collection, so a number of integrated factors are needed to comprehensive study. At the same time, the government can provide the preferential tax policies for the advanced automotive technology thus maximize the function of the tax and fee adjustments as economic levers to achieve low-carbon transport.

Acknowledgments This paper was supported in part by "National Natural Science Foundation Projects (71273023)" and "Railway Ministry Projects (2010Z012)".

References

Liu L (2010) Metropolregion: transport influencing mechanism research based on SD model, techniques and method, pp 41–45

Xing Z, Chen S, Xie J (2000) System dynamics model and its application, South China University of technology (Natural Science), pp 130–133

Zhang B, Yu Z (ed) (2010) Introduction to system dynamics and review of related software. Environment and Sustainable Development, pp 1–3

Zhang X, Liu S (2009) Tax management practices and improvement of the car sales industry [J]. Hubei State tax 11–13

Liu X (2007) Fuel tax reform pros and cons of panorama [J]. Mod Bus 16–17

Kloessn M, Müller A (2011) Simulating the impact of policy, energy prices and technological progress on the passenger car fleet in Austria—a model based analysis 2010–2050. Energy Policy 39:5045–5062

Chapter 21
Economic Evaluation of Energy Saving and Emission Reduction for ETC

Jia-hua Gan, Xiao-ming Zhang and Ze-bin Huang

Abstract We make economic evaluation for ETC energy saving and emission reduction, Firstly, we analysis the mechanism of ETC energy saving and emission reduction. Secondly, we built an index system to evaluate the effect of ETC energy saving and emission reduction. Finally, we verify the feasibility of the evaluation indexes and evaluation methods proposed in this paper through studying atoll station which using ETC.

Keywords: ETC · Energy saving · Economic evaluation · Cost-benefit analysis

21.1 Introduction

ETC can achieve automatic toll collection through e-cards or electronic tags, and it can make all ground transportation achieve automatic toll collection not need parking. Up to March 2012, 26 cities of China have been included in pilot cities of low-carbon transport system. One of major tasks is to construct ETC project, while how to assess its effectiveness and to provide support for future decision makes great difference.

J. Gan · X. Zhang (✉) · Z. Huang
Institute of Regional and Urban Transportation Economics, Chang' An University,
Middle Section of South Second Ring Road, Xi' An 710064 Shaanxi, China
e-mail: 410245855@qq.com

21.2 The Mechanism of ETC Energy Saving

There are two main charge methods for highway in China, MTC & ETC. MTC requires the vehicle to complete deceleration, idle queue, payment, and accelerating, but vehicles using ETC lanes do not need to slow down and stop, and complete the automatic payment at a faster speed, thereby reducing emissions to achieve energy saving and emission reduction (Cai 2005). According to domestic study, ordinary vehicles which using ETC lanes will emission less NO_x by 16.4 %, HC by 71.2 %, CO by 71.3 %, CO_2 by 48.9 % than using MTC lanes (Gao et al. 2009).

21.3 Evaluation Index System

According to principle of scientific, objective, operability and contradistinctive, we established evaluation model.

The input of ETC consists of 3 main parts, which are construction funds, operation and maintenance input, and user's equipment investment. According to the analysis of the mechanism of ETC in energy saving and emission reduction, the main consideration time saving, energy saving and tail gas emission output.

Based on the above analysis, the establishment of ETC on energy saving and emission reduction economic evaluation index system, shown in the Table 21.1.

21.4 Evaluation Method and Calculation

This paper uses the cost benefit analysis method to evaluate energy conservation and emissions reduction on ETC, with the net present value approach to discrimination (Zavergiu 1996).

Table 21.1 The ETC and MTC average emission factors

First class index	Second class index	Third class index
Input index	Construction fund expenses	Development costs
		Hardware cost
		Software costs
		Installation fee
	Operation maintenance cost	Staff salary expenses
		Equipment maintenance cost
	The user equipment cost	Hand-held devices cost
Output index	Time saving benefit	Save the goods in transit time benefit
		Save passenger travel time benefit
	Save energy efficiency	Fuel consumption saving benefit
	Gas emission reduction benefits	Reducing NO_x, HC, CO, reduce CO_2 emissions benefits

Investment cost computation formula is as follows:

$$C_t = C_{at} + C_{bt} + C_{ct}. \qquad (21.1)$$

In formula: C_t—the first t years total cost of ETC, Units: yuan; C_{at}—the first t years construction investment cost, Units: yuan; C_{bt}—the first t years operation maintenance into total cost, Units: yuan; C_{ct}—the first t years users equipment investment total cost, Units: yuan.

Passengers save time output can be measured by GDP, computation formula is as follows:

$$B_{Ct} = \sum_{j=1}^{3} Q_{Cj} W_j I_c \frac{TS}{3600} P \qquad (21.2)$$

In formula: B_{ct}—Passengers traveling time output value of saving, Units: yuan/h; Q_{cj}—The first j kind of bus volume of traffic in per hour, Units: car/h; W_j—The first j kind of bus take Passenger in average, Units: people/vehicles; I_c—Each passenger every hour of national income per capita, Units: yuan/people; P—transit passenger who can create value accounts for a percentage. Units: %; TS—After ETC implementation unit vehicle average save time, Units: s.

Save energy consumption with the economic output is due to the speed of travel, the number of parking and parking time reduced the energy consumption bring the economic benefit.

$$B_{Ot} = QFP_y \qquad (21.3)$$

In formula: B_{ot}—Energy consumption saving of the benefits, yuan; Q—the traffic number witch Use ETC, car; F—Unit of energy saving is an average vehicle, L/car; P_y—Energy consumption saving of the benefits, yuan/h.

Exhaust emission reduction efficiency computation formula is as follows:

$$B_{Qt} = \sum_{i=1}^{4} [Q \cdot (E_1 - E_2) \cdot H] d_i \qquad (21.4)$$

In formula: B_{Qt}—The total profit of greenhouse gas emissions, yuan; Q—Traffic flow, Car/unit time; E1—The greenhouse gases average emission factor before ETC implement, g/km; E2—The greenhouse gases average emission factor after ETC implement, g/km; H—The influence of length by ETC, km; d_i—The first i kind of greenhouse gas average prices, yuan/t.

21.5 Analysis of Cases

There is a Superhighway Toll Station located in the uncrossed position between Xian yang Airport Expressway and the 3rd Ring Road. It leads to the airport and the adjacent provinces and has made 40,000 cars driving daily, as a part of main superhighway in China, no-waiting toll collection system was brought in 2011.

21.5.1 Input Costs Analysis

Construction cost of an ETC lane is about 500,000. The wage costs, equipment replacement cost and software upgrades cost account for 20 % of the project construction cost. The overhaul costs account for 25 % of the investment in fixed assets, which is 250,000. The social discount rate is 8 %. If there is not construction cost, the conversion coefficient of other costs is 1.0. Moreover, it is assumed that the residual value of the no-waiting toll collection system is 0.

21.5.2 Output-Benefit Analysis

The future traffic volume of the toll station and ETC lanes are shown in Table 21.2 (Zhang et al. 2001).

21.5.2.1 Efficiency Savings

We can assume that the average volume of small passenger cars is 2.69 people/vehicles, and the average volume of bus is 21.38 people/vehicles (Jun 2008). Time—saving benefit can be estimated based on the above data and formulas 3.2.

21.5.2.2 Energy Saving Benefits

According to general vehicle fuel consumption, every 3 min idle oil consumption equal to the oil consumption of 1 km uniform speed, and the idle oil consumption of every second is 0.0004 L (Xiang 2001). Then we can get the fuel consumption in Table 21.3 (Zhou 2011).

Table 21.2 Toll station traffic and ETC number of transactions in future

Year	Veh/day			Veh/H					Rate (%)
				Entrance		Exit		Total	
	Entrance	Exit	Total	Car	Bus	Car	Bus		
2012	19,700	20,400	40,100	860	300	880	310	2,350	5.9
2013	21,276	22,032	43,308	1,050	366	1,074	378	2,868	6.6
2014	22,978	23,795	46,773	1,281	447	1,311	462	3,500	7.5
2015	24,816	25,698	50,514	1,563	545	1,600	563	4,271	8.5
2016	26,802	27,754	54,556	1,908	665	1,952	688	5,213	9.6
2017	28,946	29,974	58,920	2,328	812	2,382	839	6,362	10.8
2018	31,261	32,372	63,634	2,841	991	2,907	1,024	7,764	12.2
2019	33,762	34,962	68,724	3,467	1,210	3,548	1,250	9,475	13.8
2020	36,463	37,759	74,222	4,232	1,476	4,330	1,525	11,563	15.6
2021	39,380	40,780	80,160	5,164	1,802	5,284	1,862	14,112	17.6

21 Economic Evaluation of Energy Saving and Emission Reduction for ETC

Average price of gasoline in 2012 is 7.74 yuan per liter and according to formula (3), it can be calculated energy saving efficiency in the future years, as shown in Table 21.4.

21.5.2.3 Emission Reduction Benefits

Emission of NO_X, HC, CO, CO_2 are as shown in Table 21.5.

According to formula 4.4, an entrance ETC lane can reduce CO_2 emission 26.07 tons, an export ETC lane can reduce CO_2 emission 26.9 tons, and the trading price of CO_2 emission takes 8 euro per ton (about 65.79 yuan), the future ten years CO_2 emission reduction benefits as shown in Table 21.6.

21.5.3 Results of Analysis for Economic Evaluation

21.5.3.1 The Cost Flow Table for National Economic Benefit

Cost flow table for National Economic Benefit showed in Table 21.7.

Based on above analysis of results, IRR is 22.06 %, greater than Social benchmark rate of 8 %. The Net present value (the social discount rate is 8 %) is 2346.5 thousands. So using ETC will bring good effect on energy saving and emission reduction.

21.5.3.2 Sensitivity Analysis

The variation of IRR is showed in Table 21.8, given the increased input of the construction cost by 10 % and the decreased benefit by 10 %.

Table 21.3 The average fuel consumption contrast table of general vehicle

Vehicle lane		Deceleration fuel consumption	Waiting oil consumption	Service oil consumption	Leave fuel consumption	Total
ETC		0.0056	0	0.0012	0.0112	0.0180
MTC	Entrance	0.0063	0.0072	0.0024	0.0126	0.0285
	Export	0.0063	0.0168	0.0056	0.0126	0.0413

Table 21.4 Energy saving efficiency (10 thousand yuan)

Year	Energy saving benefits	Year	Energy saving benefits
2012	14.9986	2017	40.6033
2013	18.3043	2018	49.5523
2014	22.3386	2019	60.4736
2015	27.2620	2020	73.8020
2016	33.2705	2021	90.0680

Table 21.5 The average emission factor of MTC and ETC

Type	NO$_x$ (g/km)	HC (g/km)	CO (g/km)	CO$_2$ (g/km)
ETC	0.61	0.16	6.04	148.2
MTC	0.73	0.56	21.06	289.9
Emission reduction	0.12	0.4	15.02	141.7
Emission reduction percentage	16.40 %	71.20 %	71.30 %	48.90 %

Table 21.6 CO$_2$ emission reduction benefits (10 thousand yuan)

Year	Emission reduction benefits	Year	Emission reduction benefits
2012	0.3485	2017	0.9435
2013	0.4253	2018	1.1514
2014	0.5191	2019	1.4052
2015	0.6335	2020	1.7149
2016	0.7731	2021	2.0929

Table 21.7 The cost flow table for National Economic Benefit (Ten thousands yuan)

Year	Cost flow	Benefits flow	Net benefits flow	Discounted value of net benefits	Accumulative value of NPV
2011	100	0	−100	−100	−100
2012	20	28.5674	8.5674	7.9328	−92.0672
2013	20	34.8637	14.8637	12.7432	−79.324
2014	20	42.5477	22.5477	17.8991	−61.4249
2015	45	51.9252	6.9252	5.0902	−56.3347
2016	20	63.3694	43.3694	29.5165	−26.8182
2017	20	77.336	57.336	36.1314	9.3132
2018	20	94.3809	74.3809	43.4005	52.7137
2019	20	115.1824	95.1824	51.4241	104.1378
2020	20	140.5686	120.5686	60.3143	164.4521
2021	20	171.55	151.55	70.1970	234.6491
IRR	22.06 %	NPV 234.65	BCR 1.93	Payback period	6.7 years

In the most unfavorable case, IRR is 17.8, greater than social discount rate 8 %. The project has a strong ability to resist economic risk. So using ETC in this toll station is feasible, and it will bring good effect on energy saving and emission reduction.

Table 21.8 Sensitivity analysis for National Economy

IRR (%)		Utilization rate for ETC		
		0 %	−5 %	−10 %
Cost change	0	22.06	20.93	19.78
	5	20.99	19.89	18.75
	10	19.99	18.91	17.80

21.6 Conclusion

We studied the mechanism of ETC energy saving and emission reduction in this paper. Then we built an index system to evaluate the effects of ETC energy saving and emission reduction, we also discoursed the meaning and the calculation method of these index in ETC. Finally, we verify the feasibility of the evaluation indexes and evaluation methods proposed in this paper through studying a toll station which using ETC project.

References

Cai Y-H (2005) Research of OD matrix based on ITS[D]. Wuhan University of Science and Technology, Wuhan
Gao W-B, Zhang B-H, Gao Q-L (2009) Benefit analysis and Promotion measures of ETC application. China ITS J [J] (05):68–71
Jun Z-C (2008) Social economic impact evaluation of intelligent transportation system projects[M]. Tsinghua University press, Beijing
Xiang Q-J (2001) Research on vehicle fuel consumption in urban transportation system [D]. Highway (07)
Zavergiu R (1996) Intelligent transportation systems: an approach to benefit-cost studies[C]. TRIS, 102
Zhang Z-Y, Rong J, Zheng H (2001) Beijing highway toll station capacity study based on the M/G/K queuing model [J]. Highway (07):128–132
Zhou Z-B (2011) Analysis on the application benefit of ETC in Anhui Province [J]. J Highw Transp Res Dev (07)

Chapter 22
Model Calculating on Integrated Traffic Energy Consumption and Carbon Emissions in Beijing

Ying-yue Hu, Feng Chen, Wei-ming Shen and Qi-bing Wu

Abstract Based on the characteristics of Beijing traffic system, a model is built to calculate the energy consumption of urban road and rail transit and emissions of green house gas and their total amounts are calculated. At the same time, the comparison and analysis have been done to the energy consumption per capita per km and carbon dioxide emissions of different transportation means such as rail transit, public electric cars and social cars and so on.

Keywords Urban traffic · Carbon emissions · Mathematical modeling · Greenhouse gases · Low-carbon transport

22.1 Preface

With the global economic and motor vehicle ownership grew fast, traffic travel has become an important source of carbon dioxide emissions in the city. Energy consumptions per mileage for all kinds of transport modes change a lot, it is urgent

---

Y. Hu (✉) · F. Chen · W. Shen · Q. Wu
School of Civil Engineering, Beijing Jiaotong University, Beijing 100044, China
e-mail: hu-ying-yue@qq.com

F. Chen
e-mail: fengchen@bjtu.edu.cn

W. Shen
e-mail: 10121426@bjtu.edu.cn

Q. Wu
Beijing Chuangtong Infrastructure Construction Investment Company,
Beijing 100052, China
e-mail: u-qibing@163.com

to find a valid calculating method to compare the real carbon emission values of different modes. Conclusion of this paper can be used to provide decision-making for achieving a low-carbon transport and green-travelling goal for the relevant transport departments.

22.2 Models of Traffic Energy Consumption and Carbon Emissions in Beijing

22.2.1 Models of Traffic Energy Consumption

Models of urban road energy consumption. Energy consumption models of different vehicle types in Beijing are build by the classification according to the applying character, the energy difference and so on. ε, the standard coal conversion coefficient, is introduced to convert all energy consumption into standard coal for unified calculation, which can make a comparative analysis among all of them together. The specific calculation model as follows, Eqs. 22.1 and 22.2.

$$E_r = \Sigma E_i \tag{22.1}$$

$$E_i = \begin{cases} Q_{truck} \times e_{truck0} \times \varepsilon_{diesel\ oil} & \text{(If i is truck)} \\ N_i \times S_{i0} \times e_{i0} \times \varepsilon_k & \text{(If i is coach)} \\ e_{road\ lighting} \times \varepsilon_{electric\ energy} & \text{(If i is road lighting system)} \end{cases} \tag{22.2}$$

E_r is the total energy consumption; E_i is the energy consumption of the Beijing type i vehicle; Q_{truck} is the cargo turnover (in t · km); e_{truck0} is the the lorry average energy consumption intensity (in L/tkm); $\varepsilon_{diesel\ oil}$ is the standard coal conversion coefficient of diesel oil; N_i is the ownership of the type i vehicle (in vehicles); S_{i0} is the annual average mileage per car of the type i vehicle (in km); e_{i0} is the energy consumption per kilometer of the type i vehicle (in L, kg or kwh/100 km); ε_k is the standard coal conversion coefficient of each energy; $e_{road\ lighting}$ is the electric energy consumption of the urban road lighting system.

Models of urban rail transit energy consumption. The establishment of the energy consumption model as follow.

$$E_s = \sum_{k}^{n} E_{k\ vehicle} + \sum_{k}^{n} E_{k\ station} \tag{22.3}$$

$E_{kvehicle}$ and $E_{kstation}$ is the total vehicle and station energy consumption of the rail transit line k (million kwh).

22.2.2 Models of Carbon Emissions

Have a classification of the traffic energy consumption in accordance with diesel, gasoline, compressed natural gas (CNG) and carbon emissions models of different energy are set as Eq. 22.4. There is no greenhouse gases emissions using electricity, but a lot of greenhouse gases are generated within the source of the process. The model of upper reaches carbon dioxide emissions need to be constructed as Eq. 22.5.

$$M_x = E_x \times m_x \tag{22.4}$$

E_x and m_x is the consumption and carbon dioxide of energy x. (Note: diesel oil, gasoline emissions parameters directly from the IPCC (2006), the CNG emission parameters obtained through the equivalence relation between the carbon dioxide emission factor (Hao et al. 2009) of CNG buses and the energy consumption per unit distance).

$$M_e = E_e \times \zeta \times \psi \tag{22.5}$$

E_e is the traffic electricity consumption (kwh). ζ is the percentage of the national thermal power in the total generating capacity (China Electricity Council 2011). ψ is the carbon dioxide emission coefficient (Ma 2002) within the thermal power production process. The basic parameter values of all the modes can be seen in Table 22.1; the urban rail transit energy consumption shows in Table 22.2. Combined with the parameters of carbon emissions model in Table 22.3, carbon emissions of each energy can be obtained.

22.3 Estimation of Beijing Traffic Energy Consumption and Carbon Emissions

22.3.1 Models of Carbon Emissions

Estimation results of road traffic energy consumption. Access to the road lighting energy consumption from the Beijing Statistical Yearbook, and based on the established road transport energy consumption model, different cars in Beijing, such as social cars, public buses, taxis, trucks and other models as well as the road lighting energy consumption has been estimates, shown in Table 22.4. Seen from the results, social car's energy consumption is much higher than other models.

Estimation results of rail transit energy consumption. Based on the total energy consumption of vehicles and stations of the Beijing urban rail transit lines operating in 2009, urban rail transit gross energy consumption of Beijing can be seen in Table 22.5.

Table 22.1 Basic parameter values of the inertial model of energy consumption of urban road vehicles

Vehicle types	Fuel types	Freight turnover quantity in 2009	Average energy intensity	Consumption	Density	Standard coal conversion coefficient
Truck	Diesel oil	8788.87 million t · km	0.11 L/t · km		0.86 kg/L	1.46 kgce/kg
Bus	–	Operator number in 2009 (vehicles)	Annual average mileage per car	Energy loss per kilometer	Density	Standard coal conversion coefficient
	Diesel oil	15,908	6,740 thousand km	40.12 L/100 km	0.86 kg/L	1.46 kgce/kg
	Electricity	1,717	6,740 thousand km	0.96 kWh/km	—	0.13 kgce/kwh
	Natural gas	3,837	6,740 thousand km	35 kg/100 km	0.714 kg/m^3	1.21 kgce/m^3
Taxi	Gasoline	66,646	11,730 thousand km	7.58 L/100 km	0.73 kg/L	1.47 kgce/kg
Social car	Gasoline	310,340 thousand	1,530 thousand km	7.76 L/100 km	0.73 kg/L	1.47 kgce/kg

22 Model Calculating on Integrated Traffic Energy Consumption

Table 22.2 Basic parameter values of the inertial model of energy consumption of urban rail transit (million kwh)

Operating lines	Line 1	Line 2	Line 5	Line 10	Line 8	Line Batong	Line 13	Line 4	Airport line
Station energy	51.82	30.05	55.56	55.49	11.21	9.49	15.92	10.12	18.49
Vehicle energy	74.12	52.05	59.09	39.25	5.25	23.47	49.12	13.63	

Table 22.3 Basic parameter values for the calculate of the model of the carbon emissions generated by various types of energy consumption

Energy types	Diesel oil	Gasoline	Natural gas	Electricity
Carbon dioxide emissions coefficient	2.73 kg/L	2.26 kg/L	3.07 kg/kg	1.019 kg/kwh

In summary, traffic energy consumption has been classified and counted by diesel oil, gasoline, compressed natural gas, electricity, combined with the standard coal conversion coefficient (GB/T 2589-2008 2008) of each energy, the sum traffic energy consumption of Beijing in 2009 is 6.56 million tons of standard coal, shown in Table 22.6.

22.3.2 Estimation of Traffic Carbon Emissions

According to the energy consumption statistics in Table 22.6, combined with the carbon emission coefficient, carbon emissions generated from the energy consumption have been estimated respectively. It's calculated that the total amount of carbon dioxide from Beijing traffic in 2009 is 14.54 billion kg, shown in Table 22.7.

22.4 Comparison of All Kinds of Transportation

Total energy consumption are translated into standard coal quality by various modes of transport, then the contrast of the energy consumption of a variety of traffic traveling can be facilitated. Per capita energy consumption and CO_2 emissions can be calculated and comparatively analysis by combining with the annual passenger traffic of all kinds of transportation as well as the average trip distance. Seen in Tables 22.8 and 22.9.

According to the above tables, rail transit is the most energy-efficient and lowest emissions transportation models to travel. While the energy consumption required of per person per kilometer by taking the social cars is approximately 17 times of by

Table 22.4 Urban road traffic energy consumption of Beijing in 2009

Models	Social car	Bus		Taxi	Truck	Road lighting
Fuel	Gasoline	Diesel oil	CNG	Electricity	Gasoline	Diesel oil	Electricity
Energy consumption	3.69 billion L	0.43 billion L	0.09 billion kg	0.11 billion kwh	0.59 billion L	0.95 billion L	0.30 billion kwh

Table 22.5 Urban rail transit gross energy consumption of Beijing in 2009

Operating lines	Line 1	Line 2	Line 5	Line 10	Olympic line 8	Line Batong	Line 13	Line 4	Airport line	Total
Total energy (million kwh)	125.94	82.10	114.65	94.74	16.46	32.96	65.04	23.75	18.49	555.64

Table 22.6 Traffic energy consumption of Beijing in 2009

Fuel types	Diesel oil	CNG	Gasoline	Electricity
Total energy consumption	1.38 billion L	0.09 billion kg	4.29 billion L	0.97 billion kwh
Standard coal mss	1.73 million tce	0.15 million tce	4.57 million tce	0.12 million tce
Percentage	26.31	2.21	69.66	1.82

Table 22.7 Traffic carbon emissions of Beijing in 2009

Fuel types	Diesel oil	CNG	Gasoline	Electricity	Total
Carbon emissions	3.76 billion kg	0.28 billion kg	9.69 billion kg	0.81 billion kg	14.54 billion kg

Table 22.8 Calculating parameter (per capita travel) of all kinds of transportation of Beijing in 2009

2009	Total energy consumption (million tce)	Carbon emission (million t)	Passenger flow volume (million passengers)	Per capita trip distance (km)
Public bus	0.74	1.55	3861.80	7.27
Social car	3.94	8.35	3409.10	10.78
Taxi	0.63	1.35	948.74	10.78
Rail transit	0.07	0.46	1422.68	7.73

Table 22.9 Energy consumption and carbon emissions per capita of all kinds of transportation of Beijing in 2009

2009	Energy consumption of per person (gce/per person)	Carbon emission of per person (g/per person)	Energy consumption of per person km (gce/per person km)	Carbon emission of per person km (g/per person km)
Public bus	190.74	400.15	26.24	55.04
Social car	1155.29	2448.21	107.17	227.11
Taxi	666.25	1421.35	61.80	131.85
Rail transit	48.01	325.79	6.21	42.15

taking rail transport, the public electric buses is 4 times and the taxis is 10 times. Meanwhile, the carbon emissions of that are 5, 1.3 and 3 times respectively.

22.5 Conclusion

In this paper, the Beijing traffic characteristics model of energy consumption and carbon emissions is built, and the total transport energy consumption in 2009 was 6.57 million tons of standard coal, the energy consumption of bus, social cars, taxis, rail transportation were respectively 0.74, 3.94, 0.63 and 0.07 million tons of standard coal; the total amount of CO_2 generated by Beijing traffic in 2009 was 14.53 million tons, the carbon emissions of bus, social cars, taxis, rail transportation were respectively 1.54, 8.35, 1.35 and 0.46 million tons. The per capita energy consumption and the per capita carbon emissions of social cars is the highest by comparison of all kinds of transportation, were respectively 107.17 g/km of standard coal and 227.11 g/km of standard coal; while the per capita energy consumption and the per capita carbon emissions of rail transport is the lowest, were only 6.21 g/km of standard coal and 42.15 g/km of standard coal; while the carbon emissions of ordinary energy (such as gasoline, diesel) is higher than new energy (such as electricity, CNG).

With the expansion of the city, residents travel distance is increasing, the mechanization has become an inevitable trend, how to establish the dominant position of public transport and how to strengthen demand for cars controlling management is the key to the development of low carbon transport. The implementation of low carbon transport needs to promote from aspects of vehicles, energy technology and environment, the efficiency of traffic needs to be improved, new energy of low carbon transport needs to be developed, the travel environment and public transport infrastructure need to be improved, then various travel modes can be linked up and low carbon transportation of city comes true.

References

China Electricity Council (2011) Percentage of the thermal power in the total generating capacity [OL] (2011) http://tj.cec.org.cn/
GB/T 2589-2008 (2008) Comprehensive energy consumption calculation general rule[S] (2008)
Hao H, Wang HW, LI XH et al (2009) Analysis on energy conservation and emission reduction of alternative fuels for natural gas vehicle [J]. Nat Gas Ind 29(4):96–98
IPCC (Intergovernmental Panel on Climate Change) (2006) Intergovernmental panel on climate change [R] (2006)
Ma ZH (2002) Comparative evaluation research on several Chinese major coefficient of greenhouse gas emissions [D]. China Institute of Atomic Energy, Beijing

Chapter 23
Evaluation Indexes of Public Bicycle System

Yue Ma and Xiao-ning Zhu

Abstract With the concept of green traffic given public traffic get more and more attention from the people, through the survey data of the public bicycle rental of Jiangyin city it analyzes simply the characteristics of bicycles, and chooses some indexes of the public system to evaluate, such as for the whole system the indexes are analyzed by quantitative and weight quantitative, for the facilities of system consider the indexes from the utilization and the accessibility.

Keywords Traffic planning · Public bicycle system · Evaluation indexes

23.1 Introduction

The fast development of economy makes the traffic not follow its step. With the concept of green traffic given the public traffic draws more attention, bicycle because of its zero consumption, no pollution and other characteristics becomes the aid of public traffic. It helps to improve the traffic condition and reduce the pollution and energy. Bicycle system can make up the weakness of public traffic, gather and evacuate the passengers. Short distance travel takes up large proportion in medium and small cities, which creates favorable conditions for the development of slow traffic.

Y. Ma (✉) · X. Zhu
Beijing Jiaotong University, Beijing 100044, China
e-mail: 11120977@bjtu.edu.cn

X. Zhu
e-mail: xnzhu@bjtu.edu.cn

Public bicycle system (PBS) is that company/organization or two together cooperate and set lease points in some areas, such as large residential areas, commercial centers, transport junctions, tourist attractions and other people flow more intensive areas. The system provides bicycles for travelling during a certain time, and charges some fees according to the length of time. Meanwhile the system taking the related service management system and the corresponding network construction as the carrier provides the bicycle travel service for people (Gong and Zhu 2008).

23.1.1 The Advantages of Public Bicycle System

Public bicycle system as transport microcirculation to solve last mile based on the concept of "who use who borrow, publicly use". Public bicycle system has the following advantages. It is zero energy and green traffic providing convenient transport for residents and tourists. It is the trend of urban bicycle in the future. The system reflects the people-oriented sprit without time and line limited. Evacuate the passenger flow easily and efficiently. Pay more attention on the relationship between people and nature using sustainable traffic technology and concept. It has the characters of flexible operation, high accessibility and low investment. Advocate going travel by bicycle frequently due to bicycle occupying less land. It can improve the utilization ratio of road resource and ease road congestion. Widely use the information technology. Public bicycle system is the combination of high technology and traditional manufacture. It makes the traffic facilities more efficient and reasonable, which promotes the rapid development of city (Han et al. 2009).

23.1.2 The Problems Existed in the PBS

The reason of people abandoning using the bicycle is that the bicycles lack of parking garages, but the public bicycle that can solve the trouble of residents improves bicycle travelling proportion. After a period of time the public bicycle has many problems, and the main problem is that the setting scale and position cannot match with the service ability.

The locations of lease points are unreasonable. The locations generally focus on scenic and recreational areas, but in residential and public areas there are less. The way of renting or returning the bicycle is not convenient. The procedure is to be simplified. Short of bicycles' equipment, the number of bicycles is less. The speed of deployment cannot be guaranteed, so that the customers always need to wait a long time, even quit the plan to rent the bicycle. The size of bicycles should consider the difference of customers in height, sex and the people with children. Instruction should be with bilingual signs for foreigners. Propaganda and instruction both contribute to the utilization of bicycles.

23.2 The Characteristics of Bicycle Travel

Bicycle traffic is limited by terrain and climate environment greatly, and due to no protective equipment people directly are affected by environment. So in cold winter bicycle traffic may be largely limited. The distance, time and speed of bicycles are related with the people's strength and physique. Bicycle is not suitable for long distance and time trip (Lai 2007; Liang 2007). Through the survey of travel in JiangYin city analyze the data and understand more about the residents travel characters.

JiangYin whose construction planning "double speed" system—expressway and BRT, total area is about 988 km^2, coordinates the regional transport, makes sure the strategy of traffic development and promotes the development of integrated traffic. According to the survey of residents travel in 2003, travel intensity is 3.05 times. However the result of 2007 is 3.00. With the expansion of urban framework, the travel intensity reduces little, while due to the population increasing traffic demand will increase quickly (Fig. 23.1).

According to the result of survey and the situation of PBS, travel time within 30 min account for 57 %, and the corresponding travel distance is 6 km. PBS is to assist the public traffic system, it benefits travelling. Because the travel purposes are different, the requirements of traffic service are different. Such as time is important for work, while comfort for tourism. Generally speaking, in the center of city, due to the place limited in the work place and school the utilization ratio of car is low, but the public traffic is high. Time is one of the most important factors for the transport means, from origin to the destination there are various transport means, their travel time affects the choice of transports.

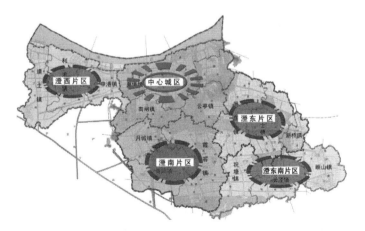

Fig. 23.1 Jiangyin group distribution

23.2.1 Travel Time and Distance

According to the survey statistics in central district bicycle travel time within 15 min takes up 18 %, 15–30 min is 39 %, 30 min–1 h is 26 %, more than 1 h is 17 %. The bicycle travel time within 30 min takes up 57 %, while within 1 h is 83 %. In Jiangyin city bicycle travel distance is within 12 km (Figs. 23.2, 23.3).

23.2.2 Characteristics on Rental One Day

Normally PBS has two peaks, namely the morning peak and evening peak. According to the statistics of implementation of PBS one week, we can see that it has three peaks. The first is the time to go to work (7:00–8:00). The second is the time after work (17:00–18:00), what is more the ratio of evening peak is higher than the morning. Besides that there is another peak (11:00–12:00), during which people maybe go out for dinner. At night rental still continues, but the rental ratio drops gradually until to 1 % during the time of 21:00 and 22:00 (Fig. 23.4).

23.2.3 Weekdays and Weekends

In weekdays there are three peaks in the chart. By contrast the weekends only have two peaks, the morning peak from 9:00 to 12:00 and the evening peak from 16:00 to 17:00. The reason of the difference between weekdays and weekends mostly is that on weekends people go out without working time limited. So the ratio is similar. What is more, the ratio of weekends is lower than weekdays'. Totally speaking the ratio of weekends is more balanced, unliking the weekdays having obvious changes. On weekends 5:00–6:00 (people going out to buy food for one week), 13:00–14:00 (some going out for dinner) and 19:00–20:00 (leisure time after dinner or returning home) all have a small peak (Figs. 23.5, 23.6).

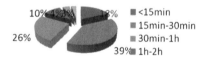

Fig. 23.2 Traffic time

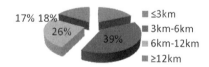

Fig. 23.3 Traffic distance

Fig. 23.4 Rental proportion at each hour one day

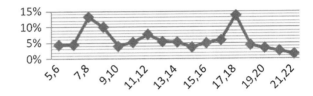

Fig. 23.5 Rental proportion in weekdays

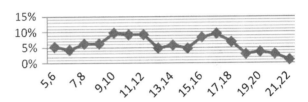

Fig. 23.6 Rental proportion at weekends

23.2.4 Rental Times

Choosing 20 lease points to analyze the ratio of rental and return bicycles, in the chart the ratio of points 1–6 and points 19, 20 are all higher and balanced. We can see that the positions and the functions are more important for the ratio. For example, the function of points 19 and 20 is commuting, where people have many changes in the aspects of living, working, studying and so on because of the emergence of PBS (Table 23.1, Fig. 23.7).

According to the difference of lease points' function choose three typical lease points to analyze. Three lease points are HuaDi department in the central section of downtown, government in the commercial district and HongQiaosan village in the common residential area. Combine with their respective characteristics to analyze (Figs. 23.8, 23.9).

Because of HuaDi department's passenger flow is big both on weekdays and weekends, the rental and return ratios are high. Government's passenger flow is balanced in weekdays, but compared with two others the ratio is lower. And on weekends the rental and return ratios decrease obviously because of personnel furlough. Compared with the two points, residential area has similar changes. During weekdays the rental and return ratios are high because of people going out for work or other reasons, but the ratio is lower than downtown. Many people go out to buy food for a week or entertain, which make the rental and return ratios increase on weekends. The following charms respectively are three points whose rental and return ratios contrast with each other on weekdays and weekends (Tables 23.2, 23.3).

Table 23.1 Lease labels and lease sites

1	2	3	4	5	6	7	8	9	10
TianHe park	HuaDi department	ZhongShan park	XingChun station	YangZi hotel	RenMin park	hospital	ChengKang bridge	ShiZheng company	JianJin street
11	12	13	14	15	16	17	18	19	20
XingGuo park Government	Examination and approval center	LiangChen furniture	HuangShanhu park	XiYuan market	JiYang village	FoQiao market	Times square	HongQiao san village	XinYi middle school

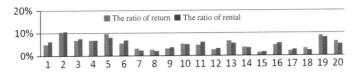

Fig. 23.7 Rental and return ratio of lease points

Fig. 23.8 Rental and return ratio in weekdays

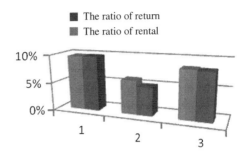

Fig. 23.9 Rental and return ratio at weekends

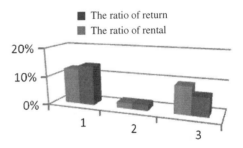

Many cities in China are single center cities. The radius of city land is within 6–10 km, which makes the bicycle travel have many advantages. According to the economical level and living conditions, single center will maintain a fairly long time. With the urban expansion it will make the bicycle demand transfer to other means of transport, meanwhile it will produce new short travel demands. In order to make PBS better implement, summarize and improve the evaluation indexes of PBS.

23.3 Evaluations of PBS

23.3.1 The View of Users

The main characteristics of bicycle travel is low consumption and zero pollution, benefiting the users' health and mental. At present the traffic congestion is serious in the cities, and people always are stuck in the way with nothing to do. People not

Table 23.2 Rental and return ratios of represented points in weekdays

	Center area of downtown	Commercial	General residential area
Site	HuaDi department (2)	Government (11)	HongQiaosan village (19)
Average rental and return ratio	10, 10 %	6, 5 %	8, 8 %
Performance characteristics	People flow is big, rent frequently all day	Activities concentrate on one day. People flow is big, rent frequently	One of big residential area, many people rent bicycles going out

Table 23.3 Rental and return ratios of represented points on weekends

	Center area of downtown	Commercial	General residential area
Site	HuaDi department (2)	Government (11)	HongQiaosan village (19)
Average rental and return ratio	12, 14 %	2, 2 %	10, 7 %
Performance characteristics	People flow is big, rent frequently all day	Activities concentrate on one day. People flow is small and rental ration is less	One of big residential area, many people rent bicycles going out

only pay attention on the rapid development of economy, but also meanwhile should protect our environment (Li et al. 2008). Bicycle traffic as a part of the green traffic meets people's expectation.

In order to make people use easily, take the characteristics and demands of users into consideration before establishment of the PBS. For example, the users maybe have the phenomenon of one way, which causes the bicycle flow unbalanced need to schedule the bicycle from the surrounding points or the storage bicycles. Make use of the system's advantages and assist the public traffic system.

23.3.2 The View of Operators

The establishment of PBS is to assist the public traffic system, so it should fully consider how to service the people well. The establishment of system is not to pursue the benefits. The main operating cost of system is from the government finance, small part from the advertisement. The establishment of system and the management personnel expenses to guarantee the system operate normally are not a small number. Although the system is a loss project in a short term, it brings many social utilities, environment utilities and so on.

23.3.3 Lease Point

Lease point distribution whether or not reasonable mainly depends on two factors, the number of people to serve (relate with the position of points in the community) and the people flow (relate with the distance to public places, crossing).

The number of bicycles in lease point whether or not reasonable also depends on two factors: static satisfaction (the satisfaction of permanent population) and dynamic satisfaction (the satisfaction of transferring or floating population).

Population density is the population in the unit of land, reflecting the degree of regional population density.

regional population density ρ = the number of regional population/regional area
$$(27.1)$$

Coverage area and the service number of lease points.

The walking distance is about 300–1,000 m. If the points are set in the center of city, the distance between the points is within 600 m advisably. Namely the service radius of point is 300 m. We can get the coverage area of every point.

$$S = \pi R^2 = 3.14 \times 0.3^2 = 0.2826 \, \text{km}^2 \tag{27.2}$$

For the service population of every point in some region P.

When the point is in the regional internal (the range of point is within the area).

$$P_1 = \rho_1 \times S \tag{27.3}$$

When the point is in the regional angle (point is in the angle, $\alpha = 0.25$).

$$P_2 = \rho_1 \times S \times \alpha + \rho_2 \times S \times (1 - \alpha) \tag{27.4}$$

When the point is on the regional edge (the distance from point to the boundary of region is less than 0.3 km, $\beta = 0.5$)

$$P_3 = \rho_1 \times S \times \beta + \rho_2 \times S \times (1 - \beta) \tag{27.5}$$

When the point is outside of region (the distance from point to the boundary of region is more than 0.3 km)

$$P_4 = \rho_2 \times S \tag{27.6}$$

23.3.4 Public Bicycle System

For the evaluation of whole PBS's planning whether or not reasonable, besides that the rationality of position and bicycle distribution, it should consider the followings, whether all the points cover the whole area; whether points distribution are relatively uniform; whether points satisfy all demands.

(1) **Quantitative score.** For $N(L_i)$, if distance is longer than 0.6 km, exceeding the distance of people accepting lead to users give up bicycle travel, the score is 0; if points locate at crossroads, commercial centers, public places and residential areas, choosing bicycle travel is convenient, the score is 10; the score criteria for L_1, L_2, L_3 and L_4.

$$N(L_i) = \begin{cases} 0 & L_1 \geq 0.6 \\ 10 - L & 0 \leq L_1 \leq 0.6 \\ 10 & L_1 = 0.6 \end{cases} \quad (27.7)$$

L_1, L_2, L_3, L_4 separately is the distance with adjacent crossroad, hotel, public places and commercial center. N_n is the number of lease points. R is the position with the adjacent community.

(2) **Qualitative analysis.** When N_n becomes bigger, it indicates that the point density is bigger. In order to benefit the users, the lease points should avoid overlap and improve the distribution rationality. For example, crossover ratio is $S_1/(2 \times S - S_1)$, when S_1 is bigger, the more population covered repeatedly, namely more score deduced.

$$N(N_n) = -10 \times S_1/(2 \times S - S_1) \quad (27.8)$$

The number of bicycles distributed by per ten thousand people in service circle

$$B = \text{the number of bicycles/the number of people in service circle} \quad (27.9)$$

This standard is used to measure the size of the static satisfaction of bicycle's number. When the score of 'position evaluation' of lease point is high, comprehensive evaluation of the location and people flow quantity are high. Namely it can slow down the urban traffic congestion and improve the residents' trips, whose utility is bigger. Therefore, in this lease point we can set more bicycles.

(3) **Weight quantitative.** Each standard with 10 scores cannot distinguish their degree effectively. To make the result of measures more reasonable, give their different weights according to actual condition and personal experiences.

The weights during L_i: because of the population flow of crossroads calculated unwell, it is slightly less than public places through the qualitative analysis and experience. Through the relevant data we can get that $Q(L_1):Q(L_2):Q(L_3):Q(L_4) = 45:12:65:8$, the sum of weights is 1, so $Q(L_1) = 0.69$, $Q(L_2) = 0.18$, $Q(L_4) = 0.12$.

23 Evaluation Indexes of Public Bicycle System

The weights between R and L_i: according to the general regulation, the connection of the residential areas with the crossroads, hotels, and public places is higher than the connection during crossroads, hotels and public places.

$$Q(\text{residential area}) = Q(\text{hotel}) + Q(\text{crossroad}) + Q(\text{public place}) + Q(\text{commercial center}) \quad (27.10)$$

Based on the population to calculate, K = population in the planning area/total population in the whole area. So the weights of R is $K \times Q$ (residential area), namely 1.99 K. N_n is the measure of overlap in service circle, for P_1 the score is 10. when $S_1 = S$, deducted score is 10. But if they intersect outside the area, the loss population only is P_4. So in order to realize the reasonability of comprehensive evaluations, the weight of N_n is $1 \times P_4/P_1$ when the weight of R is 1.

23.3.5 Index of Lease Point Facilities

(1) **Utilization**. Utilization of lease point d_1 refers to the rental times of public bicycle per hour in the working time.

Utilization of lease point d_1 = Total times of rental times/Total time in working time per day (27.11)

Utilization of bicycles d_2 refers to the rental times of one public bicycle per hour in the working time. All the bicycles in a lease point have the maximum utilization d_{max} and minimum utilization d_{min}, and utilization of bicycles d_2 can be expressed by the average of d_{max} and d_{min}.

$$d_2 = (d_{max} + d_{min})/2 \quad (27.12)$$

Utilization of parking berth d_3 is to point that the ratio of parking berth occupied in the working time. To be specific it is the ratio of the sum of time of all bicycles parking berth occupied and the working time. The formula of unit utilization of parking berth in the working time is

$$d_3 = \sum_{i=1}^{n} t_i/T \quad (27.13)$$

where T is working time(min), t_i is total time of the ith parking berth occupied(min), n is the number of parking berths in a lease point.

(2) **Accessibility**. Accessibility is to point the degree of convenience of renting a bicycle from the road network to lease points. When the resources of facilities are in short, accessibility can decide how the resources get reasonable configuration to make the planning layout satisfy the people's travel demands.

23.3.6 Management of Lease Point

The convenience from the view of managers is that generally speaking no matter the users transfer bus or subway, even to do other things, users always take or return bicycle at the nearest point which can save time. Therefore, to realize the convenience of points start when plan. Meanwhile, we can consider from the angle of users. If the formalities are simple and quick, shorten the time of people staying at the lease points, which can affect the indexes of system.

23.4 Summary

The establishment of PBS can solve some traffic problems, and improve the environment and people's physical quality with its own advantages. Through the travel demands and characteristics, analyze the characteristics of bicycles and PBS, and summarize some evaluation indexes expecting for better service for people. With the changes of bicycle transport, the characteristics of bicycles maybe have changes. The evaluation indexes need to be improved in order to make it more systematic and more reasonable.

References

Gong DJ, Zhu ZD (2008) Public bicycle system implementation mechanism. Urban traffic 6:27–33
Han HM, Zhang Y, Qiao W (2009) Leon public bicycle system. Urban traffic 7:13–22
Lai YW (2007) Bicycle traffic characteristics and facilities planning in the big cities [D]. Southeast university, Nanjing, China
Li CQ, Song R, Han Y (2008) Comprehensive evaluation index of urban bus service. In: 9th Industry economics and management (2008)
Liang CY (2007) Bicycle traffic characteristic and application. Jilin University, Changchun, China

Chapter 24
Study on Urban ITS Architecture Based on the Internet of Things

Zinan Yang, Xifu Wang and Hongsheng Sun

Abstract As an important part in the industrial chain of the Internet of Things (IOT), Intelligent Transport System (ITS) has been recognized as one of the priority sectors that IOT fells on the practical application successfully, and it is also the best starting point for government to create wisdom cities through IOT. We analyze the operation law and architecture of IOT, and put forward demands for urban ITS theory and technology on the Internet of Things. Then we combine with the existing urban ITS architecture and the IOT's architecture features, and establish urban ITS architecture based on the Internet of Things.

Keywords ITS · IOT · System architecture

24.1 Introduction

With the economic development and accelerated process of urbanization, the transportation amount increase, and the issue of urban traffic congestion and environmental pollution become worse. The rise of ITS, by increasing traffic operation management level and the utilization rate of road traffic facilities, is the indispensable effective measures to alleviate the contradiction between supply and

Z. Yang (✉) · X. Wang · H. Sun
School of Traffic and Transportation, Beijing Jiaotong University, Beijing 100044, China
e-mail: 11120905@bjtu.edu.cn

X. Wang
e-mail: xfwang1@bjtu.edu.cn

H. Sun
e-mail: hshsun@bjtu.edu.cn

demand in rapidly deteriorating traffic. The emergence of Internet of Things theory, changes the way people understand things status and control, provides a new technical and theoretical support to the development of ITS. Intelligent transportation technology applies to the entire transportation management system in order to establish real-time, accurate, and efficient transportation management and control systems.

24.2 The Principle and Architecture of the Internet of Things

The Internet of Things is showing a new idea of product tracking in the worldwide, and has a profound effect on the human life. By depth analysis of the Internet of Things' operating rules, the essence is the information exchange between the things and operating environment. "The thing" refers to entity exists of the physical world, also including the entity attributes of human being; "Human being" refers to the human will on control levels, Network and standards are the two important element(s) in the thing running environment, providing external environment support for information interaction (Internet of things and Logistics Digitalization[M] 2011). The regularity of IOT is shown in Fig. 24.1.

IOT combines the three elements organically, to ensure the free circulation of the whole operation process. The system structure of IOT can be divided into three levels: perceptron network, network communication facilities and universal application service support system, namely: the perception layer, network layer, application layer. The architecture diagram of IOT is shown as Fig. 24.2.

Specific functions of IOT three-tier architecture include: the perception layer realize comprehensive intelligent perception, and collect the physical event and data, including all kinds of physical quantities, logo, audio and video data; the network layer achieves input information management and the bearer networks of computer network and communication network; the application layer includes application support services and user application services, application support platform realize information's coordination, sharing and exchanging between cross-sector, cross-application and cross-system.

Fig. 24.1 Regularity of IOT

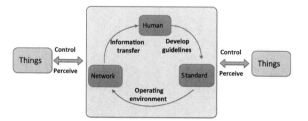

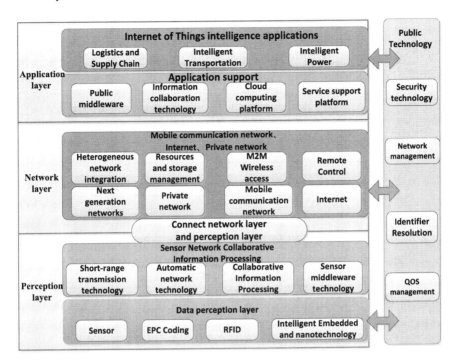

Fig. 24.2 Architecture diagram of IOT

24.3 Urban ITS Demand Analysis for the Internet of Things

Through various information marks and sensing equipment, IOT connect the things into nets to exchange and share information, and finally achieve the real-time and intelligent management network. The system structure and working principle of IOT and ITS is basically the same, based on the operation rule of the existing ITS, present the demand for ITS based on the Internet of Things.

(1) The more thoroughly perception of IOT on urban transport related elements and information management. Data is the basis for the operation of the urban ITS. The basic idea of IOT is the things connected, the road network information can be collected in time, a large number of traffic participant, including information of human and road related facilities would quickly come into IOT (Vahidi and Sayed 2003).

(2) Comprehensive interconnection and interchange on urban traffic information. The new model of transport elements information needed to build to achieve the perception and collaboration among the things in ITS, establish data input and output interface standards on traffic information subsystem, and build the traffic elements information collection system to achieve the interconnection among the information of traffic participants (Takahashi 2008).

(3) More in-depth intelligent collaborative services for urban traffic system. ITS needs to implement two-way transfer of information, not only to achieve coordination among the various intelligent transportation subsystems but also to realize the information collaboration between the uplink and downlink, to achieve the urban traffic unblocked. Decision-making and guidance information can transfer and feedback in time to urban traffic management and urban transport participants.

24.4 ITS Based on Internet of Things Structure

ITS based on Internet of Things mutually integrated by the intelligent transportation systems and the Internet of Things. According to the Internet of Things architecture and demand of ITS, this paper propose a "three layers and one network" architecture (Zhang 2011). "Three layers" refers to the data collection layer, data processing layer and application service layer, "one network" refers to the transmission network in the three layers system, as the Fig. 24.3 shown.

Data collection layer is the basic layer of the ITS based on Internet of Things, through the information collection equipment such as GPS, intelligent sensor, RFID, vehicle terminal and GIS to perceive and collect the information of the urban transportation, the coverage and accuracy of information collection directly determine the effect of the ITS based on Internet of Things. Middle layer is the

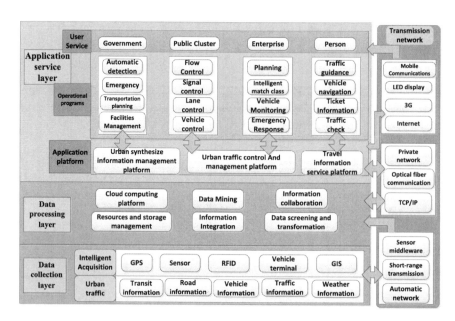

Fig. 24.3 Urban ITS architecture based on the Internet of Things

data processing layer, taking advantage of the Internet of Things technology, to analyze and use traffic information from the data processing layer. The top layer is the layer that people contact daily. Through the urban comprehensive information management platform, urban traffic control platform, urban public transport management platform and travel information service platform, manage and control the urban traffic participant.

The transmission network in ITS network can be seen as the blood vessels of system, undertake the work of traffic information transmission. The processed data transit to the management platform through a private network or optic fiber. Management information delivers to the users. The whole process of information transfer is a two-way, effectively achieve the synergy of information.

24.5 Conclusion

The Internet of Things has risen from a pure technology to an economic form, Internet of Things economic, Internet of Things industry. In the urban ITS, traditional traffic management methods develop to the bottleneck stage. The study provides the basis for the construction of the urban ITS based on Internet of Things. Make the future urban transportation become more convenient and intelligent.

References

Takahashi K (2008) Internet ITS [J]. NEC Tech J 3(1):37–41
Vahidi H, Sayed T (2003) Using the Canadian ITS architecture for evaluating the safety benefits of intelligent transportation systems [J]. Can J Civil Eng 30(6):970–980
Wang X (2011) Internet of things and Logistics Digitalization[M]. Publishing House of Electronics Industry, Beijing,
Zhang H (2011) Research on model-designing and architecture of IOT-based intelligent transportation system [D]. Beijing Jiaotong University, Beijing

Chapter 25
The Pedestrian and Cycling Planning in the Medium-Sized City: A Case Study of Xuancheng

Yiling Deng, Xiucheng Guo, Yadan Yan and Xiaohong Jiang

Abstract Walking and cycling are important component in the urban transportation, residents' life and leisure activities, especially in the medium and small cities. Improving the pedestrian and cycling environment is important, and the pedestrian and cycling planning plays an important role in the process. A pedestrian and cycling planning is introduced in this paper, taking Xuancheng, a medium-sized city as an example. Basing on the analysis of the existing conditions and issues of walking and cycling in Xuancheng, and considering the requirement of residents and the visions of upper level plannings, the planning goals including safety, priority and vibrancy are proposed as well as five objects to support these goals. Then, the planning strategies and actions including structure planning, differentiation district control, important district improvement, public transit access and advanced design guidelines are introduced in detail. At last, some suggestions are given to the planning implementation and management.

Keywords Pedestrian and cycling planning · Planning strategies and actions

---

Y. Deng (✉) · X. Guo · X. Jiang
School of Transportation, Southeast University, Nanjing, China
e-mail: acoralseu@163.com

X. Guo
e-mail: bseuguo@163.com

X. Jiang
e-mail: dxiaohongjiangseu@yahoo.com.cn

Y. Yan
School of Civil Engineering, Zhengzhou University, Zhengzhou, Henan, China
e-mail: cyadan.yan1986@gmail.com

25.1 Introduction

Walking and cycling are increasingly recognized as important forms of transportation in China. All people are pedestrians at one time or another, even those who generally use other modes of transportation, such as automobiles or transit. Walking and cycling are important in making intermodal connections as well as being travel modes of themselves. Many other benefits, including travel choice, affordability, reduced road congestion, infrastructure savings, improved health, recreation and enjoyment, environmental protection, are brought to users and nonusers alike by walking and cycling.

A good pedestrian and cycling planning in urban areas would create a more pleasant urban environment and encourage people to walk or cycle. The pedestrian and cycling planning provides a comprehensive vision for improving walking and cycling conditions. The purpose of the planning is to define an approach for the development of a safe, convenient and effective system that requires walking and bicycling as viable transportation options connecting work, shopping, residential, and recreational uses. The planning is also crucial in the plan to project delivery process. Success in construction of pedestrian and cycling facilities will only take place through good plan-to-project delivery process.

Chinese cities traditionally rely on walking and cycling daily travel, and many cities still have relatively low motorization levels despite the current surge in personal vehicle ownership. In the medium and small sized city, the pedestrian and cycling are very important in all forms of transportation, while the mode share of these two forms are always more than 60 % in the whole transportation (Deng et al. 2012). A good pedestrian and cycling planning can help the city become pedestrian and cycling friendly. Taking Xuancheng as an example, the main contents of the pedestrian and cycling planning are introduced in the paper to provide good reference to other medium or small size cities in China.

25.2 Existing Conditions and Issues

Xuancheng is a medium-sized city in the Anhui Province, which is in the middle part of China. The population was 2.76 million and the total area was 12340 km^2 in 2010. The per capita GDP was 21 thousand Yuan. In the central city which is the planning area, the population was 0.43 million. The total area was 565.5 km^2 and urban construction area was 65 km^2 in 2010.

The city recognizes that walking and cycling are currently more inconvenient than ever and dangerous than necessary, which causes unnecessary injuries, discourages non-motorized travel, and imposes economic costs on the community. The city therefore seeks to make walking and cycling safer and more convenient.

According to resident travel survey made in 2010, 39.7 % of trips in the city were made by walking, and 23.7 % were made by cycling. Walking and cycling are key forms of transportation in the city, especially through neighborhoods, around schools, and in business districts.

The planning is based on an assessment of walking and cycling network and a survey of the city's existing walking and cycling conditions. The identification of problems is fundamental in clarifying the goals and objects. The problems are always at different geographical levels, such as districts and local levels, and in various aspects such as connectivity, capacity, safety and comfort. In Xuancheng, four main problems are existed for the pedestrian and cycling:

(1) The right of way cannot be guaranteed which causes the safety of pedestrians and bikers are affected. In some roadway, the widths of sidewalks and bike lanes are too narrow to satisfy the demands for passing or activities. For another reason, the sidewalks and the bike lanes are always occupied by parking because of lacing parking space in the central city. The pedestrians have to walk with bicycles and the bikers have to ride with the automobiles;
(2) Crossing the road is difficulty and unsafe in some districts. The traffic signals in most intersections are two phase controls. The pedestrians or bicycles are in conflict with the right-turn automobiles when crossing the intersection. These phenomena are unsafe especially in some intersections around the schools or kindergartens. In addition, the crosswalks in the roadways are always not enough to meet the crossing demand for pedestrians or located in the improper places, for example, too far away from the bus station or the school;
(3) The number of bicycle parking facilities is insufficient, especially in the commercial area, around the hospital or school. Random bicycle parking always affects the passing of pedestrians and other bicycles;
(4) Xuancheng is a landscape garden city and filled with a lot of hills and rivers. But the waterfront space and the green space are not treating well to supply walking or cycling space for exercise, relaxation and recreation.

25.3 Planning Goals and Objects

When setting the goals of the pedestrian and cycling planning, the requirements of residents and the visions of the upper level planning, including the city general planning, comprehensive transportation planning and other related plannings should be considered. The pedestrian and cycling planning of Xuancheng includes three goals, i.e. safety, priority and vibrancy, to make a friendlier environment for walking and cycling. Then five objects are proposed to support the goals:

(1) Create vibrant public spaces that encourage walking and cycling;
(2) Plan, design, and build pleasant streets to meet the demand of pedestrians and bikers;
(3) Improve the connectivity of the existing pedestrian and cycling routes;
(4) Reduce the number and severity of vehicle crashes involving pedestrians and bikers;
(5) Get more people walking for transportation, recreation, and health.

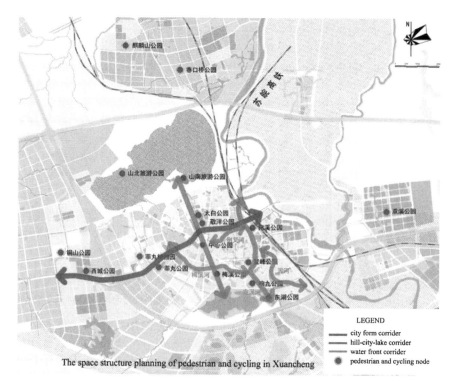

Fig. 25.1 The space structure planning of pedestrian and cycling in Xuancheng

25.4 Planning Strategies and Actions

Appropriate strategies and actions are developed to achieve these objects. The strategies and actions include space structure planning, differentiation district control, important district improvement, public transit access and advanced design guidelines.

Space Structure Planning. The macro pedestrian and cycling spaces includes corridors and nodes. The nodes are the plazas and parks. The corridors are the linear space combined with roadways, hills, rivers, and landscape green belts, which always have good landscape and high quality space for walking and cycling for travel and leisure.

In Xuancheng, twelve plazas and parks are planned with a total area of 4.19 km^2. Three kinds of pedestrians and cycling corridors (total six corridors) are planned, including four waterfront corridors, one hill-city-lake corridor and one city form corridor as Fig. 25.1 shows.

Differentiation District Control. Not all districts in Xuancheng have the same level of demand and the built environment. This strategy is to match the planning standard of pedestrian and cycling facilities to the demand in the different districts.

25 The Pedestrian and Cycling Planning in the Medium-Sized City

Table 25.1 The distance between adjacent crosswalks in the roadways (m)

Land usage/intensity	Residential, administration and public service		Commercial and business facilities		Regional transportation	Green space	Industrial, logistics and warehouse
	High	Normal	High	Normal			
Secondary road	200	250	200	300	300	300	400
Major road	250	300	250	350	350	400	500
Express way	300	350	300	400	400	500	600

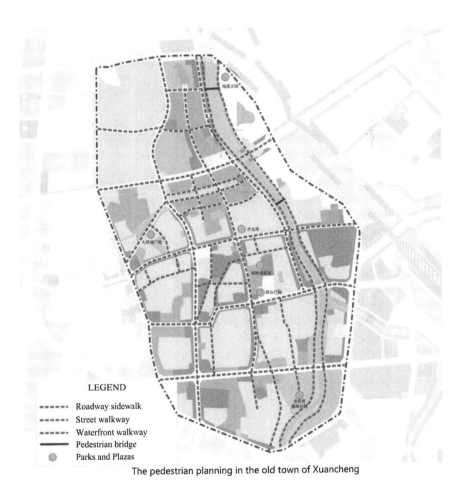

Fig. 25.2 The pedestrian planning in the old town of Xuancheng

The standards are reviewed which are used by city departments that affect walking and cycling conditions. According to the preference survey to the pedestrians and bikers, the new planning and design standards for pedestrian and

Fig. 25.3 The cycling planning in the old town of Xuancheng

bicycle facilities (e.g. bike lanes, sidewalks and crosswalks) are established. In the old city, all the roadways have already been built, the work is to reallocate the road space to satisfied the demand of pedestrian and cycling. In the new districts, minimum width standards for pedestrian and bicycle facilities (e.g. bike lanes and sidewalks) are proposed which should be obeyed in the roadway planning and design, because most of the roadways are not implemented. The standard of the distance between adjacent crosswalks is reviewed too. The standard mainly depends on the road hierarchy, the land usage and the crossing demand as the Table 25.1 shows.

Important District Improvement. The old town is chosen as the important district to do detail planning for two reasons: (1) the agglomeration of pedestrian and cycling demands; and (2) the most deficiencies of the built pedestrian and cycling facilities. The fundamental issue is how to reallocate space to meet the pedestrian and cycling needs efficiently while maintaining proper vehicular space for parking, local access and through movement, how to systematically retrofit currently deficient of sidewalks and bike lanes, and how to give the pedestrian and bicycle more routs.

In the old town pedestrian planning, except the traditionally pedestrian space, namely the roadway sidewalks, the walkways in the streets and alleys are planned and the corresponding traffic management strategies are proposed to guarantee the walking space. The waterfront walkways and pedestrian bridges are planned combing with the waterfront urban design as Fig. 25.2 shows.

In the old town cycling planning, the bike lanes are classified in isolated bike lane, mixed bike lane and street and ally bike lane. The bike lanes with high bicycle volume are physically isolated with the automobiles, and the street and ally bike lane are proposed to supply more passing space. The on road parking is also coordinated with the bike lane planning as (Fig. 25.3).

Public Transit Access. Accommodating the pedestrian and cycling with the public transit can convenient residents and increase the public transit ridership. Three actions are proposed: (1) make all bus stops conveniently accessible by walking and cycling; (2) integrate cycling into the public transit system by supporting enough bicycle racks and parks; and (3) promote smart land use development surrounding transit facilities to enhance the environment for walking and cycling.

Advanced Design Guidelines. The purpose of advanced design guidelines is to regulate the basic pedestrian and cycling facility design principle and method. The guidelines will provide technical guidance to planners and designers. The design guidelines are not only focus on the transportation function of the facilities, but also emphasize the qualities of the environment, shown in Fig. 25.4.

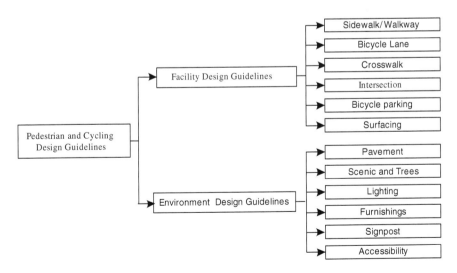

Fig. 25.4 The content of pedestrian and cycling design guidelines

25.5 Conclusions

The pedestrian and cycling planning plays an important role to improve the pedestrian and cycling environment. However, whether a plan, project or program will lead to an improvement of pedestrian and cycling environment, to a large degree, depends on the policy makers and plan executors. The city ought to prioritize the pedestrian and cycling facilities expenditures, based on the reasonable assessment of the funds, obey the standards and the schemes of the pedestrian and cycling planning in implementation.

Reference

Deng Y, Guo X, Ye M, Yan Y (2012) Building the walkable city in China: challenges and plannings. Transportation research board annual meeting (2012)

Chapter 26
A Model to Evaluate the Modal Shift Potential of Subsidy Policy in Favor of Sea-Rail Intermodal Transport

Xuezong Tao

Abstract This paper is to develop an analysis tool to assess the modal shift potential (MSP) of subsidy policy in favor of Sea-Rail Intermodal Transport (SRIT) between Ningbo Port and East Jiangxi province. Based on random utility theory and stated preference (SP) experiments, two disaggregate models, a MMNL model and a MNL model, are established and estimated. Results show that all attributes have the expected sign and the MMNL model outperforms the MNL model. The analysis of MSR under seven scenarios based on MMNL model suggests that giving a subsidy of 400 RMB/TEU to shippers who use SRIT might be the optimal choice for a cost-effective modal shift. The above methodology and results are useful for policy-makers to make freight transport policy, especially the subsidy policy in favor of SRIT.

Keywords Freight modeling · Mode shift · Sea-rail intermodal transport · Subsidy policy · Disaggregate mode choice model

26.1 Introduction

In 2011, the container throughput of Ningbo port exceeded 14.5 million, making Ningbo Port become the third largest container port in China (world's top 6). But the demand for inland transport of these containerized cargos is mainly met by road, accounting for approximately 84 % of the port container throughput. Exorbitant share of RT in the modal split resulted in ever-increasing adverse

X. Tao (✉)
Department of Transportation Engineering, Tongji University, Shanghai 201804, China
e-mail: taoxuezong@126.com

effects such as accidents, congestion, air pollution and greenhouse gases. Fortunately, Ningbo municipal government has been aware of the urgency of solving this issue and understands that a shifting mode from RT to other sustainable freight transport modes, mainly the SRIT, is a cost-effective way. According to the experience of developed countries, the achievement of modal shift needs the support of subsidy policy. Therefore, it is necessary to develop a reliable analytical tool to assess the MSP of subsidy policy, and the disaggregate mode choice models are introduced to realize this purpose.

Winston (1981) firstly developed a disaggregate model based on Random Utility Maximization (RUM) that attempts to reflect the actual behavior of freight transport decision-makers. From then on, disaggregate mode choice models were increasingly used for analyzing freight transport mode (FTM) choice decisions. Increasing efforts were made to extend the range of factors that could potentially influence the choice outcomes (i.e. transport/non-transport cost, service attributes). The most widely used model form was the multinomial logit (MNL, McGinnis et al. 1981; Nam 1997; Golias and Yannis 1998; Catalani 2001; Shinghal and Fowkes 2002; Chiara et al. 2008).

Despite the popularity of the MNL model, a number of researchers recognized that, in virtue of its independence from irrelevant alternatives property, it imposes potentially unrealistic limitations on the nature of the choice processes it could accommodate. In an attempt to address this problem, Jiang et al. (1999) proposed the use of a nested logit (NL) model. To relax another restricted assumption of the MNL model, the Heteroscedastic Extreme Value (HEV) model has been used in a number of freight mode choice applications (Holguin-Veras 2002; Norojono and Young 2003). This model allows for different scales of error to be associated with different alternatives and produced improved results compared to MNL and NL models.

Nonetheless, most of above studies have focused on competition between modes on national or international non-maritime shipments, applications addressing the mode choice determinants in the inland transport of containerized cargos via seaport being scare. Even if some researches differentiate container traffic in their analysis of intermodal traffic (Jiang and Calzada 1997; Beuthe and Bouffioux 2008), those studies do not yield specific conclusions regarding door-to-port or port-to-door traffic.

Combined with a mixed multinomial logit (MMNL) model and SP techniques, this paper presents a quantitative analysis tool to assess the MSP of different degree of subsidies in favor of SRIT between Ningbo Port and Shangrao city, which is helpful to design the efficient subsidy policy. The remainder of this paper is structured as follows. Section 26.2 presents the modeling framework. Section 26.3 describes data collection campaign and descriptive analysis of results. Section 26.4 estimates the model and discusses the findings. Finally, Sect. 26.5 concludes the paper and identifies possible future work.

26.2 Modeling Framework

According to a standard discrete choice model (McFadden 1974), an freight mode choice decision-maker (i.e., shipper or consignee) i is assumed to select the FTM m for a specified shipment with the highest utility $U_i(m)$ from amongst available freight in a choice set M_i. Considering that the decision-makers' utility cannot completely be observed, it is consequently decomposed into an observable component $V_i(m)$ and an unobservable component $\varepsilon_i(m)$, which can be expressed as:

$$U_i(m) = V_i(m) + \varepsilon_i(m) = U(x_{im}, s_i, \beta_i) = V(x_{im}, s_i, \beta_i) + \varepsilon_i(m) \quad (26.1)$$

where x_{im} is a vector of the measurable attributes characterizing FTM m given by decision-maker i, s_i is a set of socio-economic attributes of the decision-maker i that may also affect utility and β_i is a vector of parameters called tastes of the decision-maker i.

Depending on the assumptions made in regard to the distribution of errors $\varepsilon_i(m)$ and in turn, to the difference of such errors, one choice model or another will be obtained (MNL, Probit, HEV, MMNL or others). If we assume the error terms for all decision-makers i and FTM m are independently and identically distributed (IID) following a type I extreme value distribution, we can derive the most widely used model in freight transport applications, the MNL model. Then the choice probability can be expressed as:

$$P_i(m|M_i) = \frac{e^{V_i(m)}}{\sum_{h \in M_i} e^{V_i(h)}} \quad (26.2)$$

Under the IID assumption, this model displays the independence from irrelevant alternatives (IIA) property, giving rise to limitations when it comes to capturing certain patterns of behavior (Train 2003). The MMNL model (also referred to as Hybrid Logit or Kernel Logit), derived from the recognition that there would be correlation in unobserved information which can be estimated, solves the limitations of MNL. In consequence of this, the error component is partitioned into an unsystematic part $\varepsilon_i(m)$ (i.e., the random term which is IID for all decision-makers i and FTM m) and a systematic part η_i (i.e., the random term whose distribution is characterized by parameters relating to FTM m, decision-makers i or other factors y). The utility function from Eq. (26.1) can then be rewritten as:

$$U_i(m) = U(x_{im}, s_i, \beta_i) = V(x_{im}, s_i, \beta_i) + \varepsilon_i(m) + \eta_i(x_{im}, s_i, y|\phi) \quad (26.3)$$

where ϕ is a vector of parameters that describe the density of distribution, meaning that η_i can take any form of distribution. Since η_i is not given, the choice probability will be expressed as an integration of logit formula over all values of η_i weighted by the density of η_i in Eq. (26.4).

$$P_i(m|M_i) = \int_{\eta_i} L(\eta_i) f(\eta_i|\phi) d\eta_i \qquad (26.4)$$

where $L(\eta_i)$ is a logit choice probability conditional on a vector of parameters η_i that are jointly distributed with density $f(\eta_i|\phi)$.

Due to the absence of a closed-form notation for the MMNL choice probabilities, numerical techniques, typically simulation, are required in the estimation and application of this model. Details of the evolution of simulation-based maximum likelihood methods for estimating MMNL models are provided in numerous references including McFadden and Ruud (1994), Brownstone and Train (1999), Bhat (2001) and Hensher and Greene (2003).

26.3 Data Collection

The disaggregate data used in this paper are taken from a SP survey of shippers (sales managers or transport/logistics managers) in Shangrao city. A decision was made not to select a sample randomly but to interview firms: (a) who are export-orientated enterprises taking foreign countries and regions as target markets and (b) whose shipments are mainly transported in container via Ningbo port and potentially suited to SRIT. Unfortunately, for a variety of reasons, firms are often reluctant to be interviewed, so that, like in other similar studies, we were unable to gather a large sample and only 69 firms responded a telephone survey, of which 31 firms agreed to accept our interview. Given that the sample size is not large, to obtain more usable information, we request the respondents to describe two typical transport shipments and make corresponding choice under various scenarios. Even so, 57 experiments were conducted in the end. From the data gathered, five experiments have been removed due to very extreme values. Therefore, the final dataset considered in the following analysis is comprised of 52 valid experiments consisting in 832 choice observations.

Though the telephone surveys, four attributes of FTM are identified as the critical factors influencing firm's choice behavior, which are in order according their frequency mentioned by respondent as follows:

(1) Cost as out-of-pocket door-to-port transport cost, including loading and unloading, mentioned 61 times as first important factor;
(2) Time as door-to-port transport time, including loading and unloading, mentioned 52 times as second important factor;
(3) Punctuality as % of deliveries at the scheduled time, mentioned 49 times as third important factor;
(4) Safety as 100 % minus % of commercial value lost from damages, stealing and accidents, mentioned 45 times as fourth important factor.

The interviews, conducted from April 7 to May 29 in 2012. Table 26.1 reports the descriptive statistics for RT and SRIT service described by shippers.

Table 26.1 Descriptive statistics for typical transport service

Variable	RT				SRIT			
	Min	Max	Mean	SD	Min	Max	Mean	SD
Cost (RMB)	2450	2860	2672	80.779	1812	2265	2076	205.016
Time (h)	10	15	11.4	0.867	24	30	27.5	2.715
Punctuality	92 %	100 %	95.6 %	0.016	80 %	98 %	89.7 %	0.024
Safety	80 %	100 %	89.5 %	0.035	89 %	100 %	95.2 %	0.017
Current market share (%)	98.73				1.27			

In particular, the average costs is 2672 RMB/TEU for RT and 2076 RMB/TEU for SRIT respectively, while the delivery door-to-port takes on average 11.4 and 27.5 h.

26.4 Estimation and Results

One MNL and MMNL were estimated using the LIMDEP version 8.0 package, which was mostly employed to estimate the discrete choice models. The results are presented in Table 26.2.

On one hand, the coefficients of cost and time have a negative sign reflecting a decrease of the marginal utility as the values for cost and time increases. This suggests that reducing the cost or time might be an effective way to achieve the modal shift. Given that it is almost impossible to decrease the transport time with restrictions of transport infrastructure and management system in the short time, an attempt to reduce the transport cost though subsidy becomes the first choice for modal shift.

On the contrary, the more punctual and safe is the freight transport service, the more utility the shippers experience. As expected, the ASC_RT, introduced for RT, is not statistically different from zero indicating that none of the two unlabeled

Table 26.2 Estimates of MNL and MMNL models

Variable	MNL		MMNL	
	Coefficient (β_i)	t-Test	Coefficient (β_i)	t-Test
ASC_RT	−3.6378	−1.10	−8.58782	−1.16
Cost	−0.0145**	−2.11	−0.0195**	−1.97
Time	−0.1737**	−2.53	−0.1981**	−2.01
Punctuality	0.7308	1.43	0.9483	1.08
Safety	0.5624	0.81	0.7852	0.87
AIC	1025.12		1012.74	
Final LL	−527.09		−518.36	
Adjusted ρ^2	0.3518		0.3972	

Note ** → Significance at 5 % level

Table 26.3 Market share of RT, SRIT and the MSP (Unit: %)

Scenario	S_0	S_1	S_2	S_3	S_4	S_5	S_6	S_7
Market share of RT	98.73	96.98	95.07	93.19	89.36	88.34	87.41	86.59
Market share of SRIT	1.27	3.02	4.93	6.81	10.64	11.66	12.59	13.41
Total MSP	/	1.75	3.66	5.54	9.37	10.39	11.32	12.14
ΔMSP	/	1.75	1.91	1.88	3.83	1.02	0.93	0.82

Note $\Delta MSP_{i+1} = MSP(S_{i+1}) - MSP(S_i)$; $i = 0, 1, 2, 3, 4, 5, 6$; S_0—Current situation

FTM has been preferred a priori. Besides, punctuality and safety are also not statistically significant. This indicates that the shippers in Shangrao city have greater tolerance towards punctuality and safety and it will not provide enough room to alter the sensitivity of Shangrao shippers to this variable.

In terms of model fits, we note that MMNL model outperforms MNL model, with an improvement in the final LL of 8.73 points giving a statistically significant log-likelihood ratio test. The AIC statistic and the Adjusted $\rho 2$ further support MMNL model against MNL model. Therefore, we choose the MMNL model to analyze the MSP (MSP) under different subsidy policy.

According to Eq. (26.4), we calculated the Total MSP and Δ MSP under seven scenarios: S_1 means a subsidy of 100 RMB/TEU to shippers who use SRIT service; S_2 means a subsidy of 200 RMB/TEU; S_3 means a subsidy of 300 RMB/TEU; S_4 means a subsidy of 400 RMB/TEU; S_5 means a subsidy of 500 RMB/TEU; S_6 means a subsidy of 600 RMB/TEU and S_7 means a subsidy of 700 RMB/TEU. The market share of RT, SRIT and the MSP is presented in Table 26.3.

Table 26.3 shows that total MSP increase with the amount of subsidy to shippers. Specifically, a 100 RMB/TEU increase in subsidy will result in an average of 1.73 % increase in MSP. When it comes to ΔMSP_{i+1}, it presents a similar growing trend but begins to decrease from S_5 scenario. In particular, all the values of ΔMSP5, ΔMSP6 and ΔMSP7 are below the average increase value of 1.73 %. It reveals that if to achieve a cost-effective modal shift, giving a subsidy of 400 RMB/TEU to shippers who use SRIT will be a relatively optimal choice.

26.5 Conclusions

To assess the MSP of different degree of subsidies in favor of SRIT between Ningbo Port and Shangrao city, this paper established a MMNL model and a MNL model based on random utility theory and SP techniques. The results show that all attributes have the expected sign and the MMNL model outperforms the MNL model. A negative sign for cost and time indicates that the overall marginal will be reduced with the increase of cost and time, but this is opposite for punctuality and safety. In view of the restrictions of transport infrastructure and management system in the short time, making subsidy policy to indirectly reduce shipper's transport cost becomes an effective means for achieving modal shift from RT to

SRIT. The results also show that neither RT nor SRIT has been preferred a priori and there is not enough room to alter the sensitivity of Shangrao shippers to punctuality and safety. Besides, the findings suggest that giving a subsidy of 400 RMB/TEU to shippers who use SRIT might be the optimal choice for a cost-effective modal shift.

The methodology performed in this study can also be applied to other segments of the freight market or other regions. Given that the SP data may lead to potential biases, establishing a RP/SP combined model will be the future direction of our research.

References

Bhat C (2001) Quasi-random maximum simulated likelihood estimation of the mixed multinomial logit model. Transp Res Part E: methodological 35(7):677–695
Beuthe M, Bouffioux C (2008) Analysing qualitative attributes of freight transport from stated orders of preference experiment. J Transp Econ Policy 42:105–128
Brownstone D, Train K (1999) Forecasting new product penetration with flexible substitution patterns. J Econometrics 89:109–129
Catalani M (2001) A model of shipper behaviour choice in a domestic freight transport system. In: Proceedings of the 9th World Conference on Transport Research, Seoul, pp 22–27
Chiara B, Deflorio F, Spione D (2008) The rolling road between the Italian and French Alps: modeling the modal split. Transp Res Part E Logist Transp Rev 44:1162–1174
Golias J, Yannis G (1998) Determinants of combined transport's market share. Transp Logist 1(4):251–264
Hensher D, Greene W (2003) The mixed logit model: the state of practice. Transportation 30(2):133–176
Holguin-Veras J (2002) Revealed preference analysis of commercial vehicle choice process. J Transp Eng 128(4):336–346
Jiang F, Calzada C (1997) Modelling the influences of the characteristics of freight transport on the value of time and the mode choice. In: Proceedings of the 25th European transport forum: seminar E, London, pp 113–124
Jiang F, Johnson P, Calzada C (1999) Freight demand characteristics and mode choice: an analysis of the results of modeling with disaggregate revealed preference data. J of Transp Stat 2(2):149–158
McFadden D (1974) The measurement of urban travel demand. J Publics Econ 3:303–328
McFadden D, Ruud P (1994) Estimation with simulation. Rev Econ Stat 76(4):591–608
McGinnis M, Corsi T, Roberts M (1981) A multiple criteria analysis of modal choice. J Bus 217 Logist 2(2):48–68
Nam K (1997) A study on the estimation and aggregation of disaggregate models of mode choice for freight transport. Transp Res Part E Logist Transp Rev 33(3):223–231
Norojono O, Young W (2003) A stated preference freight mode choice model. Transp Plann Tech 26(2):195–212
Shinghal N, Fowkes T (2002) Freight mode choice and adaptive stated preferences. Transp Res Part E Logist Transp Rev 38(5):367–378
Train K (2003) Discrete choice methods with simulation. Cambridge University Press, Cambridge
Winston CM (1981) A disaggregate model of the demand for intercity freight transportation. Econometrica J Econometric Soc 49(4):981–1006

Chapter 27
Study of Training System Applying on Energy-Saving Driving

Haili Yuan, Bin Li and Wei Wang

Abstract In this paper, the importance of energy saving on urban railway transportation and inevitability of energy-saving training are emphasized. To improve the energy-saving performance in the process of train operation, we propose a driver training system based on CBTC simulation test platform. In this system, driver's energy-saving awareness can be strengthened. Besides, an energy-saving strategy, coasting driving, is researched and verified in this system as well. The driver training system with 3D video simulation technology and actual driving platform provides nearly practical driving experience. In this way, drivers get trained in an energy-saving form compared to the traditional form in the field.

Keywords Energy-saving driving · 3D video simulation · Energy consumption · Coasting position

27.1 Introduction

Low-carbon economy, a basic way to coordinate the development of society and economy, is gradually being paid more attention in the whole world. Moreover, it also plays a significant role in ensuring energy security and responding to climate

H. Yuan (✉) · B. Li · W. Wang
State Key Laboratory of Traffic Control and Safety, Beijing Jiaotong University, Beijing 100044, China
e-mail: 11120318@bjtu.edu.cn

B. Li
e-mail: 11120274@bjtu.edu.cn

W. Wang
e-mail: wwang6@bjtu.edu.cn

change. As a worldwide problem, it has brought new challenges to all walks of life. Urban rail transportation, as one of energy intensive consumers, must meet the challenge of Low-carbon economy and achieve the energy-saving goal.

Under normal circumstance, line conditions have effects on energy consumption from a certain extent, especially the gradient of a line. In addition, temporary speed restriction also has a significant impact on energy consumption (Wang 2008).

Besides of line conditions, the patterns of driving a train are also relevant in energy consumption (Liu and Mao 2007). So, it is necessary to build drivers' awareness of energy-saving and develop energy-saving operation habit. An efficient method for drivers to learn energy-saving driving is to operating the simulation train using 3D video simulation technology. During the driver's operation, this driver training system can compute the energy consumption, which is the criteria used to tell a driving habit is energy-saving or not. By this way, a driver can develop his energy-saving habit without consuming much energy in the field.

27.2 System Design

The driver training system introduced in this paper is based on CBTC simulation and test platform (Wang 2010), which is independently researched and developed by State Key Laboratory of Traffic control And Safety Beijing Jiaotong University.

27.2.1 Composition of Driver Training System

Driver training system consists of 3D video simulation module; energy consumptions display module and bridge module (see Fig. 27.1).

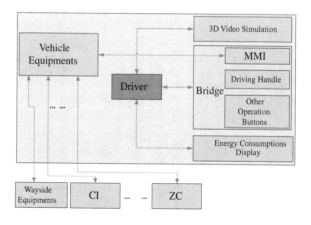

Fig. 27.1 The composition of driver training system

3D video monitor is settled in front of locomotive, and energy consumptions display monitor is settled on the bridge. During the training, with the operational handles, operational buttons and 3D video, driver can get a verisimilar feeling, just like driving a real train in a real line.

27.2.2 3D Video Module Design

The design process of 3D video module is shown in Fig. 27.2. It includes three parts: line data processing, 3D environment modeling and 3D scene displaying. Line data processing and 3D environment modeling are two preparatory stages. 3D scene displaying provides basic data for 3D environment modeling and 3D scene displaying. With the data, 3D modeling software generates railway lines, and combine the lines and 3D scenes. At last, 3D scene models and basic data are loaded to 3D scene driving program, and then the work is done.

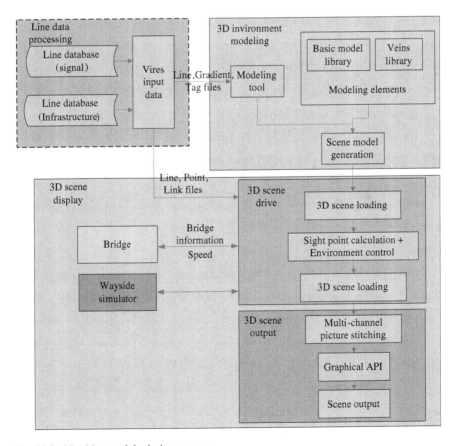

Fig. 27.2 3D video module design process

27.2.3 Energy Consumption Module Design

For simplifying the computation, this module does not take factors of infrastructure conditions and line environment into consideration. And other conditions will not change in the process of simulation, such as locomotive, line, dynamic and braking techniques. In this paper, energy consumption mainly involves three parts: start and stop energy consumption, moving energy consumption, coasting energy consumption (Xue et al. 2007).

Start and stop energy consumption E_1:

$$E_1 = 0.5nA\left(\sum Q\right)v^2 / (3.6^3 * 1000) \tag{27.1}$$

In the formula, A is coefficient of rotary inertia; Q is mass of the train; v is velocity, n is start and stop times of the train in the corridor, which is assumed to be 1.

Moving energy consumption E_2:

$$E_2 = 9.81/3600 * (\Sigma Q.w)S \tag{27.2}$$

ω is unit resistance of the train, S is running distance.

Coasting energy consumption:

$$E_3 = B * E_2 \tag{27.3}$$

Here B is control coefficient.
The total energy consumption is.

$$E = E_1 + E_2 + E_3 \tag{27.4}$$

27.3 System Function Simulation

27.3.1 3D Video Simulation

Figure 27.3 shows the development process of 3D video simulation. The left part shows daytime pattern, the right shows nighttime pattern. We can change patterns through button "L". During the operation of simulation train, we can see the scene of the train moving forward in video screen from one termination, and see the scene of the train moving backward in video screen from the other termination.

27 Study of Training System Applying on Energy-Saving Driving

Fig. 27.3 3D video simulation display

27.3.2 Energy Consumption Display

Minimum of energy consumptions produced during the training are recorded in energy consumption list. Pick the minimum energy consumption from the list, and find its corresponding coasting position, which turns to be the optimal coasting position at present, that is proposed coasting position. See in Fig. 27.4.

In our simulation, the simulation train travels from Xi'erqi station to Beiqinglu station. Energy consumption list shows energy consumptions with different coasting positions. The unit of energy consumption is 1000000 KW/h. Column in the middle shows corresponding coasting positions which is the distance from Xi'erqi. And current energy consumption shows real-time energy consumption of the train and its distance from Xi'erqi station. If there is no coasting driving,

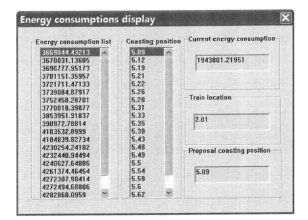

Fig. 27.4 Energy consumptions display

energy consumption is about 4.2838 KW/h, the current optimal coasting driving seen from the list consumes about 3.6698 KW/h. The difference between the two situations is obvious. Here comes the conclusion: coasting from an opportune position can contribute to saving energy.

27.4 Conclusions

This paper proposes a driver training system based on the CBTC simulation and test platform out of the view of energy saving, and it is verified on Changping Line partly. However, our simulation is based on an assumption that infrastructure conditions and line environment don't change in the process of simulation. In fact, real energy consumption involves complex factors, including traction and braking techniques, environmental change, temporary speed restriction, and the handling of emergency situations, which need to be studied later.

References

Liu H, Mao B (2007) Research on train energy conservation issues and strategy of urban rail transit. J Transp Syst Eng Inf Technol 7(5):68–73
Wang Q (2008) Ideas and methods on train energy-saving operation. Railw Locomot Car 28(4):64–67
Wang D (2010) Research on automatic test applied in CBTC system. Beijing Jiaotong University, Beijing
Xue Y, Ma D, Wang L (2007) Calculation method of train traction energy consumptions. China Railw Sci 28(3):84–87

Chapter 28
The Outlook of Low-Carbon Transport System: A Case Study of Jinan

Qiang Han and Yong Zhou

Abstract Jinan, the capital of Shandong province, is located at coastal regions in east China. It is troubled with large traffic demand as many other cities do. Transport industry has been one large energy consumer, which carbon emission accounts for nearly 10 % of all the industries. In this situation, low-carbon transport with low energy consumption, low CO_2 emission and low pollution is an increasing concern. This paper disclosed the relevance between transport and economy in Jinan by grey relational analysis, and described the causal relationship among several factors involved with low-carbon transport system. The making of low-carbon transport system in Jinan was discussed from short term and long term. The former was on the basis of energy saving potential, while the latter was based on scenario analysis. According to Jinan's target for CO_2 emission reduction, two key uncertainties, transport socialization and sustainability, are adopted as the axes to generate four future scenarios. To achieve the fourth scenario in favor of carbon reduction, much propulsive policies about structure adjustment, new energy use and public traffic travel mode have been put forward.

Keywords Low-carbon transport · Causal relationship in transport system · Scenario analysis

---

Q. Han (✉)
School of Management Science and Engineering, Shandong University of Finance and Economics, Jinan 250014, China
e-mail: qiang.han@sdufe.edu.cn

Y. Zhou
Institute of Science and Technology Development Strategy of Shandong Province, Jinan 250014, China
e-mail: bft@qq.com

28.1 Introduction

China's economy is moving onward in surprising speed. Affected by this trend, many cities of China are also developing rapidly. Transport is one fundamental, leading and service business for the development of national economy and the society. During their rapid ongoing process, they are facing critical and worsening transport problems, including traffic jam, increasing energy consumption and excessive carbon emission (Chen et al. 2009). This urban transport crisis results from continuing population growth, suburban sprawl, rising incomes and increased motorization and use (Pucher et al. 2007).

Jinan, south to Mount Tai and neighboring the Yellow River on the north, is an eastern coastal opening city and one of the 15 nation-authorized deputy provincial cities in China. It is the capital of Shandong province, and also the province's political, economic, cultural, science and technology, educational and financial center. The total area of Jinan is 8177 km^2 and the urban part occupies 3257 km^2.

The possession of motor vehicles is 1.31 million by the end of October, 2011, in which there are 0.91 million cars. Private cars have occupied vast majority share, reaching 0.76 million. As a result, Jinan is confronted with huge traffic demand. Transport industry has become a big energy consumer, which CO_2 (carbon dioxide) emissions take up 10 % of the city, and this ratio is on the increase. Traffic jam, air pollution and traffic safety it caused have greatly influenced economic efficiency and life quality of Jinan. Building up low-carbon transport system is valuable for Jinan and other cities in China.

Low-carbon transport is one transport development mode with low energy consumption, low CO_2 emission and low pollution. It aims to reduce the carbon emission of vehicles and achieve the sustainable development of transport system. A target for transport emission reduction in Jinan can be derived to help explore the likely required scales of change. In 2009, as a participant in the Copenhagen Accord, China pledged to reduce its economy's carbon intensity by 40–45 % by 2020 compared to 2005 levels. Jinan's potential transport CO_2 intensity target is based on an equivalent aspiration to the national target.

28.2 Relevance Between Transport and Economy in Jinan

As one important power to push economy forward, development of transport industry will produce a series of results, such as increase of new job opportunity, strengthening of regional self-development ability, and much possibility to increase income. So, the meaning of transport industry development is not merely to prolong the road length, but also bring positive effect for industry development, economic structure, etc. On the other hand, different factors also influence economy development in varying degree.

We adopt the Grey Relational Analysis method (Deng 1987) to seek the relevance between transport and economy in Jinan. According to the principle of authenticity, completeness and availability, we select GRP (gross regional product) to indicate economic growth, and passenger-kilometers, freight tonne-kilometers, length of highways in operation and possession of civil vehicles to describe the transport system.

The result is that Jinan's economy has strong relevance with transport. All the indices present positive correlation. Among them, the relation with possession of civil vehicles is most close, the next is freight tonne-kilometers, the following are passenger-kilometers and length of highways in operation.

Transmission mechanisms between transport industry and economic growth go as follows. First of all, output of some transport infrastructure industry constitutes part of the national economy. Second, the main utility of transport infrastructure is its function. The growth of the national economy is supported by contribution of such final production elements as labor, capital, and land. As a specific form of production elements, transportation infrastructure accommodates these final elements objectively, which plays the irreplaceable role to the entire economic growth. Third, transport infrastructure investment directly stimulates the growth of the national economy by multiplier effect, and the conduction process complies with the general investment principle. At last, from the feedback perspective, transport infrastructure keeps adjusting itself in certain sensitivity to achieve the harmonious development with national economy according to the sensitivity of economic growth.

28.3 Causal Relationship in Low-Carbon Transport System

Building low-carbon transport system is not a simple and separated task, but a systematic engineering involved with complex factors, which must be reflected comprehensively in macroscopical decision level, mesoscopical management level, and microscopic technology level. These factors include economy development, population size, ecological environment, transport infrastructure, investment, policy and technology level, etc. Their causal relationship is demonstrated in Fig. 28.1, in which the arrow shows their causal relationship, plus sign means positive effect, and minus sign expresses negative effect.

Main feedback loops are as follows:

(1) Transport Operation $\xrightarrow{+}$ Economy Development $\xrightarrow{+}$ Investment Increase $\xrightarrow{+}$ Transport Infrastructure $\xrightarrow{+}$ Transport Operation

(2) Transport Operation $\xrightarrow{-}$ Ecological Environment $\xrightarrow{+}$ Economy Development $\xrightarrow{+}$ Technology Level $\xrightarrow{+}$ Transport Operation

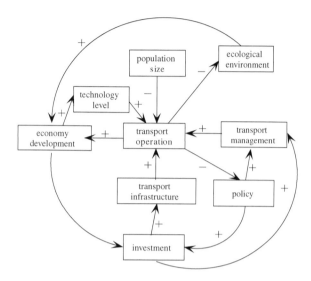

Fig. 28.1 Cause and effect diagram of transport system

(3) Transport Operation $\xrightarrow{-}$ Policy $\xrightarrow{+}$ Transport Management $\xrightarrow{+}$ Transport Operation

(4) Transport Operation $\xrightarrow{+}$ Economy Development $\xrightarrow{+}$ Technology Level $\xrightarrow{+}$ Transport Operation

(5) Transport Operation $\xrightarrow{-}$ Policy $\xrightarrow{+}$ Investment $\xrightarrow{+}$ Transport Infrastructure $\xrightarrow{+}$ Transport Operation

In addition, Jinan city is still in the rapidly developing stage, the trend of increasing population is more and more obvious. It constitutes negative polarity for transport operation.

28.4 Jinan's Short-Term Transport CO$_2$ Emission Based on Energy Saving Potential

We take international and internal advanced level as the benchmark and the time internal is from 2005 to 2015.

Energy saving potential of Jinan under efficiency convergence in 2008 is 35.1 %. This means that Jinan has a rather large space to reduce energy consumption, and so does the CO$_2$ emission.

In Jinan, transportation industry is the second petroleum product consumer, only next to manufacturing industry, and its consumption is one third of the society. Transport system is Jinan's primary industry for energy saving and carbon emission reduction.

Until now, Jinan has achieved much in energy saving of transport. From 2005 to 2010, gross production of transport industry has risen from 11.22 to 23.68 billion RMB Yuan, and annual average growth rate is 13.25 %; corresponding to this, Jinan's energy consumption has increased greatly from 2.79 million tce (ton of standard coal equivalent) in 2005 to 4.07 million tce in 2008,[1] and its annual average growth rate is 13.46 %. Energy consumption of transport accounts for 10.23 % of the total in 2005, reaching 12.61 % in 2008. Between 2005 and 2010, energy consumption per 10,000 RMB Yuan of transportation output value has reduced from 2.48 to 2.00 tce/10,000 RMB Yuan, annually 3.65 %. Learning from the historic experience of transportation industry in developed regions and combining with its basic trend in Jinan, it can be concluded that the increment speed of production and energy consumption of Jinan's transportation industry during 12th Five-Year Plan will slow down, at the rate of 11 %[2] and 8 %,[3] respectively.

According to national development and reform commission of China, burning 1 tce will produced 2.62 tonnes CO_2. So, 4.92 Mt (million tonnes) CO_2 emissions would have been reduced by 2015.

28.5 Jinan's Long-Term Transport CO_2 Emissions Based on Scenario Analysis

There has no definite CO_2 reduction goal for each industry in China. As a participant in the Copenhagen Accord, however, China has pledged to reduce its economy's carbon intensity by 40–45 % by 2020 compared to 2005 levels. Here, we assume that Jinan's potential transport CO_2 intensity goal keeps in step with the national.

Future scenarios can be generated in view of likely trends and uncertainties, and be used to help assess likely progress against aspirations. Two key uncertainties, sustainability (Banister et al. 2000) and transport socialization, are used to generate the two axes within the classic dilemma scenario matrix (Mahoney 2000). Among these scenarios, the first correspond to BAU (Business As Usual) and the fourth is the best expectancy. Scenario II and III are not unlikely to take place.

Scenario I assumes large transport input, along the road of North America. It is based on a high level of public transport, but also a lower level of sustainability. There are still so many conventional petrol cars, with little efforts to car limit (Bonsall and Young 2010) and technology innovation (Cerry et al. 2009). In this scenario, transport CO_2 emissions rise from 5.93 Mt in 2005 to 137.25 Mt in 2030. This also will result in the increase of per capita transport emissions from 0.99 tonnes in 2005 to 16.76 tonnes in 2030.

---

[1] No more latest data about energy consumption of Jinan is available from official mission.
[2] It is based on the equivalent economic growth rate of Jinan in issued 12th Five-Year Plan.
[3] It is obtained by trend extrapolation.

Scenario IV assumes much increased level of sustainability and transport socialization. It is based on a high level of public transport, as well as a high level of sustainability. There is less conventional petrol car, with much effort to car limit and public transport promotion (Hao et al. 2011). Under this scenario, transport CO_2 emissions rise from 5.93 Mt in 2005 to 91.63 Mt in 2030, and person transport emissions rise from 0.99 tonnes in 2005 to 5.02 tonnes in 2030. Although this scenario seems optimistic, the 45 % carbon reduction in intensity target would not be achieved. Compared with 2005 levels, carbon emission intensity rises by 8.97 %. So, accomplishing the task satisfactorily would require much more propulsive policies, which include advance structure adjustment of transport capacity and organizing, improve actual load rate of coaches, popularize oil consumption supervisory device and focus on vehicle maintenance (Sperling and Gordon 2009; CAI-Asia 2009), and develop convenient public traffic travel mode (Darido et al. 2009; Paulley et al. 2006).

28.6 Conclusions and Future Works

Transport is related with economy and environment. The problem before us is how to develop transport and economy while reducing carbon emission. It can be seen from the cause and effect diagram of transport system that technology and positive policy will be helpful. In short term, energy saving potential can be dug out by technology progress. In long term, however, the CO_2 emission reduction depends more on macro policies, such as structure adjustment, bus priority, etc. Then, in the next step, we should determine the contribution degree and implementation difficulties of different measures, which will be really valuable for decision makers.

Acknowledgments This work was financially supported by Humanities and Social Science Projects of the Ministry of Education of China (09YJC630141), and Shandong Province Soft Science Project (2010RKMA1001).

References

Banister D, Stead D, Steen P et al (2000) European transport policy and sustainable mobility. Spon, London
Bonsall P, Young W (2010) Is there a case for replacing parking charges by road user charges? Transp Policy 17:323–334
CAI-Asia (2009) International council on clean transportation: review of fuel efficiency standards. CAI-Asia, Manila
Cerry C, Weinert J, Yang X (2009) Comparative environmental impacts of electric bikes in China. Transp Res D 14:281–290
Chen F, Zhu DJ, Xu K (2009) Research on urban low-carbon traffic model, current situation and strategy: an empirical analysis of Shanghai. Urban Plann Forum 6:39–46

Darido G, Torres-Montoya M, Mehndiratta S (2009) Urban transport and CO_2 emissions: some evidence from Chinese cities. World Bank, ESMAP, AusAID, Washington DC

Deng J (1987) Basic method of gray system. Huazhong Polytechnic University Press, Wuhan (in Chinese)

Hao H, Wang H, Yi R (2011) Hybrid modeling of China's vehicle ownership and projection through 2050. Energy 36:1351–1361

Mahoney J (2000) Path dependence in historical sociology. Theory Soc 29:507–548

Paulley N, Balcombe R, Mackett R et al (2006) The demand for public transport: the effects of fares, quality of service, income and car ownership. Transp Policy 13:295–306

Pucher J, Peng Z-R, Mittal N et al (2007) Urban transport trends and policies in China and India: impacts of rapid economic growth. Transp Rev 27:379–410

Sperling D, Gordon D (2009) Two billion cars: driving toward sustainability. Oxford University Press, New York

Chapter 29
The Logit Model in the Urban Low Carbon Transport and Its Application

Zinan Yang and Xifu Wang

Abstract At present, low-carbon problems have become a hot spot at home and abroad, particularly in the field of transportation low-carbon transport is an important part of energy saving and emission reduction. This paper research the low-carbon factors impact on other transport modes in the city, and study the change of various transport modes share rate. A Logit model which is broadly applied in the transportation research was used for the study which selected economy, speed, convenience, comfort, safety and low-carbon as its 6 index. The share rate model of various transport modes is created according to the relevant research data and the maximum likelihood estimation method. Finally, the share rate considering the low-carbon factors are calculated and contrast the share rate value of before and after low-carbon factors.

Keywords Low-carbon · Logit model · Share rate

29.1 Introduction

The Logit model is a widely used model in the urban transport share rate. Logit model can express all aspects of travel choice influencing factors, to improve the prediction accuracy and practicality of the model (Jia et al. 2007). In this paper, the research of urban transport share rate use the Logit model especially considering

---

Z. Yang (✉) · X. Wang
School of Traffic and Transportation, Beijing Jiaotong University, Beijing 100044, China
e-mail: 11120905@bjtu.edu.cn

X. Wang
e-mail: Xfwang1@bjtu.edu.cn

the impact of share rate in the future low carbon transport. First, establish the generalized cost function of the various transport modes without considering the carbon emission factors, and calibrate the model parameters based on actual survey, and then study the share rate of the various transport modes when considering the carbon emission factors.

29.2 Model Building

According to research, in the generalized cost function, set the economy, speed, convenience, comfort, safety and low-carbon six service characteristics. The service characteristics of various transport modes quantified as follow.

(1) Economy (E_i). The public transport use various transport modes fares as measurement index, private transport use the price of fuel consumption as the measurement index of economy.
(2) Speed (F_i). Use the ratio of the average travel distance (L) and the average travel speed of all transport modes (v_i) as the measurement index of speed (F_i).

$$F_i = L/v_i \tag{29.1}$$

The travel time in the survey in fact include: pure driving time, parking time, waiting time and transfer time. According to the resident survey data and related literature (Ma et al. 2006), propose the value of the average travel speed of the various transport modes, shown as Table 29.1.
(3) Convenience (C_i). The public transport uses the average waiting time; Private transport considers the average parking time as the measurement index of convenience. Combined with the actual situation, propose the average waiting time, shown as Table 29.2.
(4) Comfort (M_i). In order to reflect the comfort features of the various transport modes in the generalized cost function, need to quantify the comfort. Comfort indicator related to travel time, expenses, psychological needs, personal space and so on, the base value take the 8 % of the various transport modes freight (Zhao et al. 2008). The specific values are shown in Table 29.2.

Table 29.1 Average travel speed of various transport modes

Transport mode	Average travel speed/(km/h^{-1})
Bicycle	10
Battery vehicle	14
Motorcycle	16
Bus	20
Taxi	25
Private cars	18

29 The Logit Model in the Urban Low Carbon Transport

Table 29.2 Various transport modes service characteristics of the Binhai New Area

Transport characteristics	Economy (E_i)	Speed (F_i)	Convenience (C_i)	Comfort (M_i)	Safety (S_i)
Bicycle	0	0.90	0.05	0.16	0.99
Battery vehicle	0	0.64	0.05	0.40	0.98
Motorcycle	2.4	0.56	0.05	0.64	0.98
Bus	2.0	0.45	0.17	0.48	0.98
Taxi	16.3	0.36	0.08	1.28	0.97
Private cars	4.0	0.50	0.08	1.44	0.98

(5) Safety (S_i). According to the number of the various transport modes vehicle accidents and security arrangement law set in the literature (Jia et al. 2007). The specific values are shown in Table 29.2.

(6) Low-carbon (R_i). The Values of low-carbon factors (R_i) use the ratio of the average travel per capita CO_2 emissions (A_i) as a reference value. The formula is (29.2).

$$R_i = 1 - \frac{A_i}{\sum_{i=1}^{n} A_i} \qquad (29.2)$$

Now actually does not consider the low-carbon factors, so take the value 1.

Due to the E_i and C_i are based on the time value, so need to multiply by the value of travel time (V(T)). M_i has subjective factors, and therefore set up separate coefficient. S_i and R_i are the characteristics can't measure by the time and price value, it should be the separate proportional, thereby establishing the generalized cost function (He et al. 2006; Zeng 2007).

$$V_i = \frac{\theta U_1 - \xi M_i}{S_i \cdot R_i} \qquad (29.3)$$

$$U_i = E_i + (F_i + C_i) \cdot V(T) \qquad (29.4)$$

A_i is the comprehensive cost of the various transport mode, $V(T)$ is the local average time value, θ and ξ are characteristic coefficient.

Use Logit model to calculate the urban various transport modes share rate. Consider the relationship with various parameters, and improve the Logit model as follow.

$$P_i = \frac{\exp(-V_i)}{\sum_{i=1}^{n} \exp(-V_i)} \qquad (29.5)$$

29.3 Example Calculation

TianJing Binhai New Area is situated in the north of the north China plain and very close to the two megalopolises of Beijing and Tianjin, with a vast area of hinterland. Binhai New Area proudly occupies the core position of the Bohai Rim economic circle, facing Japan and the Korean Peninsula across the sea, and directly encountering Northeast Asia and the rapidly rising Asian-Pacific economic circle.

29.3.1 Parameter Calibration

According to the trip research of Binhai New Area in June 2011, the he average trip distance of traveler in Binhai New Area is 8.57 km, considering the distance loss, set the travel distance at 9 km. Various transport modes service characteristics of the Binhai New Area is shown as Table 29.2.

And the average time value formula is: V(T) = GDP ÷ The local population × The average labor time. Calculated the results: V(T) = 32.61yuan/h. According to the travel choice survey, get the existing urban transport share rate. It is shown in the Table 29.3.

Suppose there are N travelers, choose M kinds of transport of modes. It can be seen as N times Bernoulli trials, using the maximum likelihood estimates.

$$f(\theta|P) = P(N_1, N_2, \ldots, N_m|\theta) = \frac{N!}{N_1! N_2! \ldots N_m!} \Pi P_i^{N_i} \tag{29.6}$$

Though calculate the partial derivative of θ and ξ, to make derivative values to 0. And input the parameter, N = 9950, M = 6, get the results. $\theta = 0.1203, \xi = 0.08769$. And then input the θ and ξ into generalized cost function, obtained Binhai New Area travel cost function model.

$$V_i = \frac{0 \cdot 1203 U_i - 0 \cdot 08769 M_i}{S_i \hbar R_i} \tag{29.7}$$

$$U_i = E_i + 32.61(F_i + C_i) \tag{29.8}$$

Table 29.3 The share rate of various transport modes

Transport mode	Bicycle	Battery vehicle	Motorcycle	Bus	Taxi	Private cars
Number	793	1647	1890	3428	750	1442
Share rate (%)	7.97	16.55	18.99	34.55	7.54	14.49

Table 29.4 The share rate of various transport modes after considering the low-carbon factors

Transport characteristics	Economy (E_i)	Speed (F_i)	Convenience (C_i)	Comfort (M_i)	Safety (S_i)	Low-carbon (R_i)	Share rate (P_i)
Bicycle	0	0.90	0.05	0.16	0.99	1	11.62
Battery vehicle	0	0.64	0.05	0.40	0.98	1	30.85
Motorcycle	2.4	0.56	0.05	0.64	0.98	0.79	14.80
Bus	2.0	0.45	0.17	0.48	0.98	0.93	37.46
Taxi	16.3	0.36	0.08	1.28	0.97	0.74	3.03
Private cars	4.0	0.50	0.08	1.44	0.98	0.54	2.24

29.3.2 The Share Rate After Considering the Low-Carbon Factors

The implementation of urban low-carbon transport operating strategy will make each properties of Binhai New Area transport improve greatly, and use Logit model to calculate the share rate of various transport modes in the future. The data is shown in Table 29.4.

29.4 Conclusion

Logit model is an effective way to research variety of transport modes share rate. In this paper, the model studies the reasonable share rate of various transport modes in the future, considering the low-carbon factors. According to the results of model calculations, the share rate of bicycle, battery vehicle and bus increase, the share rate of taxi and private cars descend. It meets the low-carbon cities policy to encourage clean energy and public transport. With the increasing attention on low-carbon economy at home and abroad, low-carbon factors will become more frequent consideration to the transport field.

References

He Y, Mao B, Chen TS, Yan J (2006) The mode share model of the high-speed passenger railway line and Its application. J China Railway Soc 28(3):18–21

Jia H, Gong B, Zong F (2007) Disaggregate modeling of traffic mode choice and its application. J Jilin Univ 37(6):1288–1293

Ma J, Bian Y, Wang W (2006) On the relative trip comfort between all traffic modes. In: Proceedings of the 6th International Conference of Transportation Professionals, pp 111–116

Zeng X, Wang C (2007) Improvement of Logit model and its application in forecasting the distribution ratios of passenger flows on Cheng-Yu intercity railroad. J Changsha Commun Univ 23(4):50–53

Zhao L, Du W, Hao G, Peng Q (2008) The transport modes lattice-order decision-making based on travel time reliability. Stat Deci 24(9):43–46

Chapter 30
Comprehensive Evaluation of Highway Traffic Modernization Based on Low-Carbon Economy Perspective

Linlin Zheng, Yongbo Lv, Li Chen and Le Huang

Abstract The comprehensive evaluation index system of highway traffic modernization is established on the basis of the long-term plan for highway energy-saving of China. The AHP-Entropy method is expounded for quantifying the weight of evaluation index. In order to make up for the deficiencies of the present highway traffic modernization evaluation method, the multi-level extensible evaluation model is developed with the extension method as the core. Jiangxi Province is studied as an example, indicating that the new method is applicable to highway traffic modernization evaluation.

Keywords Low-carbon economy · Highway traffic modernization · Index system · Multi-level extension

30.1 Introduction

In recent years, the highway traffic facilities and management facilities are improved significantly, while the situation of energy saving and environmental protection is very severe. As a huge traffic system involving multi-participation of

L. Zheng (✉) · Y. Lv · L. Chen · L. Huang
School of Traffic and Transportation, Beijing Jiaotong University, Beijing 100044, China
e-mail: 11120915@bjtu.edu.cn

Y. Lv
e-mail: yblv@bjtu.edu.cn

L. Chen
e-mail: 08121256@bjtu.edu.cn

L. Huang
e-mail: 11120882@bjtu.edu.cn

people, vehicle and road, the highway traffic is the most important field of energy saving and emission reduction that accounted for over 80 % energy consumption is concentrated in this area (Li 2011, Hu 2008). So establishing an evaluation index system with the embodiment of scientificity and applicability is imperative, through the evaluation of highway traffic development by the index system, the decision makers can recognize the gap between the current situation and the requirements of modern highway traffic system, discern the weak links and clear the focal point of development in the future.

30.2 Evaluation Index System

On the basis of analysis of influencing factors (Unified Plan Department of the Ministry of Transport of the People's Republic of China 2008, Han et al. 2011) of low-carbon highway traffic modernization, this paper establishes the highway traffic modernization index system from four aspects namely structural low-carbon, technical low-carbon, management low-carbon and policy low-carbon, as shown in Table 30.3. All of the indexes, scored by 100-point system, are divided into three grades as shown in Table 30.1.

The quantization method of each index is shown in Table 30.2. The quantization of quantitative indexes, using the planning value of highway traffic energy-saving index of our country (Unified Plan Department of the Ministry of Transport of the People's Republic of China 2008) as a standard, is determined with the comprehensive consideration of variation range and implementary difficulty; the quantization method of qualitative indexes is determined after consultation with experts.

30.3 Evaluation Method

30.3.1 Determining Weights by AHP-Entropy

The method of determining index weight in this paper is using Entropy to modify the initial weight determined by AHP, which can effectively reduce subjective

Table 30.1 Grading standard of modernization degree

Grade	Interval	Modernization degree
1	≥75	High
2	60–74	Middle
3	≤59	Low

Table 30.2 The quantization of modernization evaluation index

Evaluation index	Attribute	Quantization method
Proportion of second-class highway and above C_{11}	Quantitative	75 points when it reaches 21 %, plus(minus) 1 point per 0.5 % more(less)
Rate of road pavement C_{12}	Quantitative	75 points when it reaches 75 %, plus(minus) 1 point per 1 % more(less)
Proportion of diesel passenger cars C_{13}	Quantitative	75 points when it reaches 73 %, plus(minus) 1 point per 1 % more(less)
Proportion of diesel freight cars C_{14}	Quantitative	75 points when it reaches 90 %, plus(minus) 1 point per 0.5 % more(less)
Proportion of alternative fuel of passenger cars C_{15}	Quantitative	75 points when it reaches 6 %, plus(minus) 1 point per 0.2 % more(less)
Average tonnage of freight cars C_{16}	Quantitative	75 points when it reaches 4.5 tons, plus(minus) 1 point per 0.2 ton more(less)
Proportion of large-scale freight cars C_{17}	Quantitative	75 points when it reaches 80 %, plus(minus) 1 point per 1 % more(less)
Coverage of ETC C_{21}	Quantitative	75 points when it reaches 60 %, plus(minus) 1 point per 1 % more(less)
Technical condition of in-use vehicles C_{22}	Qualitative	Whether strengthening in-use vehicles' periodic detection, repair and maintenance or not, improving technical condition of commercial vehicles
Application of on-board energy-saving equipment C_{23}	Qualitative	Application and popularization of automobile energy saving technology and products, such as radial tire, baffle, fan clutch
Coverage of travel information system C_{24}	Quantitative	75 points when it reaches 80 %, plus(minus) 1 point per 1 % more(less)
Proportion of tractor-trailer transportation C_{31}	Quantitative	75 points when it reaches 15 %, plus(minus) 1 point per 0.5 % more(less)
Actual loading rate of passenger transportation C_{32}	Quantitative	75 points when it reaches 74 %, plus(minus) 1 point per 1 % more(less)
Degree of traffic congestion C_{33}	Qualitative	The degree of energy consumption because of congestion, whether taking effective measures to ease the traffic congestion
Utilization of freight mileage C_{34}	Quantitative	75 points when it reaches 67 %, plus(minus) 1 point per 1 % more(less)
Proportion of energy-saving driving C_{35}	Quantitative	75 points when it reaches 70 %, plus(minus) 1 point per 1 % more(less)
Standard formulation C_{41}	Qualitative	Relevant standards of highway traffic based on the perfect degree of carbon intensity and total carbon index
Low-carbon subsidies C_{42}	Qualitative	The subsidy of purchasing new energy vehicle, scrap-purchasing, etc.
Access and withdrawal institution C_{43}	Qualitative	Establishment and perfection of the access and withdrawal system based on carbon intensity

uncertainty, so as to determine the index weight more reasonably, its calculating steps as follows:

(1) Constructing the judgment matrix $\overline{A} = \{\overline{a}_{ij}\}_{n \times n}$, and we can obtain the weight coefficient of initial index $\theta_j = (\theta_1, \theta_2, \cdots, \theta_n)^T$ through AHP.

(2) After uniformization of judgment matrix $\overline{A} = \{\overline{a}_{ij}\}_{n \times n}$, we can obtain the canonical judgment matrix $A = \{a_{ij}\}_{n \times n}$, where $a_{ij} = \dfrac{\overline{a}_{ij}}{\sum_{i=1}^{n} \overline{a}_{ij}}$.

(3) The entropy E_j of index j can be got by $E_j = -\dfrac{(\ln n)^{-1}}{\sum_{i=1}^{n} a_{ij} \ln a_{ij}}$, $(0 \leq E_j \leq 1)$.

(4) Then the deviation degree d_j of index j can be calculated by $d_j = 1 - E_j$.

(5) After former step, calculate the correction coefficient $\mu_j = d_j \Big/ \sum_{j=1}^{n} d_j$.

(6) By using correction coefficient μ_j, we can correct the weight coefficient of initial index $\theta_j = (\theta_1, \theta_2, \cdots, \theta_n)^T$ from AHP method through $\theta'_j = \mu_j \theta_j \Big/ \sum_{j=1}^{n} \mu_j \theta_j$.

(7) Finally, we get the combination weight $\omega_j = (\omega_1, \omega_2, \ldots, \omega_m)$ by $\omega_j = \rho \theta_j + (1 - \rho)\theta'_j +$, where ρ is 0.5.

By using the energy saving potential of all indexes as main reference for index comparison, we can obtain the judgment matrix of each level of the evaluation index system through 1–9 comparable scale method, then calculate the weight of each level indexes relate to the higher level index, the result is shown in Table 30.3.

30.3.2 Multi-Level Extensible Evaluation Model

Extensible evaluation model is one of the main applications of extension theory. The core idea of extension theory is matter-element theory. Matter-element is an ordered three-dimensional, which is composed of matter, characters and the matter's quantity value about the characters, and denoted by R = (matter, characters, quantity value) = (U, C, V) (Pan et al. 2011). It is consistent with the thought of mapping relationship between the index system would be established for modernization evaluation and the evaluated matter. The concrete steps are as follows:

(1) The determination of classical field and segment field

Let grade field of modernization is $U = \{U_j, j = 1, 2, \cdots, m\}$; evaluation factor set of first level index is $C = \{C_i, i = 1, 2, \cdots, n\}$.

30 Comprehensive Evaluation of Highway Traffic Modernization

Table 30.3 Evaluation index of modernization and values

Target level N	First level index C_i (Weight W_i)	Secondary level index C_{ik} (Weight W_{ik})	Actual value	Score V_{ik}	Correlation degree $j=1$	$j=2$	$j=3$
Highway traffic modernization	Structural low-carbon (0.252)	Proportion of second-class highway and above (0.251)	15.42 %	64	−0.100	0.125	−0.234
		Rate of road pavement (0.179)	49.93 %	50	0.250	−0.167	−0.333
		Proportion of diesel passenger cars (0.098)	77.6 %	80	−0.500	−0.200	0.333
		Proportion of diesel freight cars (0.173)	87.9 %	71	−0.275	0.160	−0.121
		Proportion of alternative fuel of passenger cars (0.076)	4.8 %	69	−0.225	0.240	−0.162
	Technical low-carbon (0.198)	Average tonnage of freight cars (0.123)	4.05	66	−0.150	0.214	−0.209
		Proportion of large-scale freight cars (0.100)	79.2 %	74	−0.350	0.040	−0.037
		Coverage of ETC (0.121)	38 %	53	0.175	−0.130	−0.319
		Technical condition of in-use vehicles (0.403)	medium	74	−0.350	0.040	−0.037
		Application of on-board energy-saving equipment (0.319)	medium	68	−0.200	0.280	−0.179
	Management low-carbon (0.346)	Coverage of travel information system (0.157)	54 %	49	0.289	−0.183	−0.347
		Proportion of tractor-trailer transportation (0.102)	8.6 %	63	−0.075	0.088	−0.245
		Actual loading rate of passenger transportation (0.278)	79.4 %	80	−0.500	−0.200	0.333
		Degree of traffic congestion (0.233)	good	77	−0.425	−0.080	0.095
		Utilization of freight mileage (0.301)	73.1 %	81	−0.525	−0.240	0.462
		Proportion of energy-saving driving (0.086)	52 %	57	0.075	−0.065	−0.295
	Policy low-carbon (0.204)	Standard formulation (0.366)	medium	65	−0.125	0.167	−0.222
		Low-carbon subsidies (0.288)	good	75	−0.375	0.000	0.000
		Access and withdrawal institution (0.346)	medium	63	−0.075	0.088	−0.245

Tips: The relevant data of Jiangxi Province is from *The Long-term Plan for Highway and WaterCarriage Energy-saving of Jiangxi Province (2008–2020)* and *Jiangxi Statistical Yearbook*

Define

$$R_j = (U_j, C, V_j) = \begin{bmatrix} U_j & c_1 & \langle a_{j1}, b_{j1} \rangle \\ & c_2 & \langle a_{j2}, b_{j2} \rangle \\ & \vdots & \vdots \\ & c_n & \langle a_{jn}, b_{jn} \rangle \end{bmatrix}, \quad R_U = (U, C, V_U)$$

$$= \begin{bmatrix} U & c_1 & \langle a_{U1}, b_{U1} \rangle \\ & c_2 & \langle a_{U2}, b_{U2} \rangle \\ & \vdots & \vdots \\ & c_n & \langle a_{Un}, b_{Un} \rangle \end{bmatrix}$$

where V_j is a classical field and V_U is segment field of U.

(2) The establishment of correlation function and calculation of correlation degree

Then we obtain the correlation degree of the secondary level index of evaluating matter relate to modernization grade $j(j = 1, 2, \cdots, m)$ $(j = 1, 2, \cdots, m)$, which can be expressed as:

$$k_j(c_{ik}) = \begin{cases} \dfrac{\rho(v_{ik}, V_j)}{\rho(v_{ik}, V_U) - \rho(v_{ik}, V_j)} & \rho(v_{ik}, V_U) - \rho(v_{ik}, V_j) \neq 0 \\ -\rho(v_{ik}, V_j) - 1 & \rho(v_{ik}, V_U) - \rho(v_{ik}, V_j) = 0 \end{cases}$$

where $\rho(x, \langle a, b \rangle) = \left| x - \frac{a+b}{2} \right| - \frac{1}{2}(b - a)$, and $k_j(c_{ik})$ is correlation degree of the kth secondary index of the ith first level index relates to modernization grade $j(j = 1, 2, \cdots, m)$.

(3) First level evaluation

We obtain the association degree matrix of each first level index relates to each modernization grade $k(c_i)$ by

$$k(c_i) = (k_j(c_i)) = [w_{i1}, w_{i2}, \cdots, w_{ip}] \cdot \begin{bmatrix} k_1(c_{i1}) & k_2(c_{i1}) & \cdots & k_m(c_{i1}) \\ k_1(c_{i2}) & k_2(c_{i2}) & \cdots & k_m(c_{i2}) \\ \vdots & \vdots & \cdots & \vdots \\ k_1(c_{ip}) & k_2(c_{ip}) & \cdots & k_m(c_{ip}) \end{bmatrix}$$

(4) Secondary level evaluation

In the same way, we obtain the association degree matrix of the evaluating matter relates to each modernization grade $k(N)$ by

$$k(N) = [w_1, w_2, \cdots, w_n] \cdot \begin{bmatrix} k_1(c_1) & k_2(c_1) & \cdots & k_m(c_1) \\ k_1(c_2) & k_2(c_2) & \cdots & k_m(c_2) \\ \vdots & \vdots & \cdots & \vdots \\ k_1(c_n) & k_2(c_n) & \cdots & k_m(c_n) \end{bmatrix}$$

Then if it satisfies $k_{j0}(N) = \max\limits_{j=\{1,2,\cdots,m\}} k_j(N)$, the evaluating matter N is considered as belonging to modernization grade j.

Define $\tilde{k}_j(N) = \dfrac{k_j(N) - \min\limits_j k_j(N)}{\max\limits_j k_j(N) - \min\limits_j k_j(N)}$, then $j^* = \sum\limits_{j=1}^{m} j \cdot \tilde{k}_j(N) / \sum\limits_{j=1}^{m} \tilde{k}_j(N)$ is the level variable eigenvalue of highway traffic modernization, that is a degree of belonging to a certain modernization grade.

30.4 Case Study

The actual value of each index of Jiangxi Province in 2007 and the quantization score got through the method mentioned above are shown in Table 30.3. According to Tables 30.1 and 30.2, we establish the highway traffic modernization evaluation's grade field $U = \{U_1, U_2, U_3\} = \{\text{High}, \text{Middle}, \text{Low}\}$ and factor set $C = \{C_1, C_2, C_3\} = \{\text{Structural Low-carbon, Technical Low-carbon, Management Low-carbon, Policy Low-carbon}\}$ a_n, and determine the classical field and segment field of each index. Then through formulas above, we calculate the correlation degree of the first and secondary level indexes respectively, the results are shown in Tables 30.3 and 30.4.

Finally we determine the modernization grade, as shown in Table 30.4, the highway traffic modernization degree of Jiangxi Province in 2007 is second grade, indicating that the degree of modernization is middle, furthermore, we obtain the value d_1.

Table 30.4 Evaluation result of multi-level extension

	Index	Correlation degree			First level weight	Modernization grade j
		$j=1$	$j=2$	$j=3$		
First level	Structural low-carbon	−0.148	0.058	−0.148	0.252	2
	Technical low-carbon	−0.165	0.061	−0.138	0.198	2
	Management low-carbon	0.204	−0.143	−0.397	0.346	1
	Policy low-carbon	−0.166	0.092	−0.180	0.204	2
Target Level	Highway traffic modernization	−0.238	−0.003	−0.033	–	2

30.5 Conclusions

On the basis of long-term plan for highway energy-saving of China, this paper established the evaluation index system of highway traffic modernization in low-carbon economy perspective, which has the realistic significance, then further putted forward a reliable and useful method of index quantification. This paper also described the calculation steps of index weight by AHP-Entropy combination method and the multi-level extensible model's application in highway traffic modernization evaluation. Jiangxi Province highway traffic modernization evaluation of 2007 was studied as an example, indicating that the new method is applicable for highway traffic modernization evaluation which has a better realistic meaning to promote the construction of highway traffic modernization in China.

Acknowledgments This research is supported by Beijing Jiaotong University business expenses for basic scientific research (2012JBM070), Innovation work methods of Ministry of Science and Technology (2011IM040200-1106), the Specialized Research Fund for the Doctoral Program of Higher Education of China (20110009110031).

References

Jindong H (2008) Analysis for transportation energy saving potential. J Chang'an Univ (Soc Sci Ed) 10(3):39–42
Jingjie L (2011) The analysis of highway traffic management and energy saving factor. Traffic Transp 49:145–146
Ke P, Hongde W, Yanjun S (2011) Application of multi level extensible method to urban subway operation safety evaluation. J China Railw Soc 33(5):14–19
Unified Plan Department of the Ministry of Transport of the People's Republic of China (2008) The Long-term plan for highway and water carriage energy-saving
Zhi H, Deli Z, Lian L (2011) Research on demand of energy conservation and emission reduction of road traffic. Technol Highw Transp 6:135–137

Chapter 31
Traffic Congestion Measurement Method of Road Network in Large Passenger Hub Station Area

Yu Han, Xi Zhang and Lu Yu

Abstract Making correct assessment on traffic congestion intensity of road network of the large passenger hub station area, has great significance in improving the traffic condition of the road network and ensuring passengers' promptly arrival. This paper proposes a hierarchical evaluation method as section-road-network. Firstly, it determines the road section congestion intensity based on microcosmic simulation, then put the road connectivity as the weigh, to develop the weighting models for measuring traffic congestions. Finally, it conducts a case for Beijing South Railway Station by measuring congestion condition of the station area road network.

Keywords Traffic congestion measurement · Road connectivity · Traffic congestion intensity · Weighting model

31.1 Introduction

The large passenger hub station is the intersection of various transportation networks. It attracts large numbers of traffic and has a wide range radiation. It undertakes more traffic pressure than normal urban traffic road network. The main characteristics are, peak period is long and scattering, different kinds of vehicles are blended and driving in bad order. Therefore, the road network in large passenger

Y. Han (✉) · X. Zhang · L. Yu
School of Traffic and Transportation, Beijing Jiaotong University, Beijing, People's Republic of China
e-mail: 11120880@bjtu.edu.cn

X. Zhang
e-mail: xizhang@bjtu.edu.cn

hub station area is the weak point of urban traffic system. Making correct assessment on traffic congestion intensity of it has great significance in improving the traffic conditions of the road network and ensuring passengers' promptly arrival.

Presently, the majority of traffic congestion measurement methods are against the single section of the road. They select some indices and quantify them, using a certain evaluation methods such as fuzzy evaluation method (Long and Tan 2011) to calculate. Few methods were proposed to evaluate the whole road network, The research for the road network in large passenger hub station area is even nearly blank. This paper proposes a road connectivity-based weighting model for measuring traffic congestion intensity, take Beijing South Railway Station as a case to verify the model.

31.2 The Evaluation Index of Traffic Congestion Intensity

31.2.1 The Hierarchical Traffic Congestion Intensity of Road Network

The traffic congestion intensity (TCI) characterizes the congestion degree of a specific section, road, network. TCI is calculated gradually according to the order of section-road-network. The traffic congestion intensity of road section is represented by $TCI_{section}$, it is a discrete variable and its value ranges from 1 to 5. The equation is shown as follow:

$$TCI_{section} = \begin{cases} 1 \ very\ fluent \\ 2 \ fluent \\ 3 \ slight\ congestion \\ 4 \ moderate\ congestion \\ 5 \ severe congestion \end{cases} \quad (31.1)$$

By using the weighting method, we can get TCI_{road} and $TCI_{network}$. The equations are:

$$TCI_{road} = \frac{\sum_{i=1}^{n} A_i \times TCI_{section_i}}{\sum_{i=1}^{n} A_i}, \quad TCI_{network} = \frac{\sum_{j=1}^{m} B_j \times TCI_{road_j}}{\sum_{j=1}^{m} B_j} \quad (31.2)$$

In the equation, $TCI_{section_i}$ represents the traffic congestion intensity of section i in this road level, TCI_{road_j} represents the traffic congestion intensity of the road in j level, A_i and B_j are the weigh of section i and road j. n represents the section amounts of the road in a certain level, m represents road level number of the road network.

Table 31.1 Traffic congestion intensity level standard

TCI	Very fluid	Fluid	Slight congestion	Moderate congestion	Severe congestion
Value range	(1, 1.5)	(1.5, 2.5)	(2.5, 3.5)	(3.5, 4.5)	(4.5, 5)

The corresponding relation between TCI_{road}, $TCI_{network}$ and traffic congestion intensity degree is defined as follow: (Table 31.1).

31.2.2 Determination of $TCI_{section}$ Based on Microcosmic Simulation

The characteristic of section's congestion is that vehicles are in low speed and the lag time is long. In previous studies (Zhang and Yu 2008), the determination of $TCI_{section}$ needs large-scale surveys and calculations. It has low efficiency and bad application. This paper introduce a method based on microcosmic simulation, which is simple and practical.

In traffic microcosmic simulation software Vissim, by entering the traffic flow data, road geometry data and signal timing of the intersection, we can get many indices which can reflect congestion degree such as average travel speed, average lag time and so on.

We choose average travel speed, average delay time, degree of saturation three indices as the variables to determine $TCI_{section}$. The evaluation standard of these indices are shown in the following tables. The equation of $TCI_{section}$ is:

$$TCI_{section} = \frac{Index_1 + Index_2 + Index_3}{3} \tag{31.3}$$

$Index_1$ represents the $TCI_{section}$ reflected from average travel speed. $Index_2$ represents the $TCI_{section}$ reflected from average lag time. $Index_3$ represents the $TCI_{section}$ reflected from degree of saturation (Tables 31.2, 31.3, 31.4).

Table 31.2 Average travel speed evaluation reference standard (Zhang 2011)

Congestion level	Very fluid	Fluid	Light congestion	Moderate congestion	Severe congestion
Average travel speed (km/h)	>45	(35, 45)	(25, 35)	(15, 25)	(0, 15)

Table 31.3 Unit mileage average lag time reference standard (Zhang 2011)

Congestion level	Very fluid	Fluid	Light congestion	Moderate congestion	Severe congestion
Average lag Time(s)	0	(0, 23)	(23, 64)	(64, 160)	>160

Table 31.4 Degree of saturation evaluation reference standard (Zhang 2011)

Congestion level	Very fluid	Fluid	Light congestion	Moderate congestion	Severe congestion
Degree of saturation	≤0.4	(0.4, 0.6)	(0.6, 0.7)	(0.7, 0.8)	>0.8

31.3 Connectivity-Based Traffic Congestion Measurement Model

In the road network, every section is not isolated. Different sections have different extent of impact when they have different connectivity. We define the section connectivity to be the number of sections connected to the aim road section, the road connectivity to be the number of sections connected to the road in aim level. They were expressed by SC and RC.

We construct the model with $TCI_{section}$ as the index and road connectivity as the weigh. According to the three levels of traffic congestion measurement, we get the equation of TCI_{road} and $TCI_{network}$ as follows:

$$TCI_{road} = \frac{\sum_{i=1}^{n} SC_i \times TCI_{section_i}}{\sum_{i=1}^{n} SC_i} \quad (31.4)$$

$$TCI_{network} = \frac{\sum_{k=1}^{m} RC_k \times TCI_{road_k}}{\sum_{k=1}^{m} RC_k} \quad (31.5)$$

In the equation, RC_k represents the connectivity of road in k degree, m represents the amount of road degree.

31.4 The Example

We take Beijing South Railway Station as the case. The research object is the station area road network. The structure of the road network is as follow in Fig 31.1.

The red line represents fast-speed road, the blue line represents arterial road, the yellow line represents minor arterial road, the green line represents primary distributor road. According to the road network structure, we count the section connectivity and road connectivity as follow in Table 31.5.

We surveyed 15 min traffic volume of the sections above at several representative period, converted them into hourly traffic volume. According to the simulation result, we worked out $TCI_{section}$ and bring the result into the model. The variation of $TCI_{network}$ is shown as the following (Fig. 31.2).

31 Traffic Congestion Measurement Method of Road Network

Fig. 31.1 Road network of Beijing South Railway Station

Table 31.5 The connectivity of road section in road network

Section number	Section connectivity	Road connectivity
Arterial road 1	6	16
Arterial road 2	10	
Fast-speed road 1	10	17
Fast-speed road 2	7	
Minor arterial road 1	9	21
Minor arterial road 2	6	
Minor arterial road 3	7	
Primary distributor road 1	3	12
Primary distributor road 2	2	
Primary distributor road 3	6	
Primary distributor road 4	3	
Primary distributor road 5	5	
Primary distributor road 6	2	

Fig. 31.2 The variation of road network congestion

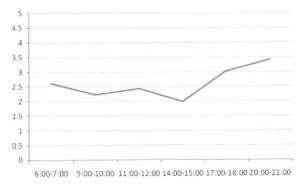

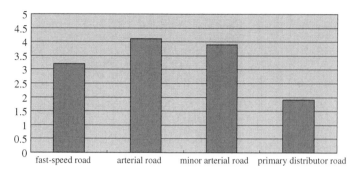

Fig. 31.3 Road congestion intensity of peak hour

From the Fig. 31.2 we can conclude that the variational regularity of traffic congestion intensity in large passenger hub station area is different to normal urban traffic road network. It has no apparent morning and evening peak period, but has great relationship with the departure and arrival of the train. 6:00–7:00 is the morning peak of urban traffic, but the station has only three trains to depart in this period. Therefore, the $TCI_{network}$ of this period is smaller than the period of 20:00–21:00, with 12 trains to depart and 5 to arrive.

We picked out the maximum of $TCI_{network}$, 20:00–21:00, to make further analysis. TCI_{road} of this period is shown as the following (Fig. 31.3).

From the Fig. 31.3 we can see that the main road and the secondary road are in moderate congestion, but the primary distributor road is fluid. To alleviate the congestion, we should guide the traffic to the unimpeded primary distributor road.

31.5 Summary

This paper proposes a road connectivity-based traffic congestion measurement model. Firstly, we determine the traffic congestion intensity of road section through microcosmic simulation results. Then put the road connectivity as the weigh, to calculate the traffic congestion intensity of the road network. This paper chooses Beijing south railway station as the example and evaluate its traffic congestion of road network. However, limited to the lack of data, this paper only makes brief analysis of the road network. More detailed works still need to do afterwards.

References

Long XQ, Tan Y (2011) Urban traffic congestion evaluation based on fuzzy comprehensive evaluation. Commun Stand 11:114–117

Zhang H (2011) The traffic congestion measurement of urban road based on microcosmic evaluation index. Urban Road Flood Prot 8:338–341

Zhang X, Yu L (2008) Traffic demand-based traffic congestion measurement models for road networks. Mod Transp Technol 12, 5(6):71–74

Chapter 32
The Governance of Urban Traffic Jam Based on System Dynamics: In Case of Beijing, China

Haoxiong Yang, Kaichun Lin, Yongsheng Zhou and Xinjian Du

Abstract This paper starts with the phenomenon of urban traffic jam, summaries the existing measures for the governance of urban traffic jam from the perspective of urban residents travel and urban logistics. On the basis of summarizing the plugging measures as well as considering the actual situation of our country, this paper establishes the system dynamics model of the management of urban traffic jam with the method of system dynamics. In the last place, this paper takes Beijing as an example, simulates the effect of the typical plugging measures in Beijing, and reaches the suggestions for China to solve urban traffic jam.

Keywords Urban traffic jam · Residents travel · Urban logistics · System dynamics

32.1 Introduction

As an important and essential condition of normal operation of city life, urban traffic can not only improve the living and travel conditions of urban residents, but also drive economic development and improvement of functional layout of the whole city. The urbanization process of China continues to accelerate under the background that economic construction developed rapid, the original city traffic system is increasingly failing to keep up with the demand of the urban development.

H. Yang (✉) · K. Lin · Y. Zhou · X. Du
Business School, Beijing Technology and Business University, Beijing 100048, China
e-mail: yanghaoxiong@126.com

The contradiction between supply and demand of urban traffic of China have become increasingly prominent under these two factors, and within some drawbacks that gradually revealed, one of the most prominent and general is urban traffic jam. The solution of this problem closely related to the improvement of urban modernization, urban economic development and urban living standards of residents. Therefore, studying urban traffic jam and finding effective measures for plugging cannot delay.

The study of traffic jam is abundant at present: In the cause of urban traffic jam, Taylor (1992) from an economic perspective, looking for the cause of traffic jam. McKnight, Eurlng (1993) thinks the control of purchase and use of private cars can solve the problem of traffic jam; In the governance of urban traffic jam, the theory of it goes through three development stages: (1) strengthen the construction of urban traffic infrastructure facilities to meet the increasing traffic demand; (2) TSM—Traffic System Management; (3) TDM—Traffic Demand Management. Nowadays, there are three mainly governance patterns of traffic jam: the increase supply mode, the demand management mode and the complete institution mode; In the studying urban traffic by application of system dynamics, the relatively famous research is the urban dynamics model of professor Forrester and the travel produce model of professor Shirazian.

This paper starts with the phenomenon of urban traffic jam, summaries the existing measures for the governance of urban traffic jam from the perspective of urban residents travel and urban logistics. On the basis of summarizing the plugging measures as well as considering the actual situation of our country, this paper establishes the system dynamics model of the management of urban traffic jam with the method of system dynamics. In the last place, this paper takes Beijing as an example, simulates the effect of the typical plugging measures in Beijing, and reaches the suggestions for our country to solve urban traffic jam.

32.2 System Dynamics Model of Urban Traffic Jam Governance

In order to alleviate the state of urban traffic jam, improve living and production environment of residents, several countries and regions are adopted some measures to govern plugging, and have achieved some effects. On the basis of combination actual condition of itself, China continues learning successful experience of other countries and looking for the effective measure.

The urban traffic system consists of two components, residents travel and urban logistics. Residents travel include urban private traffic and urban public traffic, urban logistics is the cargo professional transportation within city. In China, the measures of residents travel mainly include advocating public traffic, limiting travel policy and limiting vehicles purchase policy; the measures of urban logistics mainly include limiting freight travel policy, establishing logistics park and development of urban public truck system (urban freight taxi). At present, the plugging measures of China is various, all kinds of measures has their own different characteristics, so there is a trend to mix using some of them, but the plugging effect is the most important, which the choice and evaluation of the measures should based on, at the same time, cost and the condition of the city should be considered.

In order to analyze different plugging measures, simulate the implementation effect of the influence towards urban traffic operation, this paper puts some typical measures into the original system dynamics model, and gets the system dynamics model of urban traffic jam governance.

32.2.1 The Determination of System Boundary

The modeling purpose of this paper is to analyze the internal structure of urban traffic from the perspective of urban residents travel and urban logistics, find out the feedback mechanism of influence between the operation of urban residents travel and urban logistics and urban traffic condition, and analyze the implementation effect of different plugging measures according to the change of related policy variables, simulate the influence of these measures towards urban traffic operation, thus provide reasonable goal and feasible suggestion to optimize traffic structure.

According to the research purpose of this paper, the final scope of the research system has been determined. Residents travel system include the amount of private cars, growth rate of private cars, scrap rate of private cars, traffic volume of private cars, the amount of bus, travel rate of bus, the amount of other motor vehicles, traffic volume of other motor vehicles, total volume of urban motor vehicles, traffic volume of urban motor vehicles and the plugging measures that related to residents travel; urban logistics system include urban logistics volume, the amount of urban freight vehicles, traffic volume of urban freight vehicles, growth rate of urban logistics, increment of urban logistics and the plugging measures that related to urban logistics; the others include urban GDP, growth rate of GDP, urban environment (NO_2 stock), annual average NO_2 emissions volume of each motor vehicles and contribution rate that motor vehicles towards environment pollution.

32.2.2 The Analysis of System Causality

The urban traffic system is a complex dynamic development system that influenced by multiple comprehensive factors, the causality between factors within system of this paper is shown in Fig. 32.1.

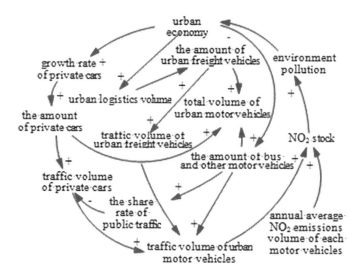

Fig. 32.1 Causality of urban traffic system

32.2.3 The Establishment of System Model

In order to analyze different plugging measures, simulate the influence of implementation effect towards urban traffic operation, this paper puts some typical measures into the stock and flow model of urban traffic system which is established by using Vensim, and gets the system dynamics model of urban traffic jam governance, as is shown in Fig. 32.2.

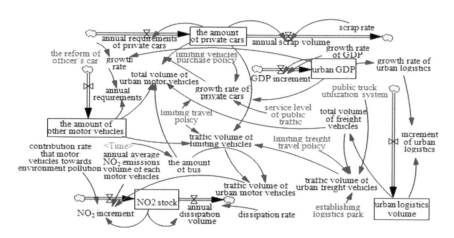

Fig. 32.2 System dynamics model of urban traffic jam governance

32.3 Model Simulation and Empirical Analysis: In Case of Beijing

Beijing is the capital of China, the study towards urban traffic jam and its governance has significant value. This part use policy optimization function of system dynamics to analyze the influence of the changes of policy variables of the model towards urban traffic status of Beijing, which is the effect of current plugging measures of Beijing towards its governance of traffic jam.

32.3.1 The Policies of Residents Travel

In Beijing, the current plugging measures that greatly influence residents travel including: limiting travel policy by tail number, limiting vehicles purchase policy by license plate lottery, advocating public traffic travel, etc. In order to ensure that the model is real and effective, this paper synthetically considers the implementation of these polices when simulate the model.

Limiting travel policy can relieve urban traffic jam, but it needs to base on the powerful urban public traffic system. If limiting travel policy is long-term implemented but the public traffic service level cannot satisfy the travel demand of residents, there will be more people choose to purchase a second car with different tail number to reply this policy. The comparison of traffic volume of urban motor vehicles between before and after the implement of limiting travel policy is as shown in Fig. 32.3.

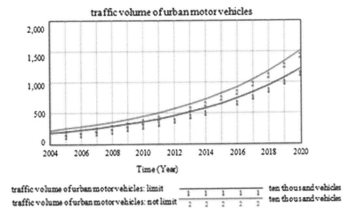

Fig. 32.3 Comparison of traffic volume of motor vehicles between before and after the implement of limiting travel policy under natural growth rate

Conclusion one: When motor vehicles grow under natural growth rate, limiting travel policy by tail number can relieve the urban traffic jam to some extent.When there are 3 % people purchase a second car, the comparison of traffic volume of urban motor vehicles between before and after the implement of limiting travel policy is as shown in Fig. 32.4.

Conclusion two: Because of the stimulation of limiting travel policy, when a small number of people purchase a second car with different tail number, the policy is short-term effective, but in the long term, it will have the opposite effect.Through the sensitivity analysis, gradually increase the proportion of purchase a second car, when there are 6.25 % people purchase a second car, the comparison of traffic volume of urban motor vehicles between before and after the implement of limiting travel policy is as shown in Fig. 32.5.

Conclusion three: When there is enough people purchase a second car, the effect of limiting travel policy will gradually disappear.The government should strive to develop public traffic, after advocating public traffic travel, the traffic volume of urban motor vehicles is as shown in Fig. 32.6

Conclusion four: Compared with limiting travel policy, public traffic priority policy can effectively relieve urban traffic jam.Considered the increasing serious traffic jam, the government of Beijing decided to take license plate lottery as a method to limit the purchase of vehicles, and take, the limiting travel policy at the same time. Therefore, according to the actual conditions, simulate the implement of these two plugging measures, and

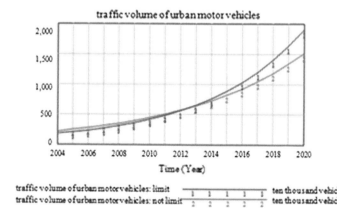

Fig. 32.4 Comparison of traffic volume of urban motor vehicles between before and after the implement of limiting travel policy when there are 3 % people purchase a second car

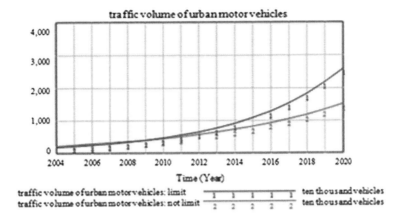

Fig. 32.5 Comparison of traffic volume of urban motor vehicles between before and after the implement of limiting travel policy when there are 6.25 % people purchase a second car

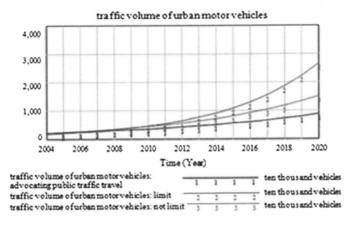

Fig. 32.6 Comparison of traffic volume of urban motor vehicles between before and after improving public traffic service level

Conclusion five: compared them with the implement effect of limiting travel policy. The simulation result is as shown in Fig. 32.7. Do not consider other discontent factors that caused by limiting vehicles purchase policy by license plate lottery, the policy relieves the urban traffic jam to some extent.

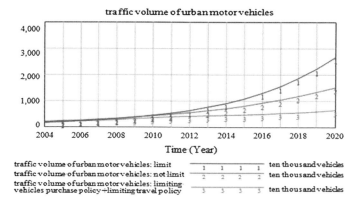

Fig. 32.7 Comparison of traffic volume of motor vehicles between before and after the implement of limiting vehicles purchase policy by license plate lottery

32.3.2 The Policies of Urban Logistics

In order to limit freight vehicles traveling in city and relieve urban traffic jam, Beijing puts forwards some relevant plugging measures to relieve urban traffic jam and improve urban environment, but which at the same time influence the normal operation of urban logistics, so whether should limit freight vehicle travel and the degree of limiting arouse much attention. The comparison of traffic volume of urban motor vehicles between before and after the implement of limiting freight vehicles travel policy is as shown in Fig. 32.8.

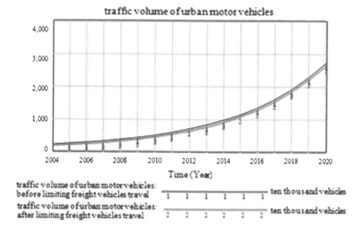

Fig. 32.8 Comparison of traffic volume of urban motor vehicles between before and after the implement of limiting freight vehicles travel policy

32 The Governance of Urban Traffic Jam Based on System Dynamics

Conclusion six: Because of the relatively small amount of freight vehicles, limiting freight vehicles travel policy has little influence on relieving urban traffic jam.Transformed passenger cars to freight vehicles to transport cargos will greatly raise the transport cost and other costs, and will bring a lot of unnecessary waste of fund and recourse. After transforming passenger cars to freight vehicles, the traffic volume of urban motor vehicles is as shown in Fig. 32.9.

Conclusion seven: Because of the appearance of transforming passenger cars to freight vehicles after limiting freight vehicles travel, limiting freight vehicles travel policy don't reduce the traffic volume of urban motor vehicles, but aggravate the traffic jam.Because of the restrictions of limiting freight vehicles travel policy in the city, Beijing introduces urban public truck utilization system to meet the fright demand of citizens. The comparison of traffic volume of urban motor vehicles between before and after the introduction of urban public truck utilization system is as shown in Fig. 32.10.

Conclusion eight: **Urban public truck utilization system can relieve urban traffic jam under the situation that meet the demand of urban logistics.**

Finally, compared implementation effects of various plugging measures is as shown in Fig. 32.11. According to Fig. 32.11, limiting vehicles purchase policy by license plate lottery combined with limiting travel policy is the most effective way to relieve urban traffic jam, but this measure will bring many negative influence; the government

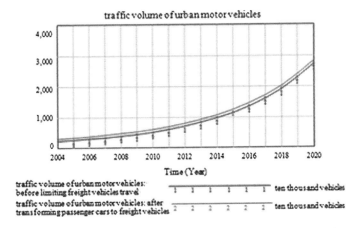

Fig. 32.9 Comparison of traffic volume of urban motor vehicles between before and after transforming passenger cars to freight vehicles

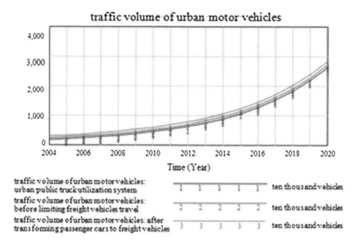

Fig. 32.10 Comparison of traffic volume of urban motor vehicles between before and after the introduction of urban public truck utilization system

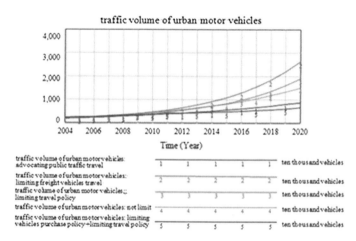

Fig. 32.11 Comparison of implementation effects of various plugging measures

advocating public traffic travel, improving service level of public traffic are effective plugging measures and will not bring other discontent; limiting travel policy by tail number is more significant to relieve urban traffic jam.

32.4 Conclusion

This paper summary the existing measures for the governance of urban traffic jam from the perspective of urban residents travel and urban logistics. On the basis of summarizing the plugging measures as well as considering the actual situation of our country, this paper establishes the system dynamics model of the management of urban traffic jam with the method of system dynamics, and takes Beijing as an example, simulates the effect of the typical plugging measures.

Acknowledgments This work is supported by the Youth Project National Fund of Social Science of China (No. 11CGL105), and supported by the Youth Project of the Humanities and Social Sciences of the Ministry of Education of the People's Republic of China (No. 10YJC630324), and supported by a Funding Project of the Academic Human Resources Development in Institutions of Higher Learning Under the Jurisdiction of Beijing Municipality (No. PHR20110877).

References

McKnight, Eurlng J (1993) Transportation 2020. Opportunities and challenges. Prof Eng 6(3):29–31

Taylor J (1992) Urban congestion and pollution. Is road pricing the answer?. In: Proceedings of the institution of Civil Engineers, Municipal Engineer, vol 93(4), pp 227–228

Chapter 33
The Design and Realization of Urban Mass Information Publishing System

Kai Yan, Li-min Jia, Jie Xu and Jian-yuan Guo

Abstract With the advocating of the low-carbon traffic and transportation concept, more and more urban residents choose urban mass as their main travel mode every day. At the same time, development in urban mass transit impels mass lines to a complicated mass network. Based on the urban mass network, this paper aims to discuss how to design and realize an Urban Mass Information Publishing System to mainly publish passenger flow guidance information including urban mass network state information, passenger flow controlling information, operation situation information of first and last train, emergency information, etc. According to the analysis of system business requirements, the system architecture is divided in three layers: presentation layer, business service layer and data layer. The function architecture is also designed based on the requirement analysis. Then we design the business flow and data organization of the system. Finally a demo system is realized as a case application.

Keywords Information publishing system · Information management system · Urban mass transportation · Intelligent transportation system · Low-carbon transportation

---

K. Yan (✉) · L. Jia (✉) · J. Guo
School of Traffic and Transportation, Beijing Jiaotong University, Beijing 100044, China
e-mail: 11121090@bjtu.edu.cn

L. Jia
e-mail: jialm@vip.sina.com

L. Jia · J. Xu
State Key Laboratory of Rail Traffic Control and Safety, Beijing 100044, China

33.1 Introduction

With advocating of the low-carbon traffic and transportation concept, more and more urban residents choose urban mass as their main travel mode in every day. The influence of the passenger flow aggregations and the emergent events such as fire or equipment fault is getting more serious. To coordinate and unify all lines in the urban mass network and realize the stable operation of the urban mass traffic work, Traffic Control Center (TCC) has been established in many cities. To provide better services for urban residents in their daily travel, the TCC needs to timely publish those information of passenger flow controlling, operation situations of first and last train, emergency events and other activities. This paper aims to discuss how to design and realize an Urban Mass Information Publishing System (UMIPS, IPS for short) base on the urban mass network.

33.2 System Requirement Analysis

As a comprehensive information management system, IPS should not only meet the demand of information publishing works, but also show some management function (Dennis et al. 2006).

The system basic requirements have been collected and filtered out, and divided into aspects as follows:

(A) **All mass network real-time publishment**

For more accurately and timely giving suggestions to all passengers riding or waiting on every train or station, and assigning tasks to all staffs working on every station and train, IPS should publish information covering all mass networks. Establishing a communication network is necessary, which can be realized by Internet or large-scale internal LAN network.

(B) **Comprehensive business publishing**

IPS aims to publish information comprehensively to meet all the business demands as follows: (1) Publishing passenger-flow-control information in high passenger flow period in every day. (2) Publishing adjustment plan of lines, stations and trains in daily operation of the urban mass transit. (3) Publishing suggestions and warnings to all passengers after emergency events happened.

(C) **Diverse publishing ways**

The ways of information publishing should be diverse and selectable, including broadcast, send messages, PIDS display and publish on website. Multiple publishing ways help passengers to receive the suggestions and warnings more quickly and comprehensively.

(D) Information templates establishment supported by preplan

For efficiently publishing information, IPS should support establishment and management of information templates base on comprehensive urban mass transit preplans which are established already or establishing by an editor of the system.

(E) System information management

To coordinate with other ITS well, IPS should support the maintenance of some basic parameter and data exchange from other systems including congestion parameter, station image and timetable of the mass network. IPS also should support some management function such as user information and authority management, and inquiry of system history operation.

33.3 System Design

33.3.1 System Architecture

(1) System Overall Architecture

Information Publishing System is a three-layer application system which adopts B/S structure mode, including data layer, service layer and presentation layer (Geisler and Quix 2012). The IPS architecture chart is as shown in Fig. 33.1.

Presentation layer is the highest layer of IPS, in which show and collect data except deal with data. It provides the interface for user operation, collect user operation information and submit request to service layer, then shown the disposal result to user according specified demand. System function is realized in this layer by two mode including GIS map and RIA. Geographic information system (GIS) map is an electric map with lots of functions like dynamic display and spatial data

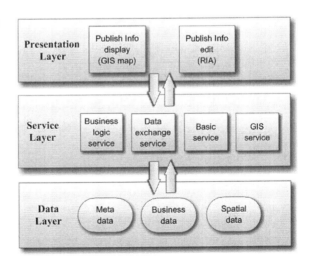

Fig. 33.1 Architecture of the IPS

exchange, etc. Rich internet application (RIA) is a more intuitive, responsive, flexible, and convenient and offline supported Web application.

Service layer is the most important layer of IPS, in which define the rules of how every logic unit can be realized and implementation of the system functions (Horsburgh and Tarboton 2009). (a) Business logic service encapsulates a lot of independent system function unit. Each business logic service corresponds to one specific function, receives presentation layer's request and returns the result. (b) The main work of data exchange service is to exchange the data between presentation layer and the data layer. It provides uniform exchange interface rules and a communication module to obtain and send message in different way. In general, it deals with data flow to be transmitted to front-end application or stored in back-end database. (c) Basic service is composed of minimum function unit that cannot be subdivided. High cohesive and low coupling is the main ability of a basic service application. In addition, every basic service application is realized with emphasis on its reusable ability and interoperation. (d) GIS service is realized on several GIS basic services and exchanges data with other service application. It is used to display the mass network states on an electric map betimes and dispose of some data from the electric map.

The data layer is the bottom of IPS and includes three data types which are metadata, business data and spatial data. All of them can be stored in a relational database or an object-oriented database with several basic data format and extensible markup language (XML) format which one can be simplicity, generality, and usability over the Internet.

(2) **System Function Architecture**

According to the requirement analyses of IPS, its function architecture is divided in four function modules that include publish editor module, publish management module, system maintenance module and GIS dynamic map module (Wang et al. 2010). The function architecture chart is as shown in Fig. 33.2.

A. **The publish editor module** is divided into two parts that include select publish template and select publish tool. The most important function of IPS is to publish guidance information to all passengers in the mass network including passengers in the stations and on the trains. Therefore, a set of complete, clear and normative guidance information template is indispensable. To publish guidance information in different ways, IPS is designed to provide a process which is used to select one or more publish tools after users confirm the content of awaiting publishing information. Such publish tools include broadcast, passenger information system (PIS), website, station screen, enquiry machine and cellphone message.

B. **The publish management module** is responsible for the management of all publishing information templates and a list of published information. The information templates can be a skeleton of a message for publishing. In the templates management module, the main function include build new information template when the user need to construct some new type information,

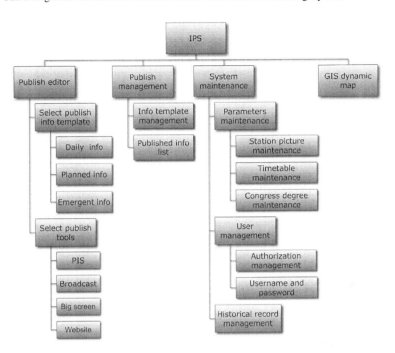

Fig. 33.2 The function architecture chart

delete and modify one or more templates. All operation on every template exchanges data with the central database. All templates can be searched in some relational datasheets of the templates.

C. **System Maintenance module** consists of user management, basic data maintenance and history inquiry. (a) The basic data maintenance function module is divided into three parts. The first is the station picture maintenance. The second part is the train timetable maintenance. The third part is the congestion degree maintenance. (b) The user management function module is responsible for the management and organization of user's information: username and password. (c) The history inquiry function module is realized with a usual history data management tool what is used to record the operating logs.

D. **GIS dynamic map module** is designed to assist all users to browse real-time state of the urban mass network better. This module consist of three sub-function module: (1) map scaling and roaming, such as zoom in, zoom out, etc.; (2) station image viewing; and (3) map rapid positioning, for instance line or station locking and positioning.

33.3.2 Data Organization

According to the system main architecture design, the data organization architecture of IPS is divided into three layers: Metadata, Business data, and Spatial

Table 33.1 Data organization

Layer	Data classification	Datasheet
Metadata	User management information	account_information, user_priority
	Data standard	data_type, storage_rule
Business data	Information templates	daily_info, planned_info, emergent_info
	Station information	station_data, station_map
	Event information	event_data
Spatial data	Mass network operation state	line_state, station_state
	Passenger flow state	pf_state

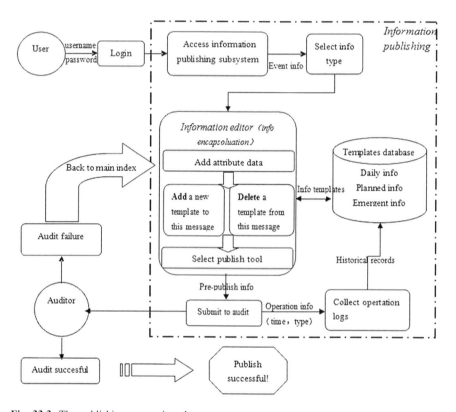

Fig. 33.3 The publishing processing chart

data. As shown in Table 33.1, the first column is Layer as a basic division, second column is Data classification in which column the data basic function type is, third column is Datasheet what show the detailed data form.

33.4 System Realization

33.4.1 Business Flow Realization

As we have detailed in the previous section, the function architecture of IPS is divided into four parts: publish editor module, publish management module, system maintenance module and GIS dynamic map module. There are different data exchanges processing in each module. The design of each processing aims to effectively realize a flexible, interactive and stable business service flow. And we now have a closer look at how the business flow is processed in the IPS. An overview of the publishing processing design in the system is depicted in Fig. 33.3.

The main function of IPS is information publishing, and it is mainly realized in three steps that including user login, information text edit and submit processed information to audit.

Step 1. User login

User who wants to login the system need input their username and password. The username and password are matching every staff that is responsible of the information publishing work.

Step 2. Information text edit

For publishing information, user has to choose the information publishing subsystem to access into an editor interface to find an information templates tree which has summarize text templates in a tree structure and classify the templates with its branches. User needs to select one template branch and input all event attribute data such as begin time, end time and location, finally, a piece of pre-published text message has been generated.

Step 3. Submit processed information to audit

User submit the processed information to audit, then the message that step 2 has generated is sent to the high level auditor. Every time that user submits to audit, IPS can store user's operation information such as submit time, information type, user's username, etc. into the datasheet of historical records. The auditor's operation information also will be stored as historical records.

Fig. 33.4 System interface

33.4.2 A UMIPS Based on Beijing Urban Mass Network

To prove the design conception of IPS, we have realized the demo system with the FLEX structure and ArcGIS service base on Beijing urban mass network. To more comprehensively display the function of BMIPS, we developed three operation interfaces as shown in the Fig. 33.4. On the left, it is a screen for watching the whole Beijing urban mass network. The middle screen is the information publishing editor interface. The right part is a screen displaying the information tables about the events state.

33.5 Conclusion

As an assistant system, the IPS tries to meet the need of its users. Therefore, the design of system business process is the major part of the overall design. A flexible, convenient, fast-response and function-complete system can not only help user save more time on daily publishing work, but also help to improve their publishing works.

As a design conception, the analysis and design of IPS is still a continuously improving and innovating process. To solve those detail problems in information publishing also need a large number of researching and exploring works and the system still have much expansion space.

Acknowledgments This work has been supported by National Natural Science Foundation of China (61074151); National Key Technology Research and Development Program (2009BAG12A10); Research Found of State Key Laboratory of Rail Traffic Control and Safety (RCS2009ZT002, RCS2010ZZ002, RCS2011ZZ004).

References

Dennis A, Wixom BH, Roth RM (2006) Systems analysis & design, 3rd edn. Wiley Publishing, USA, pp 77–78
Geisler S, Quix C (2012) An evaluation framework for traffic information systems based on data streams. Transp Res Part C 23:29–55
Horsburgh JS, Tarboton DG (2009) An integrated system for publishing environmental observations data. Environ Model Softw 24:879–888
Wang L, Qin Y, Jia L (2010) Open integration architecture of railway safety comprehensive monitor system. In: The 5th international conference on computer science & education, Hefei, China, 24–27 August 2010

Chapter 34
Research on Multiple Attribute Decision Making of BRT System Considering Low Carbon Factors

Jia-qing Wu, Rui Song and Li Zheng

Abstract Developing low carbon transportation is the inherent requirement of sustainable development of urban transportation, as well as an effective way to deal with the greenhouse effect. After treating the Bus Rapid Transit (BRT) system as the research object, this paper established an index system of the multiple attribute decision for BRT system by considering low carbon factors, and proposed methods of the BRT system multiple attribute decision making using the fuzzy comprehensive evaluation theory. The case study was conducted on the basis of BRT line 1, line 2 and line 3 in Beijing, which verified the effectiveness of the proposed index system and decision making methods.

Keywords Low carbon · BRT system · Multiple attribute decision making · Evaluation index system

34.1 Introduction

Problems of the transportation energy consumption and greenhouse gas emissions have become one of the major issues that humans have to deal with. "Low carbon transportation" is a transportation development mode with low power consumptions, low pollutions, and low emissions, which is directly on target to reduce the greenhouse gas emissions of transportation tool. As a sustainable urban

---
J. Wu (✉) · R. Song · L. Zheng
School of Traffic and Transportation, Beijing Jiaotong University, Beijing 100044, China
e-mail: jqwu1978@126.com

J. Wu
Beijing Public Transport Holdings, Ltd, Beijing 100161, China

transportation system structure, "Low carbon transportation" can effectively reflect the meaning of the sustainable urban transport system structure.

Recent years, many cities in China have built the BRT system considering the local conditions. Since the first BRT line (South-Centre Corridor BRT1 in Beijing) opened into operation, several cities such as Hangzhou, Changzhou and Jinan built BRT systems one after another. Therefore, the BRT system is becoming increasingly important in the urban low carbon passenger transportation system.

The existing low carbon transportation researches (Romilly 1999; Lia and Head 2009; Deng 2006; Hayashi et al. 2001; Cherrya et al. 2009) mostly focused on the comparison of different transportation modes instead of the BRT system multiple attributes decision making considering low carbon factors. Till now, few researches are found to carry out the case study using various BRT systems. Consequently, this paper aims to study the BRT system multiple attributes decision making by considering low carbon factors.

34.2 Multiple Attribute Decision Making Index System of BRT System Considering Low Carbon Factors

34.2.1 Construction of Index System

Multiple attribute decision is a decision problem that considers a number of sorted indices, according to certain rules, to assess the integrated status of evaluation object from some or more aspects. BRT system is a large comprehensive system including level of service, operational efficiency, economic efficiency and other factors, which should involve up to dozens of indices. The developed evaluation index system is shown in Fig. 34.1, which considers low carbon factors.

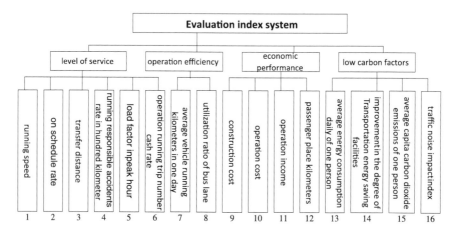

Fig. 34.1 Evaluation index system of BRT system considering low carbon factors

After confirming the index system, the method to solve the problem that each index has different dimension is to classify and determine a standard range, which are divided into four categories, positive index, negative index, segmentation processing by interval index, temporarily don't treatment index. The "positive index" refers the more the actual value the better the index is. The "negative index" refers the smaller the actual value the better the index is.

In order to facilitate the comprehensive evaluation, the index is transferred into the dimensionless form within the range of [0, 1]. For the evaluation index $u_i \in u$, its domain is set as $d_i = [m_i, M_i]$, while m_i and M_i respectively represent the minimum and maximum values of the evaluation index. Define $r_i = u_i(x_i)$, $i = 1, 2,\ldots, n$, as the dimensionless value of attribute values x_i that the decision-maker evaluate for index u_i, and $r_i \in [0, 1]$. $u_{di}(\bullet)$ is the dimensionless standard function of index u_i which is defined in the domain of d_i. Standardized conversion functions of "positive index" and "inverse index" are respectively expressed as follows:

$$r_i = u_{di}(x_i) = \begin{cases} 1 & (x_i \geq M_i) \\ \frac{x_i - m_i}{M_i - m_i} & (x_i \in d_i) \\ 0 & (x_i \leq m_i) \end{cases} \quad (34.1)$$

$$r_i = u_{di}(x_i) = \begin{cases} 1 & (x_i \leq m_i) \\ \frac{M_i - x_i}{M_i - m_i} & (x_i \in d_i) \\ 0 & (x_i \geq M_i) \end{cases} \quad (34.2)$$

34.2.2 Determination of Index Weights

The index weights are determined using the expert scoring method. Table 34.1 gives the weight value of each index scored by a number of experts based on the comprehensive, integrated, scientific, pragmatic and fair principles, as well as the index category and the value interval.

It can be found that the share of the level of service is largest in the criteria layer, accounting for 39.4 %; and the following one is that of the low carbon factor, accounting for 37.5 %. It demonstrates that the degree of BRT system meeting requirements of passengers and low carbon factors attract the most attentions of public when making an assessment of the level of development of the BRT system.

Table 34.1 Weight value of index (including index sort and interval size)

Index number	Index unit	Index sort	Interval size	Weight value relative to criteria layer	Weight value relative to target layer
	Level of service (X1)				**0.39421**
1	km/h	positive	15–30	0.25091	0.098911231
2	%	positive	0–100	0.22273	0.087802393
3	m	inverse	0–200	0.15727	0.061997407
4	Yuan/100 km	inverse	0–1000	0.11	0.0433631
5	%	inverse	30–100	0.16455	0.064867256
6	%	positive	90–100	0.09454	0.037268613
	Operation efficiency(X2)				**0.10444**
7	km	positive	100–300	0.42727	0.044624079
8	10,000 persons/km	positive	0–10	0.57273	0.059815921
	Economic performance(X3)				**0.12567**
9	10,000 Yuan/km	inverse	1000–3000	0.271362	0.034102063
10	10,000 Yuan	inverse	200–500	0.231362	0.029075263
11	10,000 Yuan	positive	500–3000	0.176824	0.022221472
12	10,000 km	positive	200–400	0.320452	0.040271203
	Low carbon factors(X4)				**0.37568**
13	KJ	inverse	0.8–1.2	0.201362	0.075647676
14	%	positive	0–100	0.331362	0.124486076
15	kg	inverse	5–8	0.236824	0.08897004
16	%	inverse	0–100	0.230452	0.086576207
Total					1.0

34.3 Multiple Attribute Decision Making Method of BRT System Considering Low Carbon Factors

Multiple Attribute Decision Method (MODM) shows particular advantages in dealing with various complex system problems that are difficult to be described using precise mathematical methods, so it has been widely used in many study areas. MODM has two different forms, as follows:

(1) Single-level fuzzy comprehensive evaluation model. Given two finite domains: $U = \{u_1, u_2, \ldots, u_n\}$ and $V = \{v_1, v_2, \ldots, v_n\}$, U is the set of all evaluation factors, and V is the set of all reviews grades.

For the ith $i(i = 1, 2, \ldots, n)$ evaluation factor u_i, if its evaluation result is $R_i = [a_1, a_2, \ldots, a_n]$, then the decision matrix for various evaluation factors is defined as R, and the fuzzy relation from U to V is shown in Eq. (34.3).

$$R = \begin{bmatrix} R_1 \\ R_2 \\ \cdots \\ R_m \end{bmatrix} = \begin{bmatrix} r_{11} & r_{12} & \cdots & r_{1n} \\ r_{21} & r_{22} & \cdots & r_{2n} \\ \cdots & \cdots & \cdots & \cdots \\ r_{m1} & r_{m2} & \cdots & r_{mn} \end{bmatrix} \quad (34.3)$$

If the weight of evaluation factors is assigned as $A = \{b_1, b_2, , b_n\}$ (obviously A is a fuzzy subset of the domain, and $0 \leq a_i \leq 1, \sum_{i=1}^{m} a_i = 1$). Through the fuzzy transformation, a fuzzy subset of the domain can be derived, which is the comprehensive evaluation results $B = A \cdot R = \{b_1, b_2, \ldots, b_n\}$.

(2) Multi-level fuzzy evaluation model. Divide the evaluation factors into several categories according to some property; and then do the comprehensive evaluation for each category; finally, do the high-level comprehensive evaluation for the evaluation results of all kinds of categories. This is the problem of multi-level fuzzy comprehensive evaluation. The model follows these steps:

Step1: The factors set U can be divided into m subsets according to some attribute c, so that they can meet $\begin{cases} \sum_{i=1}^{m} U_i = 1 \\ U_i \cap U_j = \varphi(i \neq j) \end{cases}$. Then the second stage of the evaluation factors $U/c = \{U_1, U_2, \ldots, U_m\}$ can be obtained.

In above formula, $U_i = \{u_{ik}\}(i = 1, 2, \ldots, m; k = 1, 2, \ldots, n)$ means that U_i subset contains n_k judgment factors.

Step2: n_k judgment factors in each subset U_i can be evaluated according to the single-level comprehensive evaluation model. If the weights allocation of all factors in the U_i is A and its judgment decision-making matrix is R_i, then the comprehensive evaluation result of the ith subset U_i is $B_i = A_i \cdot R_i = \{b_{i1}, b_{i2}, \ldots, b_{in}\}$

Step3: Do the comprehensive assessment to m subsets $U_i = \{i = 1, 2, \ldots, m\}$ in U/c, and the evaluation matrix is:

$$R = \begin{bmatrix} B_1 \\ B_2 \\ \cdots \\ B_m \end{bmatrix} = \begin{bmatrix} b_{11} & b_{12} & \cdots & b_{1n} \\ b_{21} & b_{22} & \cdots & b_{2n} \\ \cdots & \cdots & \cdots & \cdots \\ b_{m1} & b_{m2} & \cdots & b_{mn} \end{bmatrix}$$

If the weight of the subset of various evaluation factors is assigned to A in U/c, we can get comprehensive evaluation result $B = A \cdot R$. B is not only the comprehensive evaluation results for U/c, but also the comprehensive evaluation results of all the evaluation factors in U. There are two methods to calculate the above calculation of matrix synthesis. The first one is the main factors determination model method and the other one is the ordinary matrix model method.

34.4 Case Study

34.4.1 Case Evaluation

Until June 2012, Beijing has built three BRT lines with the total length of 55 km. Based on the proposed multiple attribute decision making index system and methods, the evaluation results of three BRT lines are illustrated in Table 34.2.

From Table 34.2 we can see, BRT1 line gets the highest score and has the best implementation effect, BRT2 line gets the lowest score and the worst implementation effect. Based on the implementation effect of the BRT1 line, the following part will comparatively analyze the existing problems of BRT2 line and judge the "gain and loss" in construction of the BRT system from different perspectives.

34.4.2 Effect Analysis

Low carbon factors: From Table 34.2, it can be seen that the index values of low carbon factor of the three lines reflect the low carbon effect of BRT system comparing with other traffic modes. But the fact that the traffic noise pollution index is close to the average level of Beijing indicates that the opening of BRT

Table 34.2 Evaluation result of three BRT lines

Index number	South-centre corridor BRT1 Operation value	Score	Chaoyang road BRT2 Operation value	Score	Anli road BRT3 Operation value	Score
1	24.4	0.062	17.55	0.0168	21.61	0.0436
2	38.47	0.0338	12.99	0.0114	64.78	0.0569
3	130.0	0.0217	120.0	0.0248	90.0	0.0341
4	414.78	0.0254	299.24	0.0304	931.15	0.003
5	71.13	0.0268	37.0	0.0584	48.07	0.0481
6	94.78	0.0178	94.95	0.0184	100.13	0.0373
7	215.82	0.0258	221.12	0.027	219.42	0.0266
8	6.5	0.0389	3.39	0.0203	4.65	0.0278
9	1801.19	0.0204	1792.56	0.0206	1443.77	0.0265
10	350.78	0.0145	396.61	0.01	386.15	0.011
11	2438.11	0.0172	1130.63	0.0056	915.78	0.0037
12	358.34	0.0319	235.29	0.0071	269.88	0.0141
13	0.98	0.0567	1.01	0.0359	0.99	0.0473
14	70.0	0.0871	75.0	0.0934	80.0	0.0996
15	6.08	0.0569	6.15	0.0549	6.13	0.0555
16	75.0	0.0216	70.0	0.026	75.0	0.0216
Total score		0.5586		0.461		0.5567

system alone cannot effectively reduce noise, it should be implemented together with other measures.

Line planning: BRT2 Line is very close to Jingtong Expressway and Subway Batong line located on its southern side. The general bus lines on Jingtong Expressway are perfect enough, so authors think that both of the two traffic modes mentioned above are more attractive. Therefore, the passengers of BRT2 Line are greatly shunted.

Right-of-way: The lane of BRT1 is nearly closed, and the length of exclusive bus lane is 15 km, occupying 94 % of the total length of the whole line. But the exclusive lane length of BRT2 line is 12.3 km, only occupying 76.8 %, the length of about 3.7 km is not the exclusive lane, and the intersections does not have been channelized. In one word, the ultimate problem of BRT2 Line is the right-of-way, which leads to low speed, more delays and low punctuality rate.

Line combination: Another current problem of BRT2 Line is that it does not deal with the general bus line combination well. When the BRT2 Line just opened, the daily passenger number was only 35,000 persons. After integrating two bus lines, as well as opening several branches, the daily passenger number reached 46,000. If the problems of vehicle type, limited capacity of the platform, and ticket system could be overcome, the daily passenger number would reach more than 80,000.

Transfer: For the three BRT lines, the layout patterns of stops include the central island and the central side turnout. The stops of general bus lines, however, are generally set outside the assistant-road. Therefore, no matter adopting what cohesive mode, the transfer distance is inevitably increased. Quite many passengers of BRT2 line give up BRT because of the long transfer distance. Of course, the "zero distance" transfer between Qianmen and the Subway line 2, as well as the vertical transfer between Muxiyuan Qiao and the Third Ring main road, are worthy of reference.

34.5 Conclusion

BRT is a new direction to promote low carbon life, but how to evaluate it holistically and scientifically is always a big problem puzzling the experts. Considering low carbon factors, this paper established an index system for BRT system multiple attribute decision making based on using the fuzzy comprehensive evaluation theory. The case study was conducted in BRT line 1, 2, 3 of Beijing, which verified the effectiveness of the proposed index system and decision making methods. The result also shows that BRT system has a good influence on low carbon.

However, this paper has not taken the waste of road resources caused by BRT exclusive lanes into consideration, which would lead high carbon emissions since more and more vehicles are gathered on the lanes beyond the BRT lanes. This is also likely to be a useful effort for further research.

References

Cherry CR, Weinert JX, Yang X (2009) Comparative environmental impacts of electric bikes in China. Transp Res Part D 14(5):281–290
Deng X (2006) Economic costs of motor vehicle emissions in China: a case study. Trans Res Part D 11(3):216–226
Hayashi Y, Kato H, Teodoro RVR (2001) A model system for the assessment of the effects of car and fuel green taxes on CO_2 emission. Transp Res Part D 6(2):123–139
Li J-Q, Head KL (2009) Sustainability provisions in the bus-scheduling problem. Transp Res Part D 14(1):50–60
Romilly P (1999) Substitution of bus for car travel in urban Britain: an economic evaluation of bus and car exhaust emission and other costs. Transp Res Part D 4(2):109–125

Chapter 35
Research on Time Cost of Urban Congestion in Beijing

Qifu He

Abstract During the past 10 years, with the geometric vehicle population growth and subsequent worse road congestion of Beijing, people have got longer travel time and more uncertainty and the speed of vehicles have got evident decrease. As road traffic efficiency drops, the cost of time delay for people increases. People often neglect time cost, but it plays an indispensable role in economics that you can hardly ignore. The externality caused by traffic congestion significantly affects the city's transportation accessibility and efficiency, ultimately having a negative impact on the sustainable development of society and economy. With the aim to present the externality with the form of currency by quantitative method from the economic point of view, the external costs of congestion are measured by the proposal and accounting of cost of urban congestion based on relatively accurate data. In this thesis, research goals are refined to time cost which is emphasized and analyzed from congestion cost and quantified. Besides, policies and advice for alleviation of road congestion are also provided.

Keywords Externality · Congestion cost · Time cost

35.1 Background

With the development of society and economy, the quickening pace of urbanization and the rapidly growing demand of urban transportation, the problem of traffic congestion becomes increasingly worse in China. Moreover, as the center of

Q. He (✉)
School of Economics and Management, Beijing Jiaotong University, Beijing 100044, China
e-mail: zhangheqifu@gmail.com

politics, economy and culture of China, the traffic demand of Beijing increases sharply with the pace of urbanization, modernization and mechanization, resulting in new records of numbers of vehicles. According to statistics of Beijing Traffic Management Bureau, the number of vehicles in Beijing has increased from 2300 in 1949 to 1,000,000 in 1997. However, it only took six years and six months to increase from 1,000,000 cars to 2,000,000 cars. Until February 17th 2012, the number of cars in Beijing amounted to more than 5,000,000. The development of transportation condition in the 12th Five-Year Plan of Beijing Transportation Research Center predicts that it is in 2016 that the number of cars in Beijing will reach 6,000,000 in accordance with the expectation of speed increase in 2011.

Traffic congestion lowers the speed of cars in urban area, increases travel time and the uncertainty and leads to external costs such as the increase of motor fuel consumption, standard contaminants and greenhouse gas emissions etc. But time cost is the most direct external cost, which directly reflects opportunity cost and profound economic meanings.

35.2 Literature Review

From the view of the definition of congestion cost, China is now at the starting stage of research on traffic congestion costs and scholars don't have clear definitions and divisions of congestion costs. Liu (2007) think that costs can be divided into private cost and social cost (See Table 35.1). Private cost refers to the part of traffic costs undertaken directly by individuals in urban transportation, which includes actual payment cost such as vehicle purchase and fuel cost etc., and invisible cost such as time cost and traffic congestion cost (traffic congestion inside cars). Social cost refers to the part of traffic costs undertaken by the society in

Table 35.1 Classification of transportation costs

Item	Classification of costs	Constitution of costs
Private cost	Actual payment cost	Vehicle purchase, fuel cost, parking charges, vehicle maintenance and traffic accidents charges
	Invisible cost	Time cost and traffic congestion cost
Social cost	Environmental pollution cost	Noise pollution cost, air pollution cost and weather change cost
	Traffic accident cost	Health loss, production loss and administration
	Traffic congestion cost	Time cost and other costs
	Urban development cost	Construction cost, urban space occupation cost and urban image decrease cost

urban transportation, such as environmental pollution cost, traffic accident cost, traffic congestion cost (road) and urban development cost.

Jing (2005) believes that urban transportation refers to the third-party influence from urban traffic users on non-traffic vehicle users. Cui (2006) maintain the external costs of urban traffic congestion includes extra trip time, environmental pollution cost caused by traffic congestion and traffic accident cost from the view of individual motors. Qi (2008) notes that travel time cost of residents is the value produced from time consumption and opportunity cost during travel.

From the view of congestion cost estimation and calculation, Cui (2006) puts forward extra time cost $C_{ET} = \bar{l} \times (\frac{1}{v_{q+1}} - \frac{1}{v_q}) \times c_{pt} \times n \times m_w$, ($\bar{l}$ is the average distance of commuter cars, c_{pt} is the value for the unit of time of car users, namely yuan/h; n is annual average times of travel of commuter cars; m_w is the average daily number of commuter cars). Jin (2007) quantifies traffic impedance and constructs toll model of environmental pollution which exceeds standard on the basis of function form of American National Highway Traffic Safety Administration.

Foreign evaluations are similar to those of domestic evaluations and they have a number of evaluation methods, which have been referred in Exteriorization of the Cost of Traffic and Society translated by Yunping etc. At present, countries in US and Western Europe divide the external effect of traffic into three spaces or time to analyze: near or short period, medium period and far or long period. The social cost of urban traffic has a characteristic of extensiveness, complexness and non-linear. Meanwhile, the nature of the effect of urban traffic and society has evident differences. Therefore, foreign evaluations of social cost currently focus on the local effect of urban traffic, namely the social costs from local air pollution, noise and traffic congestion. In recent years, in terms of the complicated relations among the speed of vehicles, time cost and motorization level and the difficulty to collect data, there are two relatively authoritative evaluation models. One is the simple concept model of the external cost of traffic congestion constructed by Maddison, $TC = g \times M = Ma + \frac{Mb}{\alpha - \beta M}$, among which a represents the actual currency cost of unit mileage (fuel cost, daily maintenance of cars and fares etc.), b represents the value of unit time, T represents the time of unit mileage, s represents the average speed of vehicles and M represents the traffic flow. The other is the evaluation model used by Institute of US Texas Traffic in its annual report. The basic assumption of Maddison model is that all the travelers and vehicles share no differences, which means that the unit travel cost and operation cost are identical and the values of unit time are equal. But the evaluation model of external cost of congestion in Institute of Texas maintains that the external cost of passenger traffic and freight should be calculated separately. In that way, it is comparatively easy to operate and understand.

35.3 Accounting of Time Cost

Congestion costs include nine sub-costs (Gao 2007), among which time cost is the most direct one. The travel time people delay has economic value and the method to measure it mainly depends on wage rate in unit time. As a consequence, time costs vary from people of different income during the same period. In terms of people with high wages, the wage per hour is higher than those of low wage in the same travel time. He probably gives more benefit (Bilbao-Ubillos 2008).

The scale of time cost has close relations with the purposes of travel, such as going to work, business trips, leisure time and going to schools etc. Generally, it can be divided into commute, non-commute and school. Different travel purposes have various affects to the value of unit time. In general, the values from commute or the period of business trips are higher than those of non-commute or the period without business trips. Considering different travel purposes and time value from congestion given up by people of various incomes, the evaluation of the cost of time should be in accordance with travel aims and the average income of various levels.

The phenomena of congestion in this study occur in the period of congestion. Firstly, calculate the total cost of time in the period of congestion. Then calculate the extra cost of time according to the percentage of the extra time of one travel to total travel time.

According to the travel volume and one travel time in the period of traffic congestion, the total cost of time can be calculated using of the values of unit time with different vehicles and their relations:

$$C_{\text{total cost of time}} = \sum_{i=1}^{3} \sum_{j=1}^{3} t_i (C_{ij} \times M_j) \times 365 \quad (35.1)$$

In the equation, $C_{\text{total cost of time}}$ refers to the total cost of time, t_i refers to one time consumption of various travel methods, C_{ij} refers to the value of unit time of purpose, M_j refers to the daily travel volumes of various purposes, i refers to buses, taxis and cars and j represents three travel methods which are commute, non-commute and school.

The calculation formula of the extra cost of time in the period of congestion is as follows.

$$C_{time} = \sum_{i=1}^{3} \frac{t_i - t_{ic}}{t_i} \times C_{i\,\text{total cost of time}} \quad (35.2)$$

In the equation, C_{time} cost refers to the total cost of time in the period of congestion, t_i refers to one time consumption of travel in the period of congestion, t_{ic} refers to the distance of one travel in the period of congestion, $C_{i\,\text{total cost of time}}$ refers to the total cost of time with various travel methods in congestion periods, i refers to buses, taxis and cars.

According to the statistics of Beijing Bureau of Statistics, the average income of employees of 2010 in Beijing is up to 48444 yuan, which reflects the average income of employees in national and collective enterprises, joint ventures, solely funded enterprises and private enterprises except the self-employed people.

In accordance with the 51th regulation of Law of The People's Republic of China, in the provided holidays, the employers should pay employees, which means they should pay the employees according to days or hours even in those holidays.

Therefore, the method to calculate the average rate of income of per hour is $x_h = \frac{I_a}{12} \times \frac{1}{D_m \times 8}$. In this equation, x_h refer to the rates of income of per hour, I_a represents the average annual rate of income and D_m refers to the days of monthly income respectively. According to the data of Beijing Bureau of Statistics, the average income of employees of 2010 in Beijing is up to 48444 yuan. In this way, the average rate of income of per hour amounts to 22.9 yuan.

In the *Annual Report of the Development of Traffic of 2011 in Beijing*, the basic travel is the most common purpose of daily travels for people, including commute and school, accounting for 30.02 % of the total travel volume. Other travels include shopping, going to gyms and leisure, visiting relatives and friends and having dinners etc. This thesis mainly focuses on three conditions in weekdays, which are commute, school and other.

Firstly, the income of people with various travel methods in 2010 can be calculated by the influence coefficients of the time cost in 2005 (Table 35.2).

In terms of the effect of the purpose of travel to the time cost, the influence coefficients recommended by the World Bank are as follows.

W = average income of per hour
The cost of per hour for commute = 1.33 W
Time cost of per hour of other non-commute travel = 0.3 W
Time cost of per hour of school = 0.15 W

The value of time of various travel methods and purposes are shown in Table 35.3.

According to *Research on Beijing Comprehensive Transportation Cost*, different modes and purposes of Beijing daily travel volume of 2005 are shown in Table 35.4.

According to Table 35.4, the percentages of three types of vehicles with different travel purposes are shown in Table 35.5.

Table 35.2 Income situations of all travel ways of 2010

	Bus	Taxi	Private car
Influence coefficient of time cost	1	1.35	1.74
Annual per capita income (Yuan)	48444	65399	84292
Per hour per capita income (Yuan)	22.9	30.9	39.84

Note Influence coefficients of time cost are from research on Beijing comprehensive transportation cost

Table 35.3 Time costs of different ways and trip purpose of 2010 (Yuan/h)

	Commute	Non-commute	School
Bus	30.4	6.87	3.43
Taxi	41.1	9.27	4.64
Private car	52.98	11.95	5.97

Table 35.4 Different modes and purposes of Beijing daily travel volume of 2005 (Million people/day)

	Commute	Non-commute	School
Bus (include subway)	351.5	434.2	84.4
Taxi	89.7	110.7	21.5
Other	351.5	434.2	84.4

Table 35.5 Different ways and purposes of Beijing travel daily proportion of 2005

	Commute	Non-commute	School
Bus (include subway)	1	1.24	0.24
Taxi	1	1.23	0.24
Other	1	1.23	0.24

Table 35.6 Different models and purposes of Beijing daily travel volume of 2010 (Million people/day)

	Bus	Taxi	Other
Total daily travel volume	1317	188	882
Business	531.1	76.1	357.1
Non-business	658.5	93.6	439.2
School	127.5	18.3	85.7

In the *Annual Report of the Development of Traffic of 2011 in Beijing*, daily travel volume of various travel methods in 2010 can be found. With Table 35.5, daily travel volume of various models and purposes are shown in Table 35.6.

In this thesis, the calculated cost is the cost of traffic congestion, thus this calculation chooses the period of congestion and all the costs are costs from vehicle congestion in the period of traffic congestion. The periods are set from 7 a.m. to 9:30 a.m. and 4:30 p.m. to 7 p.m. From Fig. 35.2, it can be seen clearly that during the two periods, the traffic volumes reach the peak and those of other periods are quite small, thus not included in congestion periods.

Figure 35.1 shows the rate of travel areas of private cars during congestion periods to the daily areas. By this, the total travel volume of cars during congestion periods can be calculated and so do other travel modes, which are shown in Table 35.7.

The average numbers of the average travel time for peaks in the morning and evening of three types of vehicles are taken as the average travel time of those three types of cars during congestion periods. According to the traffic research, parts of the data are shown in Table 35.8 (except taxis).

Combine different travel figure out all kinds of car travel for the purposes of the ratio of their total trip different travel purposes as shown in table 35.9.

35 Research on Time Cost of Urban Congestion in Beijing 231

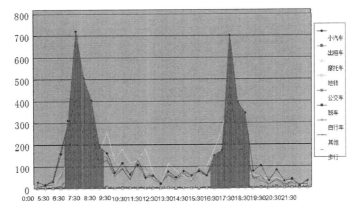

Fig. 35.1 Relationship between travel volume of private cars during congestion periods and total travel volume

Table 35.7 Different modes of Beijing daily travel volume for residents during congestion periods (Million people/day)

	Bus	Taxi	Other
Ratio of travel volume during congestion periods to total daily volume	0.63	0.38	0.69
Total daily travel volume	1317	188	882
Travel volume during congestion periods	829.71	71.44	608.58

Note Total daily travel volume is from Beijing annual report of transportation development of 2011

Table 35.8 Average travel time of vehicles in rush hour on (Min)

Vehicle	2010 rush hour in the morning	2010 rush hour in the evening
Car	36.43	35.78
Subway	68.31	62.72
Bus	54.4	55.41
Bike	23.04	24.87
Walk	12.14	11.34

Note Data from Beijing annual report of transportation development of 2011

Table 35.9 Different modes and purposes of Beijing travel volume for residents during congestion periods of 2010 (Million people/day)

	Total travel volume during congestion periods	Commute	Non-commute	School
Bus	829.71	334.56	411.51	80.29
Taxi	71.44	28.92	35.57	6.94
Other	608.58	246.38	303.05	59.13

Table 35.10 Taxi trip distance during different periods

Trip distance	12 p.m.–7 a.m.	7 a.m.–9 a.m.	9 a.m.–5 p.m.	5 p.m.–7 p.m.	7 p.m.–12 p.m.	All day
Distance (km/car)	10.26	8.23	7.63	8.35	8.89	8.22

Due to the lack of data of taxis in Table 35.9, the travel time of taxis in peak periods is calculated separately. According to the third traffic comprehensive survey in Beijing, the travel distance of one time for taxis of various periods in 2005 can be figured out (details in Table 35.10). Since there is no big change in travel distances of one time, one travel distance of 2005 is regarded as the travel distance of 2010 and is dealt with the speed conditions of road network in peak periods of 2010.

The average travel distance of every time for taxis in peak periods is (8.23 + 8.35)/2 = 8.29 km.

In the beginning period of data preparation, receipts of taxis should be extracted and collected randomly, figuring out the travel speed of taxi in congestion periods reaches 20.63 km/h, thus the travel time of every time in that periods should be 8.29 km/(20.6 km/h) = 24.1 (min).

Furthermore, related survey shows that the rates of bus time, walking time and waiting time amount to 64:36:13, which is shown in Fig. 35.2.

According to the data of traffic survey, one travel time of people by bus in congestion periods is $\frac{54.4+55.41}{2} = 54.9$ (min). After subtracting the waiting time and transferring time, the average consumption of time on bus is $\frac{(54.4+55.41)\times 64\%}{2} = 35.14$ (min). Therefore, the average consumption of time by bus in congestion periods amounts to $\frac{36.43+35.78}{2} = 36.1$ (min) (Table 35.11).

1. **Total time cost of three types of vehicles in congestion periods**:

$$C_{i\,\mathrm{totaltimecostofvehicle}} = t_{i\,\mathrm{everytimeconsumption}}$$
$$\times \sum \left(C_{\mathrm{thevalueofunittime\,for\,certainpurposes}} \times M_{\mathrm{travelvolumeofcertainpurposes}} \right)$$

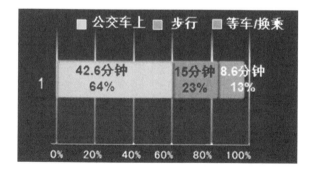

Fig. 35.2 A public transport travel time

35 Research on Time Cost of Urban Congestion in Beijing

Table 35.11 Time consumption and travel distance of three types of cars in congestion periods

	Bus	Taxi	Car
One travel time consumption in congestion periods (min)	35.14	24.1	36.1
One travel distance in congestion periods (km)	11.1	8.29	10.78

a. During the congestion periods in 2010, the time cost of all the residents by buses is:

$$C_{\text{totaltimecostofbus}} = t_{\text{everytimeconsumptionofbus}}$$
$$\times \sum (C_{\text{thevalueofunittime for certainpurposes}} \text{travelvolumeofcertainpurposes}) \times 365$$
$$= \frac{35.14}{60} \times (334.56 \times 30.4 + 411.51 \times 6.87 + 80.29 \times 3.43) \times 10^4 \times 365$$
$$= 283.74 \text{ (billion yuan)}$$

b. During the congestion periods in 2010, the time cost of all the residents by taxis is:

$$C_{\text{totaltimecostoftaxi}} = t_{\text{everytimeconsumptionoftaxi}}$$
$$\times \sum (C_{\text{thevalueofunittime for certainpurposes}} \times M_{\text{travelvolumeofcertainpurposes}}) \times 365$$
$$= \frac{24.1}{60} \times (28.92 \times 41.1 + 35.57 \times 9.27 + 6.94 \times 4.64) \times 10^4 \times 365$$
$$= 22.73 \text{ (billion yuan)}$$

During the congestion periods in 2010, the time cost of all the residents by taxis is:

$$C_{\text{totaltimecostofcar}} = t_{\text{everytimeconsumptionofcar}}$$
$$\times \sum (C_{\text{thevalueofunittime for certainpurposes}} \text{travelvolumeofcertainpurposes}) \times 365$$
$$= \frac{36.1}{60} \times (246.38 \times 52.98 + 303.05 \times 11.95 + 59.13 \times 5.97) \times 10^4 \times 365$$
$$= 373.94 \text{ (billion yuan)}$$

Consequently, the total time cost of residents during congestion periods in 2010 is 68.041 billion yuan (Table 35.12).

By the sampling survey of taxis in Beijing, the average speed of taxis during non-peak periods (6:30 a.m. to 7 a.m. and 9 a.m. to 5 a.m.) is 25.39 km/h. As for the speed of road network of Beijing, when the congestion coefficient reaches four, the average speed is 24.5 km/h. Since the speed in the evaluation system of congestion coefficients includes buses, it is difficult to identify them. Therefore,

Table 35.12 The total time, extra congestion time and total time cost of different ways of travel in 2010 in Beijing during congestion periods

	Bus	Taxi	Car
Total congestion time (ten thousands hours)	177365.7	10473.7	133649.2
Total cost during congestion periods (billion yuan)	283.74	22.73	373.94

this study adopts 25.4 km/h as the critical speed of congestion and flow for taxis and cars.

Due to the lack of travel volume of buses in different periods, the weighted average speed of buses cannot be calculated. Thus, this thesis takes the average speed of critical congestion periods as the critical speed of congestion and flow for buses.

Table 35.13 shows the critical flow speed and congestion speed of buses, taxis and cars in the road network.

The passages above figure out one time consumption and distance of different travel modes for residents during congestion periods. By combining the standard of flow speed, the time difference between congestion travel and flow travel can be calculated and the total cost of extra congestion can be drawn from the proportional relation.

2. One travel time consumption of flow speed for three types of travel modes

$$t_{\text{flow speed of bus}} = \frac{11.1}{20.4} \times 60 = 32.6 \,(\text{min})$$

$$t_{\text{flow speed of taxi}} = \frac{8.29}{25.4} \times 60 = 19.6 \,(\text{min})$$

$$t_{\text{flow speed of car}} = \frac{10.78}{25.4} \times 60 = 25.46 \,(\text{min})$$

3. Total cost of extra congestion time of Beijing vehicles in 2010

$$C_{\text{itimecostofextracongestion}} = \frac{t_{\text{itimeconsumptionofextracongestion}}}{t_{\text{itotaltimeconsumptionofcongestion}}} \times C_{\text{totaltimeconsumptionofcongestion}}$$

Table 35.13 Standards of flow speed and congestion speed of Beijing urban road network

	Bus	Taxi	Car
Flow speed (km/h)	20.4	25.4	25.4
Congestion speed (km/h)	18.8	20.6	20.6

$$C_{\text{timecostofextracongestionofbuses}} = \frac{35.14 - 32.6}{35.14} \times 283.74 = 20.51 \text{ (billion yuan)}$$

$$C_{\text{timecostofextracongestionoftaxis}} = \frac{24.1 - 19.6}{24.1} \times 22.73 = 4.24 \text{ (billion yuan)}$$

$$C_{\text{timecostofextracongestionofcars}} = \frac{36.1 - 25.46}{36.1} \times 373.94 = 110.2 \text{ (billion yuan)}$$

Thus, the total cost of extra congestion time of Beijing vehicles in 2010 is:

$$C_{\text{costofextracongestiontimeofBeijing}} = 20.51 + 4.24 + 110.2 = 134.9 \text{ (billion yuan)}$$

35.4 Conclusion

In the above analysis, it can be found that the time cost caused by congestion is enormous and that of private cars account for a proportion of the total. Furthermore, it can be judged that a large number of private cars lead to traffic congestion. As a result, the time cost of private cars increases a lot and affects the time cost of taxis and passengers of buses. Consequently, two basic suggestions are offered here.

Firstly, we need to solve the preference of people when they go out and lead them to travel with less emissions of carbon dioxide. We need to change their blind pursuit for private cars, enhance the guidance of modern traffic awareness, improve the concept of traffic consumption, use cars reasonably and moderately, strengthen the awareness of modern traffic safety and drive safely and scientifically.

Secondly, with the characteristic of low carbon, green traffic is the component of urban sustainable traffic system. According to the conditions of traffic congestion in Beijing, great efforts should be made to construct a comprehensive urban traffic system of bus, walk and bike, which is a healthy development mode for the actual circumstance of Beijing. For instance, we can give a plan of traffic system rules for walk and bike, make regulations and rules of example streets and districts for walk and bike in the western part of Zhongguancun and central business districts and gradually promote it. We can also encourage and support the development of public bike systems, set up about 1,000 spots for bike rent, which will make the ratio of travelling bikes in Beijing at about 20 % and make the structure of traffic travel more reasonable. Therefore, the integrity of traffic and land is the technical key of the construction of sustainable traffic system and the government is responsible for making policy structure of the coordination and development of traffic and land.

References

Bilbao-Ubillos J (2008) The costs of urban congestion: estimation of welfare losses arising from congestion on cross-town link roads. Transp Res A 42(8):1098–1108

Cui Z (2006) An analysis of external cost of urban traffic congestion. Acad J Wuhan Univ Techonol 2:61–63

Gao T (2007) Quantization study of urban traffic external cost. J Tianjin Univ Technol Educ 3:26–29

Jing W (2005) Assessment on traffic social cost of Beijing. Traffic Environ Prot 2:35–37

Jin L (2007) Municipal transportation exterior cost analysis and internalization quantitative method. Technol Econ Areas Commun 5:27–31

Liu L (2007) The analysis of external cost of road congestion in Beijing. Beijing Technol Bus Univ 4:19–23

Tongyan Qi (2008) A study of the traveling time cost of Beijing residents. J Highw Transp Res Dev 6:12–14

Chapter 36
The Development and Application of Transport Energy Consumption and Greenhouse Gas Emission Calculation Software Based on the Beijing Low-Carbon Transport Research

Weiming Shen, Feng Chen and Zijia Wang

Abstract According to the established calculation model for energy consumption and greenhouse gas (GHG) emissions from the Beijing low-carbon transport research, a general calculating software visualizing the whole calculation and providing flexible data input interface was developed. The user can calculate the values of energy consumption and GHG emission of urban road transport and rail transit, and do some scenario analysis through the succinct interface of the software and rich parameters to input. This tool allows the transport agencies to analyze transport policies and evaluate the implication of these policies, the analysis can improve the scientific and rationality of the traffic decision-making and ensure the implementation of green transport goals.

Keywords Transport energy consumption · Greenhouse gas emission · Calculating and predicting · Scenario analysis

36.1 Introduction

Fossil fuels, the global energy supply and greenhouse gas (GHG) emissions have increased rapidly because of the development of industrialization. Global energy crisis and climate change make energy saving receive worldwide consensus. According to the established calculation model for energy consumption and GHG emissions, a general calculating tool visualizing the whole calculation and providing flexible data input interface was developed, which allows the transport

---

W. Shen (✉) · F. Chen · Z. Wang
Urban Rail Transit Research Center, Beijing Jiaotong University, Beijing 100044, China
e-mail: swimming.1988@live.cn

agencies to analyze transport policies and evaluate the implication of these policies, The analysis can improve the scientific and rationality of the decision and ensure the implementation of green transport goals.

With this software, the user can input the values of parameters included among the urban road transport and urban rail transit, through the calculation of the computer, the energy consumption and GHG values can be obtained. Especially, the user can utilize the energy consumption model formulated and the calculating program developed according to the basic data and traffic policies, the future annual traffic energy policies and trend of traffic energy consumption and GHG emissions be analyzed, and the efficacy of the energy saving policies be predicted. These analyses can provide reference for traffic policy making.

36.2 Basic Framework and Design Goals

The transport energy consumption and GHG emission calculation software is built based on the establishment of Beijing low-carbon transport quantitative model. According to the calculating tool built, the user can calculate the values of energy consumption and GHG emission and do some scenario analysis through the succinct interface of the software and rich parameters to input. The software aims at the visualization of the calculation and analysis procedure, which can help to improve the decision-making more rationally and scientifically, so that to ensure the implementation of green transport. The basic framework of the software is as shown below.

Software could be divided into two parts: calculation and predicting of the energy consumption and GHG emissions and scenario analysis. The calculation and predicting part is composed of three modules: input modules, calculation module and Output module. Calculation module is actually the visualization of calculation model developed above which generates the calculation process with the data input by users, and provides results, Therefore, this module bear little linkage with users. Parameter input module and the result output module are more important to the users.

(1) **Data input module**

Parameter input interface is divided into two parts: urban road parameters input and urban rail transit parameters Input. Urban road traffic includes vehicle amount of different types, VKT of different types, vehicle fuel economy, passenger volume and PKT of different vehicle types, average trip distance and other parameters. The urban rail transit involves line length of different alignment types, numbers of station of different alignment, annual train departure, passenger volume, average trip distance and so on Beijing (2007). Energy consumption per load per kilometer is user-selectable input parameters. The two parts of parameter input bear a relationship through passenger attribution. The figure below shows the input parameters of public diesel vehicles and parts of the urban rail transit (Fig 36.1).

Fig. 36.1 Input parameters of public diesel vehicles and parts of urban rail transit

(2) **Results Output Module**

The results of the software include the transport energy and greenhouse gas emissions can also be output by the corresponding mode of transport and energy consumption and vehicle emissions inventory, and the form of a chart to show users.

The scenario part includes two main interfaces, the parameter input interface and the result interface, and the latter one is same as the result output module of the calculation part. Entering the interface of Scenario analysis, Firstly the user chooses target year he wants to calculate or predict, and the right parameter table can be chosen to change the corresponding parameter values. If the user cannot know some of the parameter values, the software set them as the default values automatically.

There are 4 scenario situations settled in the software: Reduction of average trip distance and promotion of fuel economy of social car, limitation on purchase of social cars, reduction of energy economy of rail traffic and modes shift of passenger traffic. In the next chapter, the first scenario will be mainly interpreted.

Except for the three modules, there are shortcuts for all the types of calculation and the macro indexes (such as GDP per capita and the number of year) on the left of the interface, so the user can enter the interface which he wants to calculate directly without click in one by one (Fig 36.2).

36.3 Sample Application

There are two examples to show how the software operates: The first one is to calculate the values of Beijing energy consumption and GHG emission in 2009 which are already analyzed by the relevant departments, and this process can be seen as a test of the calculating accuracy of the software. The other is to analyze the scenario "Reduction of average trip distance and promotion of fuel economy of social car".

Fig. 36.2 Parameter input of scenario analysis

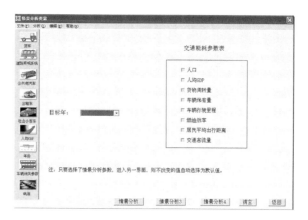

(1) Calculation of Beijing energy consumption and GHG emission in 2009

After choosing the urban road transport on the main interface, firstly setting the parameter values of lorry: Freight turnover: 8543719901 t · km Beijing (2010) and the average energy intensity value as the default value, 0.1079 L/(t · km) IEA (2004) and click on the calculate button, the energy consumption value is 921867377L. And do the same procedure in the Public electric vehicle, taxi, public passenger vehicles and road lighting system. In every page there is a energy and GHG result of the corresponding vehicle type.

About the urban rail transit, the parameters are divided into 2 parts. For the metro vehicles, the energy consumption for unit of freight traffic turnover for both underground and ground lines is needed to input. For the metro stations, the average energy consumption for the action and lighting per month for 3 different kinds of stations (underground station with platform screen doors (PSD), underground station without PSD and ground one without PSD), the number of each kind of station and turnover volume for the underground and ground lines should be obtained by the user. However, if the user cannot get any of the parameter value, he can use the default value set in the software which is related to urban rail transit in Beijing. After calculation, the total result of energy consumption and GHG emissions of Beijing in 2009 are 6.50 billion kgce and 42.19 billion kg.

In the urban rail transit part, if the user wants to calculate the energy consumption for unit-load of the new-bulit metro line, the figure below is available and the user just needs to input the corresponding parameter values, and the values of unit-load for every new-bulit lines will be shown in the table and the total energy consumption for metro vehicles are calculated out.

(2) Scenario analysis

In the scenario "Reduction of average trip distance and promotion of fuel economy of social car", considering average resident trip distance gradually decreases with the rationalization of urban distribution of Beijing. Because of the increasing of the proportion of Euro V vehicles Wang (2009), the average fuel

Table 36.1 Energy consumption and GHG emissions in scenario of travel distance decrease and fuel economy promotion

Target year	Social car energy consumption (million L)		Social car GHG emissions (ten thousand t)	
	Default value	Value of scenario	Default value	Value of scenario
2015	5761	3904	57.61	39.04
2020	9623	37.80	96.23	37.80

efficiency of Beijing increases from $7.76 \text{ L} \times 100^{-1} \text{ km}^{-1}$ to 7.09 and 6.57 in 2015 and 2020. The average trip distances of social cars change from 10.78 km to 8 and 5 in 2015 and 2020. Using the default value in inertial model for other parameters, the results are drawn from the software, as shown in table. According to the result, Reduction of average trip distance and promotion of fuel economy of social car can make the values of energy consumption and GHG emission drop by 60 % at most (Table 36.1).

36.4 Conclusion

Various and flexible parameter input, enable the software having relatively strong function and general usage to some degree.

Software can be used to establish transport energy consumption and GHG emissions inventory. The inventory involves energy consumption and GHG emissions from on-road and rail transportation, and different vehicle types. As long as the basic transport data is input, the corresponding energy consumption and emissions inventory could be drawn and presented to the user in the form of chart.

Software can be applied for the scenario analysis of a variety of policies and the implication analysis of energy saving policies. There are many types of input parameters, including parameters of vehicle amount, passenger volume, average trip distance and other factors, so the transport policies or management measures such as control on vehicle purchase, clean fuel technology and promoting the traffic sharing of urban rail transit, which can impact transport energy consumption and GHG emissions in Beijing, could be fully analyzed, and conduct comparative analysis between the different scenarios could be launched.

References

Beijing municipal third traffic comprehensive survey [R], 2007
Beijing traffic development annual report [R], Beijing Transportation Research Center 2010
IEA. Energy Statistics and Balances [M]. International Energy Agency, 2004
Wang Q (2009) Energy Data- China's Sustainable Energy Project Reference Material [R], Energy Foundation

Fig. 37.1 SoC-EMF curve of battery (20 °C)

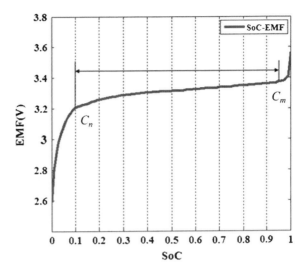

Fig. 37.2 SoC of electric busover operational time

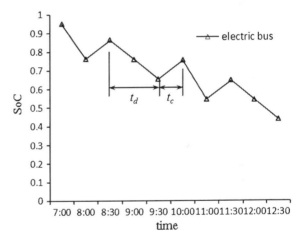

schedule after the fast charge. Once the charging strategy of operating-charging-operating is carried out, SoC of electric bus would be changed over operational time showed as Fig. 37.2, where t_d is discharging time of battery and t_c is charging time of battery.

37.2.3 Vehicle Scheduling Model

The study of bus operating and planning with electric bus can start with a single depot and single route vehicle scheduling problem with route and charging time constraints. The vehicle scheduling model is based on followed assumption:

(1) the style of electric bus is same,
(2) the average round-trip time T_r is invariable,
(3) the average round-trip consumption of the battery power C_r is invariable,
(4) there are enough charging equipments to serve the operational electric bus with $n_c C$ charge in the charging station.

Let N be the fleet size of electric buses with a single route, related to F that defined as the sum of the departure trips, which can be calculated if the timetable is fixed. The vehicle scheduling model of electric bus with a single route is aiming to solve the problem about the minimum fleet size of electric buses when the timetable is fixed, that is to promote the maximum trips of a electric bus according to the charging strategy which can meet the service of the fixed timetable. Using variable f defined as the maximum trips of electric bus with the fixed timetable, the vehicle scheduling model of electric bus with a single route can be formulated as follows:

$$\min N = \frac{F}{f} \tag{37.1}$$

$$C_m + fC_r + n_c T_c \geq C_n \tag{37.2}$$

$$C_m + fC_r + n_c T_c \leq C_m \tag{37.3}$$

$$T_d + T_c \leq T \tag{37.4}$$

$$T_d = fT_r \tag{37.5}$$

Equation (37.1) is the objective function, Eqs. (37.2) and (37.3) assure that SoC of a electric bus is in the working range each trip, and Eq. (37.4) define a bus operational time T where T_d is the sum of t_d, T_c is the sum of t_c.

37.3 Vehicle Route Test

The Entertainment and Food Special Line (EFSL), which is located in Dongguan Songshan Lake National High-tech Industrial Development Zone, is used as test operational line. As Fig. 37.3 showed, EFSL is connecting the industrial, commercial and education areas, in order to increasing the accessibility of each area. EFSL starts from A towards J and returns along the same path to A as the terminal of the bus line. The single-way distance of the line is 12 km and the round-trip is 24 km.

Higer KLQ6702EV is chosen as the operation bus of EFSL, major parameters are showed in Table 37.1, and the results showed in Table 37.2 are testing parameters and running indicators on EFSL according to the test mode without charging.

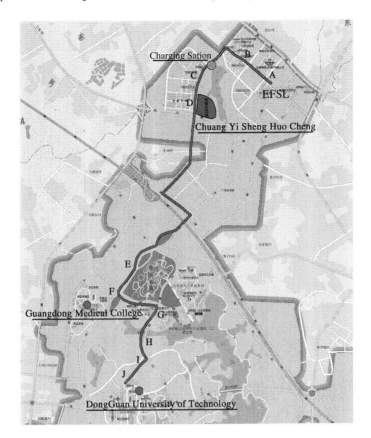

Fig. 37.3 Entertainment and food special line

Table 37.1 Major parameters of higer KLQ6702EV electric bus

Size (mm)	Curb weight	Capacity	Max speed	Battery pack
7020 × 2040 × 2790	6000 kg	13 person	90 km/h	336 V/250 Ah

Table 37.2 Test result of higer KLQ6702EV electric bus

Round-trips	Distance	SoC	AVS	Energyconsumption
4	98 km	95–10 %	26.3 km/h	0.65 kWh/km

37.4 Scheme of Operational Planning

The operational time of the EFSL is 7:00–22:00. The buses departs every hour and reducing the departure interval to 30 min at traffic peaks, as shown in Table 37.3.

Fixed timetable given, the operational time of electric bus T is 16 h, and the sum departure trips F are 19 trips. The battery of electric bus has a capacity of

Table 37.3 Fixed timetable of EFSL

7:00	10:00	13:00	17:00	20:00
8:00	11:00	14:00	18:00	21:00
8:30	11:30	15:00	18:30	22:00
9:00	12:00	16:00	19:00	–

Table 37.4 Process of vehicle-chain construction

Number	Vehicle-chain construction
1	1 → charge → 3 → charge → 5 → charge → 7 → charge → 9 → charge → 11 → charge → 13 → charge → 15 → charge → 17 → charge → 19
2	2 → charge → 4 → charge → 6 → charge → 8 → charge → 10 → charge → 12 → charge → 14 → charge → 16 → charge → 18

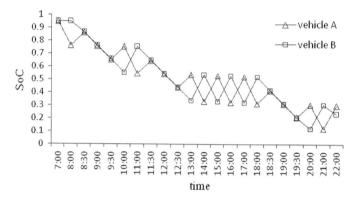

Fig. 37.4 Appraisal curve of real-time SoC of electric bus

75 kWin EFSL, and its working range is from $C_n = 10\%$ to $C_m = 95\%$ by test, which charge at the rate of 0.2 C in the charging station. It's based on the test data of Higer KLQ6702EV electric bus that T_r is 0.91 h, C_r is 15.6 kWh/km. According to Eqs. (37.2), (37.4), and (37.5), the result is $f \leq 10.3$ trips, so the minimum fleet size of electric bus N are 2 vehicles by Eq. (37.1). The vehicle-chain of electric bus in EFSL is constructed showed in Table 37.4.

According to the data and operational vehicle-chain described above, the appraisal curve of real-time SoC is calculated and shown in the Fig. 37.4, which indicates that the constraints of route and charging time could be fulfilled by the vehicle schedule, that is, the vehicle scheduling model of electric bus with a single route is effective.

37.5 Conclusions

Considering the battery state of charge, a model is built to solve a single depot and single route vehicle scheduling problem with route and charging time constraints. In the case study of Dongguan Songshan Lake National High-tech Industrial Development Zone, there is an effective operational Planning scheme of electric bus based on the vehicle scheduling model. However it hasn't been taken into account that the number of charging is finite for a battery, which will be considered in the further research aimming at lengthenning the battery life and reducing the operational costs of electric bus.

Acknowledgments This research was funded by the National 863 Project (No. 2011AA110305).

References

Francfort J (2004) TH! NK city-electric vehicle demonstration program: second annual report 2002–2003
Han X, Jiang J (2011) Simulation of operating plan of E-bus charge station. Microprocessors 2:88–91
Heber CU (1997) Testing electric vehicles of the latest generation on the Island Rugen. Automobiltechnische Zeitschrift 99(9)
Liu H (2008) Building a good environment and facilitating the industrialization of electric vehicle. Shantou Technol 2:19–20
Shi W, Jiang J (2010) Research on SOC estimation for $LiFePO_4$ Li-ionbatteries. J Electron Meas Instrum 8:769–774
Teng L (2009) Power supply modes for electrical vehicles and impacts on grid operation. East China Electr Power 10:1675–1677
Wang N, Gong Z (2012) Prospect analysis for hybrid electric and battery electric city buses based on lifecycle cost and emission. China Soft Sci 12:57–65
Xu F, Yu G (2009) Analysis for the layout of electric vehicle charging station. East China Electr Power 10:1678–1682

Chapter 38
Scheme Research of Urban Vehicle Restriction Measures According to Synthesis Criterion

Long Chen, Ming-jiang Shen and Xing-yi Zhu

Abstract Transportation demand management is an effective way to realize urban low-carbon transport. The urban vehicle restriction is one of the modes of transportation demand management. However, why, when, and how to start the vehicle restriction are the problems the traffic administrators will meet with. Therefore, in this paper, three criterions are presented, and then are verified by the case of vehicle restriction schemes adopted in Hangzhou to show the significance and effectiveness.

Keywords Vehicle restriction · Synthesis criterion · Transportation demand management

38.1 Introduction

The vehicle restriction is one of the modes of transportation demand management (TDM). Compared with the district congestion charge mode mature adopted in England, Singapore, and other developed countries, the district vehicle restriction

---

L. Chen · M. Shen
Municipal Department, Architectural Design and Research Institute of Zhejiang University, Hangzhou 310012, People's Republic of China

X. Zhu (✉)
Shanghai Institute of Applied Mathematics and Mechanics, Shanghai University, Shanghai 200072, People's Republic of China
e-mail: zhuxingyi66@yahoo.com.cn

X. Zhu
Key Laboratory of Roadand Traffic Engineering of Ministry of Education, TongjiUniversity, Shanghai 200092, People's Republic of China

is a method quite suitable to resolve traffic problems in China, where the quantity of vehicles is rapidly increasing, however, the transportation manage measurements still exist many limitations (Cynthia et al. 2010).

The district vehicle restriction scheme has been carried out in San Diego, Bogota, Beijing, etc. In china, this scheme is developed from the countrywide activity of "No vehicle day", which obtained public praise. Then, different categories and scope of vehicle restriction schemes were adopted accompany with the important pageants, such as Olympic Games in Beijing (Cynthia et al. 2010; Mahendra 2008) Asian Games in Guangzhou, the World University Student Summer Sports Games in Shenzhen. After these three successful attempts, the regular district vehicle restriction policy was unveiled in Beijing in Oct. 2008, which implemented by three stages. From then on, Nanchang, Lanzhou, and other cities in China started to adopt this scheme.

The regular vehicle restriction policy is an intrepidity and effective attempt of TDM, however, it will seriously affect all the citizen's daily life. Therefore, the two questions, namely, how to accurately judge whether the vehicle restriction policy is the best choice to resolve the urban transportation problem, and which kind of vehicle restriction schemes are more suitable to the city's characteristic, are the principal problems the traffic administrators will meet with during the decision-making.

In this paper, combined with the research experience of urban vehicle restriction policy in Hangzhou, three criterions, namely, starting criterion, causation criterion, choice criterion, are suggested to improve the design of vehicle restriction schemes. Finally, these three judging criterions are used and verified by the implement of vehicle restriction schemes in Hangzhou on Oct. 2011.

38.2 Starting Criterion

The objective of vehicle restriction is car, and the direct intention of restriction is to control such rapid increase of the quantity of cars as well as to decrease the trip frequency, therefore, to change the develop mode of motorization in the future. Hence, the thousand people car-occupancy's rate (P) is used as the starting threshold. Since by using P as the division criterion of the develop stages of the quantity of city's cars, the develop stages can be separated into the following three stages (Li 2006): primary stage ($P = 0–40$ cars per thousand people); intermediate stage ($P = 40–200$ cars per thousand people); mature stage ($P > 200$ cars per thousand people);

The increasing trend curves of P in different developed countries are plotted in Fig. 38.1. It is shown that the measures have been taken in order to relieve the increasing trend after the value of P comes into the mature stage, such as $P = 357$ cars per thousand people for France, $P = 213$ cars per thousand people for England, and $P = 202$ cars per thousand people for Japan. Take England as an example, in 1970, when the value of P in London comes into mature stage

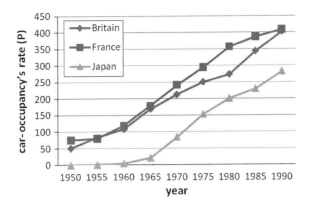

Fig. 38.1 Thousand people posse's rate curves for the developed countries

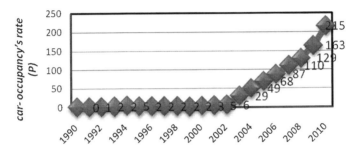

Fig. 38.2 Thousand people car-occupancy's rate curves for Hangzhou

($P = 213$ cars per thousand people), the congestion charge measurement (charged two dollars for each vehicle driving into city center) was adopted and gained great successful.

As for Hangzhou, from 1990 to 2010, the value of P increases rapidly (see Fig. 38.2). In 2010, the value of P in the main city zone reaches to 215 cars per thousand people, which means Hangzhou has already goes into the mature stage. Therefore, according to the starting criterion, Hangzhou city has been provided with the starting reason to implement the vehicle restriction policy.

38.3 Causation Criterion

The main reason for serious traffic congestion in China is that the capacity of road network can not satisfy the demand of increase of motor vehicles. Recently, with the increasingly strained resource of urban land and the basic formation of city road network, the sustainability of increase of capacity of road network is decreasing rapidly. Under this background, the main reason to choose the vehicle restriction policy is that through this measure we can effective prevent the demand of too fast increase of motor vehicles. Therefore, we choose the degree of space–time saturation (S_{at}) as the causation criterion. When the value of S_{at} is larger than 1.05, the vehicle restriction measures can be considered to be adopted.

The degree of space–time saturation can be defined as the ratio of total trip volume of motor vehicles in peak hour (D_n) with the actual capacity of vehicle road network (C_{apr}) (Chen 2002), namely,

$$S_{at} = \frac{D_n}{C_{apr}} \tag{38.1}$$

where D_n can be calculated by

$$D_n = \frac{Q_h s_h + Q_c s_c + Q_b s_b + Q_t s_t}{(1 - s_w)} \tag{38.2}$$

in which Q_h, Q_c, Q_b, and Q_t are the after-convert number of freight carriers, taxis, buses, and the other vehicles, respectively, in the observation district on hand; s_h, s_c, s_b, and s_t are the trip proportion of freight carriers, taxis, buses, and the other vehicles during the peak hour, respectively,; s_w is the proportion of the non-local vehicles, which is as one part of the traffic flow.

The actual capacity of vehicle road network (C_{apr}) can be calculated according to the space–time consumption method, which is based on a certain level of service. The formulation of C_{apr} is given as follows:

$$C_{apr} = \frac{\sum_{i=1}^{4}(l_i \times \eta_{1i} \times \eta_{2i} \times \eta_{3i}) \times (1 - S_w) \times \alpha \times c}{l_p \times s_t} - (\frac{v_t}{l_p \times s_c} - 1) \times Q_C - (\frac{k_b \times v_b \times s_b}{} - s_b) \times Q_b \tag{38.3}$$

where l_p is average trip distance for vehicles; l_i is the equivalent length for the i-th level road, in which the road level includes expressway, main road, secondary road, and collector road; η_{1i} is the coefficient of effective length for all levels road; η_{2i} is the utilization factor of junction for all levels road; η_{3i} is the number of lanes for all levels road; α is the coefficient of the level of service for total road network; c is the travel traffic in a single lane (is equal to the road section capacity).

Here, we take Hangzhou city as an example. Firstly, we set down the research area considered to carry out the vehicle restriction policy. Then, According to the observation data of vehicle flow, all vehicles' trip coefficients in Eqs. (38.2) (38.3) can be obtained. The coefficient of the level of service for total road network is 0.75–0.8. Finally, according to the Eq. (38.1), the degree of space–time saturation in peak hour of the research area is 1.08, which is large than the causation threshold. Hence, it can be concluded that under the present traffic network conditions, the research area in Hangzhou already has satisfied the causation criterion to adopt the vehicle restriction policy. However, the degree of space–time saturation in peace hour of the research area is 0.95, which is less than the causation threshold. Therefore, the final decision of vehicle restricted only in peak hours was made. Besides, stage estimate is needed during the implement of the vehicle restriction, especially when Railway No.1 is put into operation.

38.4 Option Criterion

During the design of vehicle restriction schemes, three important variables are needed to be opted for, they are time, range, and type (namely, the type of vehicle to be restricted). The diverted traffic volume is a parameter closely related to the variables of time, range, and type. Besides, it is also a determinate parameter for judging the effect of vehicle restriction and judging if the other supporting system is perfect enough. The estimate of the diverted traffic volume is able to offer the basis for decision-making of the selection of the design schemes. Therefore, the diverted traffic volume is regarded as the option threshold for these three variables.

Here, we still take Hangzhou city as an example. Three factors are considered as follows.

(1) Firstly, the public transportation includes railway and bus. However, the railway in Hangzhou is still in construction. Therefore, the affordable diverted traffic volume of buses can be regarded as the threshold in the public transportation level (Chen et al. 2006).

The restricted area is involved 387 bus lines, which is 63.8 % of the total lines, and is involved 3476 buses, which is 46.7 % of the total amount. The passenger flow volume is 280.67 million per day during the peak load shifting period, and the passenger flow volume is 119.73 million per day during the morning and evening peak hour period. The average load factor of bus line during the morning and evening peak hour period is 90–92 %. Therefore, based on the actual traffic conditions in Hangzhou, a more conservative scheme is adopted, namely, the vehicles are restricted according to the tail number rather than the odd or even number, and the vehicles are restricted during the peak hour rather than all day around.

Based on the above measures, the diverted traffic volume will reach to 12–15 million trips. Referenced by the data from the "No vehicle day" over the year, the average convey velocity of bus in the restricted area will increase from 12.5 to 17.61 km/h, that is to say the velocity would be increase 40.88 %. Therefore, if we only intend to increase the average convey velocity of bus by 8–10 %, we can increase the turnover rate of bus. Then, the diverted traffic volume can reach to 8–10 million trips correspondingly. According to the above idea, the exclusive bus lane, whose length reaches to 60 km, was set up, and 170 buses, whose body length is 12 m, were added before the implement of vehicle restriction measures, which satisfy the bus diverted traffic volume.

(2) Secondly, because of the large amount of the ancient architectures, the resource of parking berth in the old city zone of Hangzhou is quite shortage. Therefore, the affordable diverted traffic volume of static parking berth can be regarded as the threshold in the static transportation level.

If the static parking berth cannot satisfy the requirement of the diverted traffic volume, the phenomena of no place to park and no space to move will occur, which will significantly affect the effect of the vehicle restriction measures. Therefore, in Hangzhou, the restricted area does not include the old city zone. And more parking berths are added. Finally, the number of the parking berth in the restricted area is 150246, including 31536 public parking berths inside road and 118710 parking berths outside road. Considering that the vehicles are restricted only during the peak hour, the requirements of "park and ride" or "park and wait" will occur. Besides, the temporary public parking berths and the corresponding guidance sign will set up. (Table 38.1).

(3) Finally, Hangzhou is the provincial capital of Zhejiang, there are 134 provincial administration units in the restricted area in Hangzhou. Besides, it is also a famous tourist city, including the most famous scenic spots, such as West Lake, canal, Xixi wetland, etc. Hence, the number of nonlocal vehicles is huge. Therefore, the affordable diverted traffic volume of nonlocal cars can be regarded as the threshold in the function level (Xiao (2008)).

Here, we take Raocheng Highway entrance as an example to analysis the diverted traffic volume of nonlocal cars in Hangzhou. Figure 38.3 presents the distribution chart of the directions and numbers of nonlocal cars driving into Hangzhou during the peak load shifting period. The observation time is morning peak hour (from 7 am to 8:30 am) and evening peak hour (from 17 am to 18:30 am). The numbers marked in the figure is the nonlocal cars number driving in and out of the interchange entrance. Figure 38.4 gives the comparison of the numbers of nonlocal cars and total cars driving in and out of different interchanges. Figure 38.5 plots the proportion of nonlocal cars number to total cars number driving

Table 38.1 Present status of parking berths in Hangzhou city

District	Total number	Number of parking berths (auxiliary construction)		Number of public parking berths (inside road)		Number of public parking berths (outside road)	
		Number	Percentage of total (%)	Number	Percentage of total (%)	Number	Percentage of total (%)
Shangcheng	41529	30461	73.3	9573	23.1	1495	3.6
Xiacheng	38580	27673	71.7	10194	26.4	713	1.8
Xihu	80134	59438	74.2	17537	21.9	3159	3.9
Jianggan	39913	33352	83.6	6423	16.1	138	0.3
Gongshu	33099	25030	75.6	7999	24.2	70	0.2
Total	233255	175954	75.4	51726	22.2	5575	2.4

*Data are from special planning book of parking facility in Hangzhou city

38 Scheme Research of Urban Vehicle Restriction 257

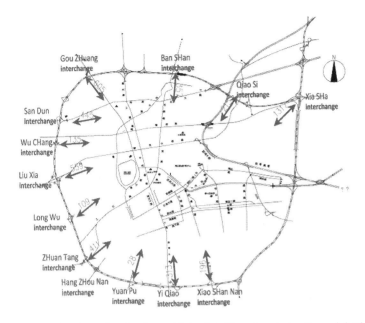

Fig. 38.3 Distribution chart of nonlocal cars driving into Hangzhou during the peak load shifting period

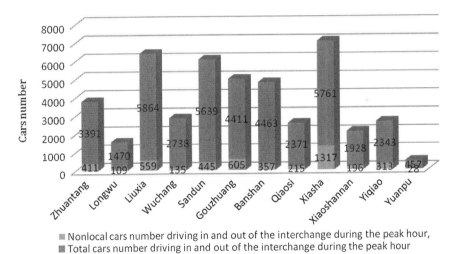

Fig. 38.4 Nonlocal cars number and total cars number in interchange peak

in and out of the interchange of Raocheng Highway entrance during the morning and evening peak hours in the work day. It can be seen from these three figures that the districts most affected by the nonlocal vehicles are Gouzhuang, Sandun, Wuchang, Liuxia, and Banshan. Specially, the entrances in Liuxia and Gouzhuang are quite close to the vehicle restricted area. Therefore, the detour scheme is

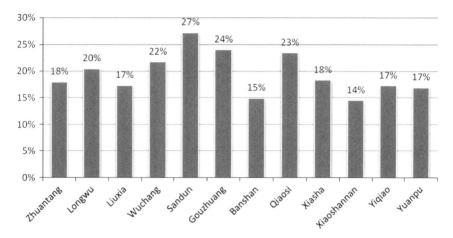

Fig. 38.5 Proportion of nonlocal cars number to total cars number in the interchange entrance during the morning and evening peak hours in the work day

suggested and the places for "park and ride" should be set up during the design of the scheme of vehicle restriction measures in Hangzhou.

38.5 Conclusions

In this paper, the synthetic criterions are put forward, which include starting criterion, causation criterion, and operation criterion. Then, these criterions are used and verified. Actually, the proposals of these three synthetic criterions are the significant supports to make plans of the vehicle restriction measures. In Oct. 2011, the vehicle restriction measures were started and were implemented during the peak load shifting period. After one month, according to the statistics, the average trip velocity during the morning and evening peak hours increases 21.6 %. Meanwhile, the trip is orderly, and the citizens show their support to the vehicle restriction measures. Furthermore, more and more citizens begin to trip during non-peak hours and are willing to park and ride, which helps to realize the demand management of Hangzhou's transportation.

Acknowledgments This research wassupported by the National Natural Science Foundationof China (No. 11102104)

References

Chen CM (2002) Research on capacity of road network. (in Chinese) Ph.D., Dissertation of Beijing Jiaotong University

Chen Z, Liu YS, Shi F (2006) Study on distance diversion curve of inhabitant trips modal. (in Chinese) Commun Stand 8:157–159.

Cynthia CY, Zhang W, Umanskaya VI (2010) The effects of driving restrictions on air quality: São Paulo, Bogotá, Beijing, and Tianjin. University of California

Li ZY (2006) Transportation planning for modern urban road. (in Chinese) Shanghai Jiaotong University Publication, Shanghai

Mahendra A (2008) Institutional perspectives on road pricing: essays on implementation, response, and adaptation. Essay submitted to the Department of Urban Studies and Planning in partial fulfillment of the requirement for the degree of Doctor of Philosophy in Urban and Regional Planning at the Massachusetts Institute of Technology

Xiao L (2008) Whether the conditions of vehicle restriction are ripe in Hangzhou? (in Chinese)

Chapter 39
The Research of Low-Carbon Transportation Management System

Wen-shuai Guo and Chao-he Rong

Abstract Energy constraints and climate change are being more and more concerned. It is the objective requirement of following the world trend to develop the low-carbon economy vigorously. As the transportation industry is an important section of energy-saving, the low-carbonization of it is inevitable. However, resource-wasting still exists in the construction of traffic infrastructure at present in China. The irrationality of transportation structure results in inefficient use of resources, which opposite to the idea of low-carbon transportation. These problems caused by the discordance between the traffic management system and traffic development limit the advance of low-carbon transportation. This paper argues that it is necessary to build integrated transportation management system to realize the advance of low-carbon transportation, and also put forward ideas and specific suggestions on the establishment of integrated transportation management system.

Keywords Low-Carbon transportation · Integrated transportation · Management system

39.1 Introduction

39.1.1 The Connotation and Necessity of Low-Carbon Transportation

Low-carbon transportation is a main component of energy-saving economy. It is a mode of transportation development which is characterized by high-efficiency, low-energy, low-pollution, and low-emission. Its core lies in the realization of

W. Guo (✉) · C. Rong
School of Economics and Management, Beijing Jiaotong University, 100044 Beijing, China
e-mail: 10113097@bjtu.edu.cn

shifting the mode of transportation development from an extensive way to an intensive one, the overall arrangement of transportation infrastructure construction, the improvement of transportation structure, and the advancement of energy efficiency.

Energy is the foundation of the development of our national economy, and the environment is one of the necessary conditions of sustainable development of the society. On one hand, as a major energy consumption, transportation industry consumes one third of the world energy, and more than half of the oil in the world was put into use of transportation. In recent years, China's energy consumption of transportation industry is with an average annual growth at the rate of more than 10 %, which accounts for 8 % or so in the total energy consumption. On the other hand, a lot of traffic energy consumption caused severe environmental pollution.

39.1.2 The Existing Problem of Low-Efficiency of Energy Usage and Resource-Wasting Phenomenon in the Transportation Filed

39.1.2.1 The low-Energy Efficiency Caused by Unreasonable Transportation Structure

In the domestic and foreign material, aviation has the highest energy consumption, highway is the second and the third is railway, waters has the lowest energy consumption. However, the development of railway and inland waters transportation in China are relatively slow. The imbalanced development of each mode of transportation results in the unreasonable transportation phenomenon, such as the expensive long-distance transportation of bulk goods. This kind of problem is serious in coal and ore area and other area of large items. In some of the important coal transportation, the capacity of railway transportation is in severe shortage. The ability of main railway lines is very limited, which causes that some main highways assume a lot of coal transportation tasks. Because of the poor construction of the inland river, it was left to the road to take over much low value-added coal, ore, chemical fertilizer, grain and other commodities in the long distance transportation in some parts of the channel along. However, the excessive use of road transportation is bound to bring huge uneconomic energy consumption and large amounts of carbon emissions.

39.1.2.2 The Lack of Overall Plan of Transportation Infrastructure Results in Waste of Resources

Some parts of important transportation area in China have highway, railway, and the inter-city rail lines densely, where lack of overall arrangement of specific position of the line, and technical standards. Thus, to some extent, it led to the waste of resources. A typical example is that the Jingjin channel which has three expressways, 2 first class roads, and 2 high speed railways, 1 common railway with 3 branches. These roads are basically parallel. It is hard to get effective use of the land between them. The urbanization will be affected. Meanwhile, there still exists the problem of surplus of transportation capacity.

On the use of resources of the river-crossing channel in China remains the same problem, i.e. lack of overall arrangement. At present, there are a total of 86 blocks of bridges (excluding Yangtze River Tunnel) across the Yangtze River from Shanghai to Yibin, in which only seven bridges can be used both by cars and trains. In addition, there are the Hangzhou Bay Bridge, Zhoushan Islands Bridge, and most of these river-crossing bridges are highway bridges. Along with the rapid growth of railway network construction, the river-crossing channel resources will be more deficient.

39.2 The Systematic Reason that Restricts the Development of Low-Carbon Transportation

Low-carbon transportation requires the cohesion among every part of the system. This is mainly a multi-modal transportation style, and it requires the coordinated state between transportation system and its external environment. This mainly refers to the state of coordinated and sustainable development between transportation and land exploitation, environmental protection and energy usage. However, in the condition of the present level of productivity force of transportation section, the existing management system has not been able to well adapt to the condition of productivity force, and has begun to restrict its development, which mainly displays in that low carbon traffic demand to the development of energy conservation and environmental protection mode of transportation, however, the traditional system make the railway and water transportation become a shortcoming of transportation system. Urbanization and urban agglomeration required efficient urban transportation, especially urban rail traffic with the characteristics of low carbon and environmental protection. But in the traditional system the examination and approval for urban rail traffic is always after when people bought cars. Resources and environment restriction require that transportation industry would change the development mode, but preempted resources, waste resources and the destruction of the environment always characterize the traditional system.

Rectifying the systematic reason, it mainly includes the division among different transportation managements, the lack of cooperation and coordination mechanism among high efficient departments, and the lag of legal systematic construction and social governance. In a nutshell, there hasn't been set up an integrated transportation management system. If we do not adjust current transportation management system promptly, there would be less and less harmony between the transportation system and the environment. This will not only cause bigger resistance but destruction to the development of the integrated transportation. What's more, because the transport construction has the characteristics of path dependence, it will bring greater cost to the transportation industry and systematic reform in the future. For example it can lead to huge waste of resources and even damage, excessive emissions, the deterioration of the environment, inefficient transportation and so on (Rong 2005).

39.3 Reference from the Experience of Foreign Transportation Management System

Taking the international experience for reference, during the speeding up stage of industrialization, urbanization and transportation, it is an important time to reconstruct the transportation infrastructure and basic structure. To construct a high efficient, comprehensive and sustainable development mode of integrated transportation, all sorts of technologies, policies, plans, legal systems should jointly work together. Thus, we will get better results. Each mode of transportation has the support of inherent advantages and interest group, and they develop according to their respective system. It may not get favorable results for the overall efficiency of integrated transportation and the whole social economy, if the plan, construction and operation of integrated transport hub was ignored. However, the influence of transportation facilities and transportation structure would last long and become difficult to adjust to some extent. From this point of view, the establishment of integrated transportation management system should be realized the sooner the better in the condition that every mode of transportation have got development to some degree (Rong 2008).

On the construction of integrated transportation management system, the developed countries already have some very good researches. According to the research of Dominic Stead, since a lot of transportation policies cross various departments, different departments need cooperate with each other during the drafting. During the lateral management of the department policies, according to the close degree of various departments' cooperation these policies can be divided into: Policy cooperation, which contains dialogues and informative communication among various departments, and Policy coordination, which refers to further transparency on the basis of policy cooperation among various departments, and avoiding the conflict of policies through the coordination of interest between each other, as well as Policy integration, which means joint work and unifying the goal to formulate policy on the basis of policy cooperation and coordination (Dominic Stead 2008). At present most countries in the world, choose the integrated transportation management system to rule over all kinds of transportation, even including the management of land development, urban construction, which are closely tied to transportation development, so as to improve the degree of integration on policy formulation and implementation.

In practice, among the 126 countries in the world that establish railways, there are 119 that had implement the integrated transportation management system, including 40 countries which established the integrated transportation departments ruling over all kinds of transportation (such as the Department of Transportation), while, the rest of 79 countries established more comprehensive management department including integrated transportation management system. Such as Department for Transport in England, Ministry of Land, Infrastructure, Transport and Tourism in Japan and so on (review and enlightenment on the strategy of Foreign integrated transportation development 2011; Wang and Yang 2008; U.S.department of transportation 2012; Japan website traffic department of transportation 2012).

39.4 The Construction of Integrated Transportation Management System

Through the analysis of the management systematic problems restricting low carbon traffic development in China at present, and using the successful experience of the developed countries on the management system reform for reference, this paper argues that transportation management system reform in China should abide by the following directions.

39.4.1 Take the Opportunity to Complete the Super-Ministries Reform of Transportation

Realizing integrated transportation must be guaranteed by organizations and systems, the establishment of the integrated transportation authority, to a great extent, is the prerequisite condition to form integrated transportation system in China and be put into positive operation as soon as possible. At present the separation of government and enterprise on roads, water transportation and civil aviation has already achieved. The conditions of establishing the super-ministry of transportation is mature. While the main obstacle is that the separation of government and enterprise on railway has not been achieved in China. We should separate the function of governance from the Ministry of Railways, and merger it into the super-ministry of transportation. The further reform and reorganization of the railway industry can be researched and realized in the framework of the super-ministry of transportation.

After the establishment of the super-ministry of transportation, we should set up professional bureaus like: Railway Bureaus, civil aviation Bureaus, Highway

Bureaus and Shipping Bureaus. After straighten out the set-up of organizations in super-ministry of transportation, we should classify, integrate and comb the various departments in it. It is the formulation of laws and regulations, policies, plans, as well as constructive and administrative construction.

39.4.2 To Set Up Efficient Cooperation and Coordination Mechanism Between Different Departments

Integrated transportation of administrative management involves the various aspects of the social and economic life, besides the relationship among a few transportation industries, it also involves many other administrative departments. Even if the Ministry of Transport covered all the administrative managements of transportation industry, there will still exist overlapping responsibilities between this department and others, which mainly includes but not be limited to: The balance of transportation plan and national economic and social development plan with the National Development and Reform Commission, and the examination and approval of transportation infrastructure constructive project, and the management and coordination of relevant freight rate or the charge, as well as the plan and management on urban road, subway, and rail transportation with the City Construction Department; the coordination of overall urban planning and integrated transportation system planning of city with Ministry of Housing and Urban–Rural Development; the coordination of land development, utilization planning, examination and approval and regulatory, with the Ministry of Land and Resources; the coordination of the energy supply, the energy conservation, the environmental protection areas, with the National Energy and Environmental Protection Departments.

In recent years, there are some affairs such as the governance of overload vehicles on highway transportation and the promotion of the development of the logistics industry, have connected more than ten ministries to cooperate with each other. This proved that the requirements of coordination in terms with transportation is obvious. To some extent, it is apparently not enough just to merge a few transportation industry management departments in order to solve all problems. Some countries continued to implement the super-ministry management in a greater scope after realizing the super-ministry of transportation, to further integrate various administrative institutions and resources. This done really has its immanent rationalization, which can improve the administrative efficiency to a certain extent. However, there is no country which is likely to integrate all administrative affairs related to transportation into one government department. No matter how large the integrated transportation department is, the cooperation and coordination among different departments in necessary fields is inevitable.

Recently the reform of integrated transportation management system in China can only be limited to a few direct transportation management departments. We should gradually create conditions for further integration. Therefore we must make great efforts to establish efficient cooperation and coordination mechanism between different departments. For example, we should establish related mechanism based on information sharing and policy-making dialogues among Ministry of Transport, Ministry of Housing and Urban–Rural Development, Ministry of Land and Resources and Ministry of Environmental Protection. We can also organize joint working group or hold up conference among Ministries, and establish inter-departmental training mechanism for civil servants so as to avoid the shortcomings such as confusion about responsibilities, non-uniform goals and so on. This can also help to overcome the barriers of lacking knowledge of such aspects as law, organization, financial and engineering technology. In a word, we must ensure the system and mechanism to conform to the important policies, laws and regulations, strategic planning, rules and standards of integrated transportation in order to meet the requirements of comprehensiveness, continuity and collaboration.

39.4.3 Strengthen the Construction of Legal System, Promote Responsible Administration, Introducing Social Management

The super-ministry reform is related to the innovation of governmental system and mechanism as well as the readjustment of the power interests. Because of the influence of the planned economy and the department legislation, in China, to a quite extent, there exist different sets of legal systems in different departments. After the super-ministry reform, there may be some problems in the aspect of relevant law enforcers, litigation jurisdiction and legal application. So we should strengthen the construction of the relevant laws and regulations system and promote the administrative organization setting, function orientation as well as the legalization operation mechanism in order to solve the problem of overlapping functions, decentralizing functions, confusion of responsibilities, and shuffles over responsibilities in the government affairs from a system point of view.

After integrating the different modes of transportation management organization, the super-ministries reform of transportation realized the different modes of transportation management, which objectively caused the problem of the construction of power. We also need to build supervision mechanism to match the degree of power centralization, and to realize reasonable power restriction. Generally speaking, the division of interests is up to the policies, the realization of interests is up to execution, and the rectification of interests is up to supervision. Under the current transportation management system, the configuration of decision-making power, executive power, and supervision power of relevant departments are not reasonable, which results in department interests override the public interests. This leads to the deformation and distort of public policies, such as unreasonable transportation structure, lack of comprehensive utilization of the resources, inconvenient transportation of both passengers and goods, and even causes power departmentalization, departmental benefits, the interest collectivization. Therefore, we should strive to promote scientific decision democratization, executive specialization, supervision independence, and explore the mechanism in which decision-making power, executive power, and supervision mutually restrict and coordinate.

The premise of effective super-ministry system is "limited government". In the transportation administrative area, we must also change the main concern from the previous "effective administration" to the "responsible administration". That is, from excessive one-sided emphasis on concentrating resources to accomplish large undertakings, i.e. the so-called "efficient", to the equivalence of administrative responsibilities and power. What's more, through the administration according to law, restrict accountability and the supervision of the general public to achieve the constraints of government power. We should make the formulation and implementation of our integrated transportation policy can reflect the public interest, as well as show normativity and operability, and avoid the dislocation of government, which could cause the problem of "department with too much power", "power benefit" and even "benefit legalization" situation. It's very significant to bring public policy program and social management in this mechanism.

References

Dominic stead (2008) institutional aspects of integrating transport, environment and health policies. Transp Policy 15:139–148
Japan website traffic department of transportation: http://www.mlit.go.jp/en/index.html. July 2012
review and enlightenment on the strategy of Foreign integrated transportation development, Integrated transportation development strategy study of China, (2011)
Rong C (2005) The thinking that build the integrated transportation management system as soon as possible in China. China Soft Sci Mag 2:10–16
Rong C (2008) The reform direction of transportation management system are horizontal integration and vertical breakthrough. Financial
U.S.department of transportation website: http://www.dot.gov/. July 2012
Wang X, Yang X (2008) Foreign transportation management system, China Communications Press, Beijing

Chapter 40
Urban Low-Carbon Transport System

Peng Xing and Tianjun Hu

Abstract Establishment of urban low-carbon transportation system is the main means to achieve urban low-carbon transport. Learn from the advanced experience of foreign countries, the construction of urban low-carbon transport system mainly consists of the layout of urban traffic, the public transport system construction, the main optimization of the traveling population, the development of the transport of low-carbon technologies and the improvement of traffic management policies. As to how to realize low-carbon transport, this paper first analyzes the energy consumption and carbon emissions; then it illustrates the concept and connotation of low-carbon transport; then it raises the above five aspects of the construction of urban low-carbon transport system. Finally, based on china national conditions, the paper proposes some suggestions for establishing green transportation system characterized by low-carbon and promoting sustainable urban development.

Keywords Urban traffic · Energy consumption · Carbon emissions · Low-carbon transport system

40.1 Introduction

Urban transportation is an important carrier for carrying urban development, but also an important source of energy consumption and CO_2 emissions. With the rapid process of urbanization and motorization, urban traffic has been rapid development. As a result, the problems of traffic jams, environmental pollution and

P. Xing · T. Hu (✉)
School of Traffic and Transportation, Beijing Jiaotong University, Beijing 100044, China
e-mail: 11120902@bjtu.edu.cn

energy consumption have become increasingly prominent. Transport development should be coordinated with economic and environment is a worldwide consensus.

In recent years, many concepts and theories of urban low-carbon transportation have been proposed, and some strategies have been applied to the practice process. Our new research topic is mainly to clarify the relationship between them, and establish a scientific, reasonable, comprehensive urban low-carbon transport system to promote the sustainable development of cities.

40.2 Energy Consumption and Carbon Emissions

40.2.1 Energy Consumption

With the rapid growth of urban population and the rapid expansion of urban space, a large number of residents travel and the flow of goods have led to the rapid increase of energy consumption in urban traffic.

As shown in Table 40.1 Zhang et al. (2011), energy consumption of China's urban transportation is increasing rapidly every year. Energy consumption (per person per kilometer or energy consumption of per ton -kilometer) is different for different modes of transport. The energy consumption of the bus as a benchmark, the energy consumption of bicycle, bus rapid transit, streetcars, light rail, subway, trolley buses, motorcycles, cars are 0,0.3,0.4,0.5,0.8,5.6,8.1.Therefore,in the various transport modes, energy consumption per unit the biggest is cars, the smallest is public transport such as light rail, subway, tram, bus, and so on.

40.2.2 Carbon Emissions

Carbon intensity is defined to carbon dioxide emissions per unit of GNP growth. This indicator is used to measure the relationship between a country's economic growth and carbon emissions growth. If a country has a growing economic, while carbon dioxide emissions per unit of GNP are declining, it indicates that the country has a low-carbon development model. The carbon intensity of the transport is usually expressed as the carbon dioxide emissions from energy consumption of person km.

Table 40.1 2004–2009, China's urban transportation energy consumption (million tons of coal)

	2004	2005	2006	2007	2008	2009
Gasoline	2191	2717	2908	3205	3646	4177
Diesel fuel	2641	3297	4096	4857	5779	6645

Table 40.2 2004–2009 urban transport CO_2 emissions (Unit: million tons)

	2004	2005	2006	2007	2008	2009
CO_2 emissions	1.06	1.28	1.46	1.65	1.83	2.14

Table 40.3 Carbon intensity of several modes of transportation

Way to travel	Rail transport	Bus	Cars	Electric cars	Bicycle	Walk
Carbon emissions kg/person km	0.049	0.075	0.135	0.017	0	0

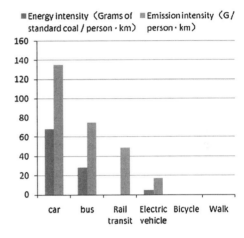

Fig. 40.1 Different modes of transportation carbon emissions intensity comparison

Table 40.2 shows that the city's carbon dioxide emissions present the trend of rapid growth during these years. As shown in Table 40.3, the carbon emissions have a significant difference for different modes of transportation. If we want to reduce CO_2 emissions, we must choose low-emissions transports, reduce the use of high-emission vehicles, and provide a reasonable urban transport system (Fig. 40.1).

40.3 Low-Carbon Transport

40.3.1 The Origin of the Low-Carbon Transport

Low-carbon transport is actually an important integral part of a low-carbon economy. "Low-carbon economy" first appeared on Energy White Paper published by the British Government in the 2003 for our energy future: Creating a low-carbon economy". Low-carbon economy is to reduce greenhouse gas emissions as a starting point, bases on low-carbon energy systems, low-carbon technologies systems and low-carbon industrial structure to adapt to climate change, and build an ecological civilization economic model as the main content.

The construction of low-carbon transport system is an important part of energy conservation and low-carbon economy, the strategic plan of the national response to climate warming, the inherent requirements of the changing patterns of development of urban transport. An important part of building a low-carbon city is exploring more efficient, more energy, more low-carbon and cleaner transportation modes, promoting green travel, creating low-carbon transport.

40.3.2 The Connotation of Low-Carbon Transport

Low-carbon transport is a new concept of sustainable development and practical goals. We have basically reached a consensus on the low-carbon development, but because of the different national positions and stages of development, it leads to different metrics. There is no a clear and unified concept in the low-carbon transport, it can be summarized as: In the context of a growing awareness of climate change and its serious impact on the survival of mankind, for the purpose of energy saving, in order to achieve the environmental, social, economic sustainable development, by using the system adjustment and technological innovation and other means, we can achieve transport way efficiency to enhance, the transport structure optimization, the effective regulation of traffic demand, transportation organization and management innovation for ultimately achieving the full cycle of the whole industry chain of low-carbon development, promoting the transition of low-carbon social and economic development Yu et al. (2011).

40.4 Urban Low-Carbon Transport System

The core of the low-carbon transport is to use different means to reduce energy consumption and reduce carbon emissions from transport travel. In order to achieve the sustainable development of urban transport, it is necessary to establish the urban low-carbon transport system. From a macro, urban low-carbon transport system includes the layout of urban transport, public transport-based system construction, the optimization of the travel groups, low-carbon technology development and the improvement of traffic management policies (Fig. 40.2). These five respects can realize the traffic congestion minimum, traffic demand minimum, motor vehicles use minimum, motor vehicles lowest carbon emissions, inefficient transportation minimum, so as to achieve the goal of traffic emission minimum.

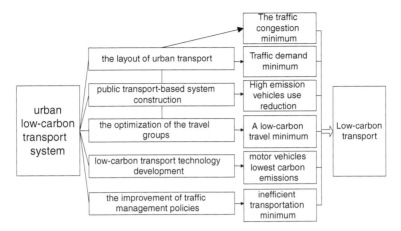

Fig. 40.2 Urban low-carbon transport system Wang (2011)

40.4.1 The Layout of Urban Transport

The layout of urban space and urban transport modes are mutually reinforcing. Multi-center layout is the basis of the urban space for the low-carbon transport. The spatial structure of the multi-center is the best form, which can coordinate the aggregation efficiency and transport costs for big cities. The spatial structure of the multi-center can reduce traffic demand and balance distribution of traffic. We can use the rail transit which is the high transport capacity connecting the centers; within the centers, we mainly use the public transport, private transport as an auxiliary; it not only can reduce the traffic total, but also reduce traffic congestion, thereby greatly reducing carbon emissions.

Consider land use and transportation planning. Build a balanced distribution of traffic demand compact and networking of urban spatial form, mixed development of high-density of land use and public transport to reduce travel distances and car use, and reduce the traffic demand from the source.

40.4.2 Public Transport-Based System Construction

In a variety of urban transportation system, the carbon emission of the system which is based on urban public transport is far less than car-based transport system. Among them, the rail transit public transport system is the lowest carbon emissions.

We must vigorously develop intelligent traffic information systems to provide transport options, the path to guide, transfer, real-time traffic and other traffic travel information services for the traveler. Transport services efficiency improvement can significantly reduce carbon emissions.

40.4.3 The Optimization of the Travel Groups

Implement bus priority strategy and change resident travel mode. Improve the operating environment for public transport and slow traffic, enhance passenger transport hub functions and transfer facilities, and improve related safeguards to enhance public transport attractiveness and level of service in order to guide the public shift to low carbon modes of transportation.

We should enhance residents' knowledge of low-carbon trip and sense of responsibility to promote and encourage the people to support low-carbon travel.

40.4.4 Low-Carbon Transport Technology Development

The low-carbon of transport depends on the development and popularization of low-carbon technologies. The increasing of the low-carbon motorized transport and non-motorized transport is a strong guarantee for the construction of low-carbon transport system. Transportation technologies need to complete the dual task of energy saving and low-carbon in China. We should try to catch up with the technical level of developed countries, while we enter to the new low-carbon technology field.

Low-carbon technologies of motorized transport mainly refers to the utilization of new energy technologies, In addition to using alternative energy or alternative fuels and develop new energy vehicles, we can develop lightweight materials to reduce motor vehicle weight in order to reduce fuel consumption and emissions. Non-motorized transport is carbon-free, so it should be strongly innovation and upgrading. With the development of low-carbon transport technologies, non-motorized transport has been put on the agenda to improve the performance of the bicycle and other non-motorized transport, comfort and safety, so that the bike fit and attract more people to use. Of course, we can also look for other ways, liking the invention of better non-motorized transport.

40.4.5 The Improvement of Traffic Management Policies

Traffic areas are conducing management and innovation by freight logistics, intelligent transportation, information systems, and efficient work to improve the transport organization, management and service level, thus we can achieve intensive use of transport resources, reduce energy consumption and greenhouse gas emissions, and implement low-carbon transport.

The traffic order is a guarantee of the urban traffic. The traffic rules are essential to traffic order. Traffic management is the basic way to maintain traffic order. Traffic management in urban low-carbon transport system is necessary to maintain traffic order, also to supervise the transportation low carbonization. Good traffic order can reduce congestion and emissions; eliminating the motor vehicle on the road is the guarantee for low-carbon transport.

The construction of traffic management should comply with the requirements of the low-carbon transport. We can build from the following three areas Yu and Wei (2011). First, build the low-carbon transportation laws and regulations system. We can prohibit effectively the acts which isn't meeting the requirements of the low-carbon based on legal. Second, establish the intelligent traffic management system. Intelligent traffic management system is the integrated use of advanced information technology, electronic technology and computer processing technology to the traffic management system, in order to establish a real-time, accurate and efficient, all-round playing a role in traffic management systems. It is internationally recognized as the best way to solve urban traffic congestion, improve operational efficiency, improve traffic safety, etc. For example, it can make the road capacity increased by two to three times; vehicles traveling on intelligent road, the number of stops can be reduced by 30 %, the parking time is reduced by 13–45 %; traffic accidents can be greatly reduced. The last one, strengthen motor vehicle emissions inspection management. The arbitrary use of the vehicles of non-standard emissions will inevitably lead to low-carbon transport lost by the wayside, so we must vigorously develop and improve motor vehicle emissions testing technologies and management methods. We can prohibit the non-standard emission vehicles on the road at any time, and ensure that all the motor vehicle on the road reach the low-carbon criteria.

40.5 Conclusion

Low carbon development is an important strategic measure which can achieve economic development to shift from the traditional extensive economy to intensive economic. The city as a center of human production and life plays an important role in the economic and social development. Urbanization is a low carbon way that the human society pursues material and spiritual civilization. High-carbon urbanization development model is not desirable; low-carbon urbanization is the inevitable choice.

Low carbon transport system and the construction of low-carbon cities interact and complement each other. Development of low-carbon economy must conduct the city's low-carbon construction. Low-carbon transport system as an important part of the urban construction is essential in the urban modernization. Low-carbon transport system can reduce environmental pollution and carbon emissions Liu and Wei (2011). It can effectively alleviate the traffic pressure on urban centers and change people's lifestyle and urban development model. It will surely promote the city to develop low-carbon economy, promote low-carbon life, so as to promote the city's low-carbon transition, and promote the sustainable development of cities.

References

Liu Q-l, Wei C (2011) Research on construction of urban low carbon transport system. Materials for renewable energy and environment (ICMREE), 2011

Wang G (2011) The construction of urban low-carbon transport system. Forw Position (13)

Yu S, Mu L, Jl B (2011) On green transport and low carton transport. ICTE 2011-proceedings of the 3rd international conference on transportation engineering, 2011

Yu D, Wei D (2011) The planning strategy of low carbon urban transport based on the vision of low impact development. Remote sensing, environment and transportation engineering (RSETE)

Zhang T, Zhou Y, Zhao X (2011) Study on the current situation and approaches of urban low carbon transport. Urban Stud (1)

Chapter 41
Study on the Control of AC Dynamometer System for Hybrid Electrical Vehicle Test Bench

Ying Tian, Zhenhua Jing, Keli Wang, Shengfang Nie and Qingchun Lu

Abstract This paper evaluate the testing bench of AC dynamometer system, the induction machine can run in electro motion and generate electricity status under speed and torque control by direct torque control (DTC). It develops the AC dynamometer system controller based on PXI, and study on the AC dynamometer steady-state control strategy under the rapid prototype method in LabVIEW RT software. It effectively enable to n/P, M/P, n/M and M/n control mode, get ideal control quality by adjusting the PID control parameter, each control mode charge smoothness. Experimental results show that the controller is excellent, and the control performances for engine testing completely satisfied the demand of quality control. It provides the basis for further study on dynamic control of AC dynamometer.

Keywords AC Dynamometer · Control strategy · Hybrid electrical vehicle

41.1 Introduction

Dynamometer is a key equipment in motor performance test platform, mechanical transmission test bench and engine test-bed etc. Alternating current dynamometer use three-phase alternating current engine as load equipment, converter affords

Y. Tian · K. Wang (✉)
School of Mechanical and Electrical Control Engineering, Beijing Jiaotong University, Beijing 100044, China
e-mail: 11125680@bjtu.edu.cn

Z. Jing · S. Nie · Q. Lu
State Key Laboratory of Automobile Safety and Energy, Tsinghua University, Beijing 100084, China

drive power of variable frequency for asynchronous motor as well as accurately control the rotation speed and torque. Compared with other dynamometer technology, alternating current dynamometer has many characteristics such as easily control power, high control accuracy of torque and rotation speed, short dynamic response time, flexibility and variety of the structure, high efficiency energy saving and high reliability. In the application of engine bench or motor test, target motor's torque and rotational speed which are main control objects and measurement parameters of alternating current dynamometer system. In this paper the control of target motor system adopts direct torque control technology.

41.2 Direct Torque Control

Direct torque control compare set torque value with actual torque of motor, through a simple hysteresis control to get control signal of torque. Then combining stator flux control signals, and choosing a suitable stator voltage space vector. Thus making asynchronous motor's electromagnetic torque quickly tracks the set torque to realize direct torque control. During the process of controlling torque, it also achieve flux control to make sure that stator flux change in a error prescribed scope. The whole control system includes flux control and torque control (Gao 2005; Takahashi and Ohmori 1989).

41.2.1 Flux Control

The foremost task of direct torque control is to accurately obtain stator flux, because during the process of control, comparing actual amplitude of flux with the set value of flux to get flux control signal, then the state of inverter switch is confirmed based on flux location. At the same time, electromagnetic torque is obtained according to calculating flux and stator currents. From the above whether flux is good or not which surely affect whole properties of control system.

U-I Model of stator flux can be expressed as

$$\psi_s = \int (u_s - i_s R_s) dt \qquad (41.1)$$

This model is simple, which is only used stator resistance of motor parameter, It can be calculated the value of stator flux through stator's current and voltage from sensor sampling. But when the rotational speed is much lower, voltage drops from stator resistor (Rs) is occupied a large part which will affect the observer exactitude of flux, so this model is used in high speed.

In low speed, it makes U-I model has large error in flux observer due to the influence of the stator resistance. In order to make sure that flux observer is accurate, we adopt stator currents and rotor speed of flux observer model which is I–N model.

According to the equation

$$\psi_s = \frac{1}{1+\frac{L_\sigma}{L}}(i_s L_\sigma + \psi_r) \tag{41.2}$$

$$\frac{d}{dt}\psi_r = \frac{R_r}{L_\sigma}(\psi_s - \psi_r) + j\omega\psi_r \tag{41.3}$$

$$\psi_s = \psi_{s\alpha} + j\psi_{s\beta} \tag{41.4}$$

$$\psi_r = \psi_{r\alpha} + j\psi_{r\beta} \tag{41.5}$$

$$i_s = i_{s\alpha} + ji_{s\beta} \tag{41.6}$$

We could get,

$$\begin{cases} \psi_{s\alpha} = \dfrac{1}{1+\frac{L_\sigma}{L}}(i_{s\alpha}L_\sigma + \psi_{r\alpha}) \\ \dfrac{d}{dt}\psi_{r\alpha} = \dfrac{R_r}{L_\sigma}(\psi_{s\alpha} - \psi_{r\alpha}) - \omega_r \psi_{r\beta} \end{cases} \tag{41.7}$$

$$\begin{cases} \psi_{s\beta} = \dfrac{1}{1+\frac{L_\sigma}{L}}(i_{s\beta}L_\sigma + \psi_{r\beta}) \\ \dfrac{d}{dt}\psi_{r\beta} = \dfrac{R_r}{L_\sigma}(\psi_{s\beta} - \psi_{r\beta}) - \omega_r \psi_{r\alpha} \end{cases} \tag{41.8}$$

Rotor resistance (Rr) inductance (L_σ) and rotational speed are used during the process of calculation in this model. In high speed, measurement error of rotational speed will bring certain effects to the model, so it has high precision in low speed. Therefore, we may combine two flux observers, adopting I–N model in low speed, but taking U-I model in high speed.

41.2.2 Torque Control

According to motor mathematical mode, electromagnetic torque

$$T_e = \frac{3}{2}P_n(\psi_s \otimes i_s) \tag{41.9}$$

derivation to both sides, and at the same time multiplied by L_σ, get

$$L_\sigma \frac{d}{dt}T_e = \frac{3}{2}P_n L_\sigma \left(\frac{d}{dt}\psi_s \otimes i_s + \psi_s \otimes \frac{d}{dt}i_s\right). \tag{41.10}$$

and

$$L_\sigma \frac{d}{dt} i_s = \frac{L + L_\sigma}{L} \frac{d}{dt} \psi_s - \frac{d}{dt} \psi_r .\tag{41.11}$$

though the deduce get

$$L_\sigma \frac{d}{dt} T_e = \frac{3}{2} P_n \psi_r \otimes u_s - \left[\left(1 + \frac{L_\sigma}{L}\right) R_s + R_r\right] T_e - \frac{3}{2} P_n \omega \psi_s \psi_r .\tag{41.12}$$

It can be seen in a control cycle of delta t, stator flux ψ_s and rotor flux ψ_r's change is very small. The main factor is u_s that affect instant torque change, once u_s has apparent change in a cycle which will induce motor's torque to produce rapid change. It will realize a high-performance control of torque through improve DC voltage U_d and shorten the control cycle time under inverter promised condition.

By formula (41.12) we can see that ψ_r and u_s are cross product relationship. The response of torque is fastest when they are perpendicularity, but the relationship between stator voltage vector and rotor flux linkage is not easy to know, approximate using stator flux to express the rotor flux, namely $\psi_r \approx \psi_s$. Formula (41.12) expressed as:

$$L_\sigma \frac{d}{dt} T_e \approx \frac{3}{2} P_n (\psi_r \otimes u_s) - \left[\left(1 + \frac{L_\sigma}{L}\right) R_s + R_r\right] T_e - \frac{3}{2} P_n \omega \psi_s^2 .\tag{41.13}$$

By formula (41.13) we can see, choosing a voltage vector vertical with the current stator flux linkage in direct torque control, it can get the maximum torque response, but inverter offers alternative voltage vector is limited, only voltage vector's direction similar to vertical could be choose to get high torque.

From above analysis, when voltage vector applied in advance of the stator flux, the change rate of torque is above zero ($dT_e/dt > 0$), electromagnetic torque increase. When voltage vector applied behind of the stator flux, the change rate of torque is below zero ($dT_e/dt < 0$), and electromagnetic torque reduction. This feature can be use to control torque.

In physical concept, electromagnetic torque is decided by cross product between the stator flux and rotor flux, written as:

$$T_e = \frac{1}{L_\sigma} \frac{3}{2} P_n (\psi_r \otimes u_s) = \frac{1}{L_\sigma} \frac{3}{2n} P_n |\psi_r| * |\psi_s| \sin\theta .\tag{41.14}$$

From formula (41.14) we can see that in direct torque control, keeping the stator flux changeless, but rotor flux's amplitude is decided by load. Change stator flux angle θ to alter electromagnetic torque. The rotation speed of rotor flux changes slowly, it is mainly through the control of stator flux's rotating speed to realize torque control during control process.

41.3 Control Effect

This paper builds alternating dynamometer system which include motor control integration, energy processing system, throttle actuator control integration, rotation speed and torque measurement devices, and hardware real-time controller etc. The whole control system belongs to complex, large inertia, time-variant nonlinear system, accurate mathematical model is difficult to establish (Sandholdt et al. 1996; Bunker et al. 1997; Diana 1998). Therefore each control loop controller is adopted control strategy based on PID controller.

In the test bench, the rated power of Toyota 8A engine is 63 kW, highest rotating speed is 6,000 r/min, and maximum torque in 5,200 r/min is 110 Nm. A serial of experiments of n/M mode are schemed out, and test results are given.

The rotating speed step and torque curves of n/M mode are shown in Fig. 41.1. The torque keeps in 40 N·m during the process of rotating speed from 2,000 r/min step to 2,500 r/min. The analysis reveals that the whole process of adjustment is quickly and stabilization, rotating speed without overshooting, and the adjusted time is about 3 s. At the beginning of the process, for the AC dynamometer control speed is faster, the torque decreases first and then steadily rising. The whole transition process approximately 5 s then the system reaches steady state.

The torque step from 40 to 60 N·m curves of n/M mode are shown in Fig. 41.2. The rotating speed is 2,500 r/min and torque step from 40 N·m to 60 N·m. It can be concluded that the whole transition process approximately 3 s, the torque curve is smoothly and less fluctuation. The rotating speed fluctuate 1 r/min during the throttle begin to change, and the subsequent speed fluctuations are in scope of ±1 r/min.

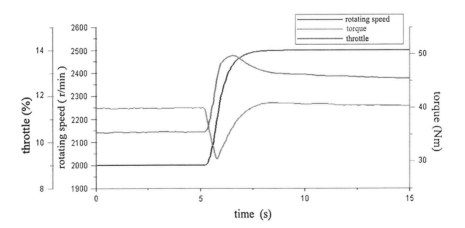

Fig. 41.1 n/M mode rotating speed step curve

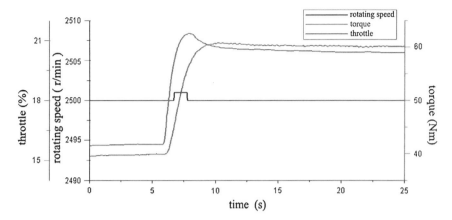

Fig. 41.2 n/M mode torque from 40 to 60 N · m step curve

41.4 Conclusions

AC dynamometer system controller development experiment platform is based on PXI. It adopts rapid prototype method to study steady control algorithm for AC dynamometer in LabVIEW RT, which the n/P, M/P, n/M and M/n four control modes, Through test adjusting that control parameters achieve an ideal control quality, various states switch smoothly, an ideal control effect of system steady control, which is satisfied and realized the requirements of measurement and control system for engine experiment bed.

Acknowledgments This research was support by the Fundamental Research Funds for the Central Universities.

References

Byron JB, Matthew AF, Bruce ET (1997) Robust multivariable control of an engine-dynamometer system. IEEE Trans Syst Technol 5:189–199
Diana Y (1998) Adaptive control of diesel engine-dynamometer systems. IEEE Conference on Decision and Control Tampa, pp 1530–1534
Gao J (2005) Analysis of AC motor and system[M]. Tsinghua University Press, BeiJing
Sandholdt P, Ritchie E, Pedersen JK, Betz RE (1996) A dynamometer performing dynamical emulation of loads with nonlinear friction. Proc IEEE Int Symp Ind Electron 2:873–878
Takahashi I, Ohmori Y (1989) High-performance direct torque control of an induction machine. IEEE Trans Ind Appl 25(2):257–264

Chapter 42
Low-Carbon Transport System by Bicycle, in Malmö, Sweden

Yingdong Hu and Xiaobei Li

Abstract Low-carbon transportation is the trend of development of the sustainable urban transport. With the features as lightweight, flexible, environmentally friendly, and comfortable, bicycle transport has become an important part of the urban low-carbon transport. The article describes the background of low-carbon transportation in Malmö, Sweden, analyzes the status and strategies taken for bicycle travel and proposes problems to be solved and solutions.

Keywords Malmö · Low-carbon transportation · Bicycle travel

42.1 The Background of the Low-Carbon Transportation Development in Malmö

Malmö is Sweden's third largest city. In recent years, Malmö is in economic transition—transforming from traditional and declined shipbuilding and textile industry to new high-tech information industry and biotechnology industry. How to shape a city image of cultural and environmental sustainability in the post-industrial period is the strategic objective of Malmö. To create an ecological livable urban environment, Malmö has made groundbreaking contributions to the combination of old city reconstruction, urbanization process and sustainable development. The

---

Y. Hu (✉) · X. Li
School of Architecture and Design of Beijing Jiaotong University, No.3, Shangyuancun, Haidian District, Beijing 100044, China
e-mail: ydhu@bjtu.edu.cn

X. Li
e-mail: 11125931@bjtu.edu.cn

project of "City of Tomorrow—Bo01" has achieved the goal of more than 1000 residential units 100 % relying on renewable energy and self-sufficiency.

Low-carbon transport refers to the low power, low emission and low pollution modes of transport, which is the trend of development of the sustainable urban transport. With features of lightweight, flexible, environmentally friendly, comfortable, bicycle transport has become a zero-energy green way to travel. And encouraging non-motor vehicle travel has a positive effect on the reduction of urban transportation energy consumption. Within the range of 5 km, bicycle travel is generally no more than 20 min, and its travel cost is about one twentieth of the car. Compared to walking, it is of higher speed and a wide range of travel. The development model of public bicycle system in Malmö has provided the ideas for solving the problem of urban traffic, reducing impact on the climate and environment and establishing low-carbon transport system and livable city.

42.2 The Status of Bicycle Travel in Malmö

Malmö has mild seasonal climate and flat terrain, one of the most suitable cities for bicycle travel. Bicycle travel accounts for about 25 % of the total travel. The municipal government has committed to improve the driving conditions of bicycles and gradually built a bicycle path network covering the entire urban area. At present, Malmö has about 2100 km sidewalks, 420 km bike paths and 1125 km motor vehicle lanes. There are obvious bicycle paths on urban main roads, the city has many free bicycle parking, and bicycles and public transport achieve seamless. During 8 years from 2002 to 2010, the number of citizens using public transport has increased by 30 %. And 40 % of 280,000 urban citizens choose to ride bicycles to school or work. As shown in Fig. 42.1, with city center square (Stortorget) in Malmö and Central Station and other public facilities in the center, cycling within 5 min can cover the center city in Malmö and within 30 min, the whole urban area of Malmö.

42.3 Strategies Taken for Bicycle Travel in Malmö

42.3.1 Transport Policy

In order to create a better transport, Malmö has developed a series of programs, objectives and plans.[1] The programs have laid the foundation for further enhancing the status of citizen cycling and walking and made the two travelling ways more attractive.

---

[1] Including "Overall planning in 2000", "Environmental programs and transportation policy of Malmö in 2003–2008", "Traffic environment in 2005–2009" and its new bicycle travel program and the first walking program.

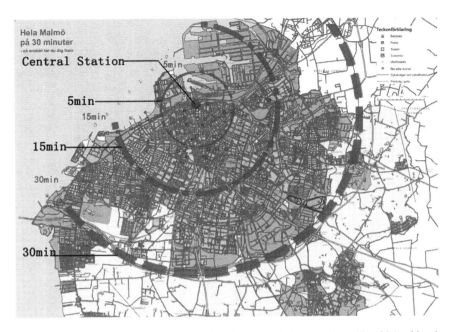

Fig. 42.1 Bicycle paths coverage and travel radius in Malmö, the *red road* is 420 km bicycle path. (*Source* Organized and drawn according to the information, CykelkartaMalmö 2009 slutversion (Malmo bicycle map in 2009—http://www.malmo.se/)

Fig. 42.2 Contrast of the proportion of travel volumes in different travel ways in 2003 and 2008 in Malmö. (*Source* Self-drawn)

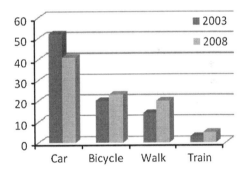

(1) **The survey of travel habits**. Malmö conducts a large-scale survey of travel habits every 5 years and most recently in 2008. As the data shows, in the premise of per capita travels in 2008 and 2003 basically the same, the proportion of a short-distance travel by car has decreased and the proportion of walking, cycling travel and a long-distance travel by train has increased[2] (Fig. 42.2).

---

[2] Organized and drawn according to the relevant data of "Improving Malmo's traffic environment" (http://www.malmo.se/).

(2) **Travel guidelines**. The pedestrian can receive a free map indicating all bicycle paths from City Hall and the Tourist Information Office. The map updates every 2 years to label the newly constructed bicycle paths. Besides, the website of Skånetrafiken (Transportation network in Skane) provides online bicycle trip planning assistant to check the best route for cycling trips in Malmö conveniently.

42.3.2 The Set of Bicycle (Dedicated) Paths

There are two bicycle paths. One is a dedicated lane of independently set without the interference of motor vehicles and traffic lights, and the other is a side-by-side arrangement with the motor vehicle lanes. In order to ensure the safety of the rider, most of bicycle paths are located between the sidewalk and street parking area. So there is protection and buffering of the parking vehicles between riders and fast moving car flow. Besides, there are dedicated signal lights on bicycle paths, which give bicycles the same status as motor vehicles.

In order to improve the conditions of bicycle travel, Malmö has taken a series of improvements on bicycle paths. Firstly, bicycle paths pass through urban residential areas, railway stations, subway exits and other important urban public areas, combined with the riverbank, beach, public green, squares and other open space to form scenic humanized walking and bicycle paths. Secondly, to carry out alterations to the road in the central areas of the city, turn part roadways of originally bi-directional four-lane into bi-directional two-lane and reduce the width of roadways to increase or widen bicycle paths and sidewalks (Fig. 42.3). Thirdly, reduce the scale of motor vehicle parking and encourage public and bicycle travel.

Fig. 42.3 Bicycle paths combined with seaside landscape, located in Bo01 residential district in Malmö, east coast of the Öresund (*Source* Self-shot)

42.3.3 The Design of Bicycle Paths and Bicycle Parking Space

(1) The relation between bicycle paths, parking space and roadways

In theory, addition of bicycle lanes in the limited street space is bound to compress the motor vehicle lanes and brings the risk of traffic jams. But the setting of bicycle paths will help improve traffic efficiency and occupy a smaller area of the road to meet the travel needs of large groups of people. So it is very important to layout bicycle paths reasonably and deal with the relationship between parking space and roadway. The following chart (Table 42.1) reflects the settings of different situations between bicycle parking areas, roadways and sidewalks.

(2) The design of bicycle racks

Malmö currently has about 45,000 bicycle racks. Freely parked bicycles may cause troubles to pedestrians, vehicles, the blind and other disabled. Therefore,

Table 42.1 The settings of bicycle parking and the roadway and sidewalk

Traffic	Width of the street	Width of sidewalks	Characteristics	Icon
Heavy	Narrow	Narrow	Bicycle parking parallel to the road, combined with car parking. The width of bicycle parking is 2.0 m, bicycle paths and roadways are separated	
	Wide	Narrow	Bicycle parking parallel to the road, combined with green. Bicycle paths and roadways are mixed	
	Wide	Wide	Bicycle parking is combined with roadways: 1. have an angle of not less than 45° with building facades; 2. parallel to the road	
			Bicycle parking is combined with sidewalks: 1. Vertical to the road, combined with green; 2. Vertical to the road, combined with bus shelters and the distance between is no less than 1.5 m	
Light	Narrow	Narrow	The angle between bicycle parking and the road edge is not less than 45°. The width of bicycle parking is 1.5 m	

This table is organized and drawn according to the information and research data, Cykelparkeringshandbok for Malmö (Bicycle Parking Guide in Malmö—http://www.malmo.se/

Fig. 42.4 Bicycle parking areas on one side of the bus station, Malmö Central Station (*Source* Self-shot)

encourage people to use bicycle racks as many as possible. Bicycles in sur-

Fig. 42.5 The design of new bicycle rack (*Source* The official website of Malmö—http://www.malmo.se/)

rounding areas of Malmö Central Station must be parked on bicycle racks to keep the pavement clean (Fig. 42.4). If incorrectly parked, they will be towed away.

Figure 42.5 is the design of Malmö's new bicycle rack. The color is striking orange and the overall appears like the car outline, in order to trigger people's reflection: choose 2 cars or 20 bicycles? This is a new symbol of proposing the sustainable development of urban bicycles.

(3) **The design of indoor bicycle parking**

Near public areas such as city squares, railway stations and subway stations, it is often necessary to set the indoor bicycle parking to meet transfer needs of the public.

Fig. 42.6 The transfer between bicycles and public transport (*Source* Drawn based on existing information, CykelP(Bicycle parking——http://www.malmo.se/)

42.3.4 The Connection and Transfer Between Bicycles and the Public Transport

Another key to popularize bicycles is whether the use of bicycles achieves convenient transfer with other public transport. The general practice is to set up the bicycle parking on the side of the road near the train station or subway station entrances and exits and have good convergence with the train station (Fig. 42.6). Only near Malmö Central Station, there are about 3,000 people parking bicycles every day.

The Three-dimensional bicycle garage near Västra station in Lund adopts double-layer trailers. The area of approximately 200 m^2 in two floors can accommodate about 460 bicycles, average 0.434 m^2/vehicle, much lower than domestic standards for cars 1.5–1.8 m^2 (Fig. 42.7).

42.3.5 Innovative Technology Solutions

In order to make travel by bicycle faster, safer and more comfortable, Malmö has tried to set up a variety of facilities of more innovative sense.[3]

---

[3] Organized according to relevant information of "Improving Malmo's traffic environment" (http://www.malmo.se/).

Fig. 42.7 Bicycle garage near Västra station (*Source* The official website of Malmö http://www.malmo.se/)

(1) **Bicycle counting license recording cycling number**. Malmö has equipped with "bicycle counting license" at Kaptensgatan and Södervärn to record the number of passing bicycles. Values indicate the current situation of bicycle riding, which make cycling traveler have the sense of pride on contributing to urban low carbon transport (Fig. 42.8).

(2) **Bicycle first**. Radar sensors have been installed in nearly thirty intersections to detect bicycles passing. The traffic signal will automatically transform into green when bicycles pass. This can ensure the smooth bicycle travel.

(3) **Bicycle service station**. Malmö has placed the inflatable pump at six locations (Kaptensgatan, Södervärn, Värnhem, GamlaIdrottsplatsen, Hovrätten and Erikslust), which is convenient for bicycle inflation (also applies to baby carriages and wheelchairs). Three pump stations are equipped with simple vehicle maintenance tools to be small service stations.

(4) **Other facilities**. In addition to the above measures, Malmö has also installed railings at the traffic lights for cyclists to rely and eliminate the trouble of on and off bicycles, placed several large mirrors at intersections of narrow perspective to make it convenient for people to see road conditions at corners. Besides, try to set up a variety of lighting facilities to improve the visibility of the night along the road.

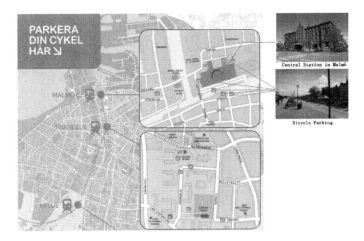

Fig. 42.8 Bicycle counting license (*Source* The official website of Malmö—http://www.malmo.se/)

42.3.6 Promotion to the Public

In addition to rehabilitation measures on the transport policy and hardware facilities, the propaganda work to guide the public to participate can not be ignored. The most successful one is "travel not by bicycle within 5 km is shameful". Through campaigns close to the life, the use of private cars in Malmö has decreased by about 10 % in 10 years.

42.4 Conclusion

Although Malmö has achieved certain results on the bicycle low-carbon travel, still need to make efforts in the following areas:

(1) **Improve the bicycle path network**

In the past 5 years in Malmö, among citizens riding bicycles, 105 people out of an average of 420 will hurt.[4] This may be due to the set of some bike lanes is not coherent, lacking the necessary isolation between bike lanes and motor vehicle lanes, which have increased traffic accident risks. At this point, the experience of bicycle lanes setting in Copenhagen is worth learning. Bicycle lanes and driveways, sidewalks are clearly differentiated in elevation, clear at a glance and without disturbing each other.

---

[4] Cykelkarta Malmo 2009 slutversion (Malmo bicycle map in 2009——http://www.malmo.se/).

(2) **Realize seamless between bicycles and other travel modes**

As a green transport, bicycles should be combined with other public travel modes, forming the green travel pattern of bicycles and public transport (B + R), to better achieve short distance travel advantages of bicycles. Only by reaching the integration between transfer and convergence on space and time, realizing seamless and zero transfer, the radiation range of the bicycle traffic can be extended and realize in the end the sustainable development of low carbon transport.

Acknowledgments This research is supported by National Natural Science Foundation of China (51178038); longitudinal research projects funded projects of Beijing Jiaotong University (2011JBM180, 2010RC030, 2012JBM120).

References

The official website of Malmö. http://www.malmo.se/
Improving Malmo's traffic environment. http://www.malmo.se/sustainablecity

Chapter 43
A Study on Low-Carbon Transportation Strategy Based on Urban Complex: Taking Shenzhen and Hong Kong as Examples

Yezi Dai

Abstract Public transportation is an important way to realize the mode of urban low-carbon transportation. Reasonable design and planning of urban complex will effectively organize as well as integrate regional public transportation. And it could also achieve low-carbon travel of urban residents on the premise of comfortableness and convenience. The article took cities of Shenzhen and HongKong as examples, and discussed planning and design of urban complex oriented by low-carbon transportation.

Keywords Urban complex · Low-carbon transportation · Shenzhen · Hong Kong · Multidimensional transportation hub · Regional walking system

43.1 Introduction

From a worldwide perspective, transportation mode based on traditional fuels is one of the main areas that emitting greenhouse gas. To alleviate the trend of global warming low-carbon transportation mode which aims to reduce carbon emissions is definitely imperative. In general, there are several ways to achieve this: improving the carbon emissions performance of transports, optimizing the way of travel and the model traffic management. Optimizing the way of travel means reducing use of private cars by promoting public transportation so as to achieve the goal of low-carbon transportation. The essay will take this as a starting point to

Y. Dai (✉)
Gold Mantis School of Architecture and Urban Environment, Soochow University, Soochow, China
e-mail: Leaf568@hotmail.com

inquiry into planning and design of urban low-carbon transportation mode which take urban complex as the core.

Urban complex is a kind of building or building complex composed by two or more space with different functions. It is now one of the effective measures to handle increasingly lack of land resources in cities. And it also becomes an important linkage of urban public transport network for its characteristic of collecting and distributing a large number of people and its countless ties with urban space and urban functions. Reasonable layout and design of urban complex will be able to improve use efficiency of public transportation and relieve urban traffic pressure. As economically developed SAR cities, Shenzhen and Hong Kong enjoy high level of municipal construction, favorable conditions of road infrastructure and convenient public transportation. However, as immigrant cities, they inevitably encounter extremely high population density and relatively scarce available land resources for construction. In view of this, using land efficiently through urban complex and integrating surrounding public transport resources by means of planning or architectural design become important ways for the two areas to promote green travel and develop low-carbon transportation.

43.1.1 Combining with Subway Stations

The aim of multidimensional comprehensive development of urban complex is to obtain the maximum economic efficiency with the least urban land. However, comprehensive development only considered from the perspective of commercial area without matchable traffic conditions and urban facilities also contraries to the original intention of sustainable development. As its rich commercial and recreational functions which bring about public and open character embracing different ages and industries will unavoidably gather lots of people and vehicles. Absence of efficient transport modes and convenient diverting mechanism will definitely further increase the regional traffic burden and affect normal use and benefits of the complex. But one of the important solutions for low-carbon transportation is to encourage efficient and fast public transportation. Consequently, high-frequently distributing of urban complex has an inner connection with high-efficiently diverting of public transportation.

Currently, the most common public transports include conventional ground bus and capacious rail transit (subway). Subway is second to none in terms of transportation efficiency, safety and comfort. Its characteristics of gathering sustainable and stable crowd and a series of deeper transforms aroused by which in underground space development and urban business model further agree with the spatial characteristics of complex. Therefore, combination of urban complex and subway stations embody the priori necessity. Underground commercial space link up with subway station and the connection become the gateway and transfer space of subway station. The business and transport functions of complex integrate into urban public transportation and it not only achieve the goal of saving land and investment but also make complex an indispensable part in the urban space system. In Shenzhen Special Economic Zone, most of commercial complexes link directly with subway stations through underground passages. (Figure 43.1) In fact, development and site selection of the complexes matched with planning and construction of urban subways from the very beginning. In Hong Kong, the MTR Corporation even led or participated in exploitation of several shopping malls along subways. Linkage development between subway construction and shopping centers has maximized utilization of public transports, brought great convenience to people's life, saved investment cost and realized maximization of economic and social benefits.

43.1.2 Setting Up Multidimensional Transportation Hub

When combining complex with subway stations, we can not ignore the important role of traditional ground transportation at the same time which is irreplaceable in popularity and coverage. So setting up public transport hubs such as bus starting stations, subway stations, and intercity terminals on or under the ground of complex buildings is benefit for overcoming traffic bottlenecks between bus stations and public place, reducing traffic pressure of street, and diverting flow quickly. Based on statistics about number of average daily passengers made by Hong Kong Transport Department in September, 2011, subway passengers accounted 39.7 % of all public transport passengers while bus passengers reached 50.7 %.(http://www.td.gov.hk/filemanager/en/content_4494/chart27.pdf) It can

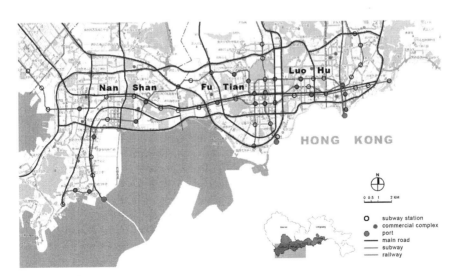

Fig. 43.1 The location of main ShenZhen commercial complexes and their relation to the subway station

clearly be seen that bus which combines with bus transfer stations or terminus is still an important supplementary for subway. The integration becomes a major feature for Hong Kong urban complex in terms of fully use of public transport. And buses successfully bind with subway stations and bus stops. Complex has become intersection of flow ground and underground and inadvertently appeared as transfer station for a variety of transports. These bus stations are often located on the ground floor of shopping centers and passengers will be able to enter directly into shopping malls as soon as getting off buses. It is convenient and fast and people can avoid inconveniences from sun and rain.

Taking Elements as an example, it is a super large urban complex consisted by business, houses, offices and hotels. And the three-floor shopping mall constitutes the main body of the complex podium. As it is located in the "bridgehead position" where Kowloon Peninsula links with Hong Kong Island, the complex becomes a traffic artery converging a number of important public transport routes including MTR Tung Chung Line, Airport Express Kowloon Station, Shenzhen Airport Express Bus, West Kowloon Underground Terminus of Guangzhou-Shenzhen-Hong Kong Express Rail link. (Figure 43.2) The ground floor and basement of the shopping center collect lots of transport routes and terminus of bus lines, (Fig. 43.3) and its first floor becomes a shared lobby for the shopping center and these transportation lines. Commercial space and transportation hubs are in perfect harmony. Broad perspective, good lighting and clear identity maximized the public character of space. Large numbers of people gathered here frequently

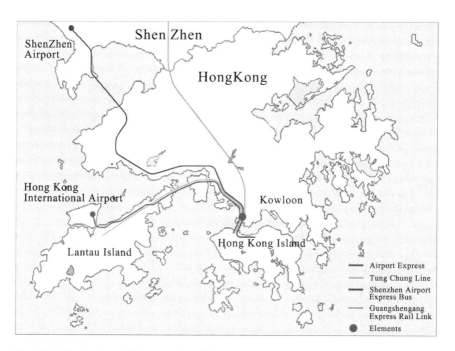

Fig. 43.2 The location of Elements in public

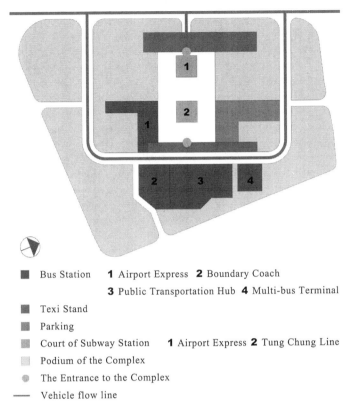

■ Bus Station **1** Airport Express **2** Boundary Coach
             3 Public Transportation Hub **4** Multi-bus Terminal
■ Texi Stand
■ Parking
■ Court of Subway Station **1** Airport Express **2** Tung Chung Line
▫ Podium of the Complex
● The Entrance to the Complex
— Vehicle flow line

Fig. 43.3 Functional analysis on transportation the ground floor of Elements

and passed through quickly. Commercial space provides a platform of buffering and conversion for urban public transportation system here, and high transport accessibility, in turn, promotes appreciation of real estate and commercial benefits.

43.1.3 Organizing Regional Walking System

Low-carbon transportation mode also includes planning of urban public walking system. Streets are narrow and ground transports are busy in Luohu and Futian district of Shenzhen, then separating people from vehicles becomes a basic solution in solving issues of urban low-carbon transportation and integration of urban architecture.

Central Walk, for example, located on the central axis green belt of CBD of Futian district in Shenzhen is a business complex integrating shopping, entertainment and dining. Its roof, a part of green pedestrian system penetrating the whole CBD from north to south, connects with the north public green space and

Fig. 43.4 Central Walk on the central axis green belt of ShenZhen CBD Commercial Area

the roof green space of the south under-construction business complex through overpass. Meanwhile, the west and east municipal green land has been sunk to solve the lighting problem of the shopping center's basement, and the space composition of the complex has been extended to the south building. Therefore, a multi-layered multidimensional green space has been formed surrounding the complex. It is extremely distinguishing as it is not only self-contained but also runs through the walking system in the whole area. (Figure 43.4) The other kind of walking system opposite to the aforementioned is a typical renewed old business district. Taking South Renmin business district of Shenzhen as an example, relying on the favorable location of the Luohu port, it once was the central and symbolical area of Shenzhen. As Hong Kong and Shenzhen port moved to west and the surrounding business districts sprung up, business of the area declined gradually these years. Thus, a renovation program of South Renmin business district which is dominated by government and actively joined by the developers came up. The main outcome of the program is the pedestrian overpass which connects the LuoHu Port and the Dongmen business district connects the lined main commercial complexes. (Figure 43.5) The difference compared to the former one is that this channel is much more pleasant and the space is more lively and familiar as it is located in an old business district with relatively cramped space and freer layout, moreover, most of the sections are located within the commercial space.

43.2 Summary

With continuous development of China's economic construction and unceasing advance of urbanization, conflict between urban development and environmental protection has become increasingly prominent. Expression in urban transportation is the contradiction between the significant increase of private cars and travel efficiency and urban environment. Maximizing the use of public transports by

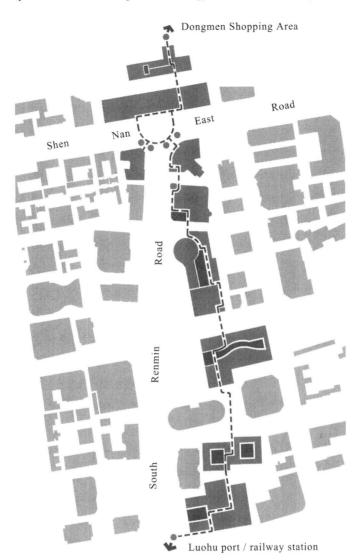

Fig. 43.5 The pedestrian overpass system of South Renmin

adroitly guiding action according to circumstances, meeting low-carbon development trends of future urban transportation and satisfying humane, comfortable and convenient travel requirements of public are not only the subjects placed in front of city governors, but also social responsibilities which the real estate developers should assume. Both Shenzhen and Hong Kong are economically developed areas, they enjoy some valuable experiences in building urban complexes based on low-carbon transport modes. And they should be able to provide us with lots of useful inspiration and reference.

Chapter 44
Study on Data Storage Particle Size Optimization of Traffic Information Database for Floating Car Systems Based on Minimum Description Length Principle

Rui Zhao, Enjian Yao, Xin Li, Yuanyuan Song and Ting Zuo

Abstract Particle size partition plays a key role in the optimization of historical database precision and data storage space. When establishing the historical database of traffic information for floating car systems, proper size of data storage particle can optimize data precision and storage space simultaneously and gives minimum comprehensive cost. This paper proposes a data storage particle size optimization model, which tries to balance data precision as well as data storage space for floating car systems. Furthermore, the proposed data storage particle size optimization model is executed in Beijing case study. The results show that the data storage particle size is 35 min at night while 10 min in the day under the given constraints of minimum cost of data precision and storage space, which is consistent with the real traffic condition and application requirement.

Keywords Floating car · Traffic information · Particle size optimization · Data precision · Data storage space

R. Zhao (✉) · E. Yao · X. Li · Y. Song · T. Zuo
School of Traffic and Transportation, Beijing Jiaotong University, Beijing, China
e-mail: 11121030@bjtu.edu.cn

E. Yao
e-mail: enjyao@bjtu.edu.cn

X. Li
e-mail: 10120909@bjtu.edu.cn

Y. Song
e-mail: 11120986@bjtu.edu.cn

T. Zuo
e-mail: 11121051@bjtu.edu.cn

44.1 Introduction

Due to rapid urbanization and ever-increasing number of vehicles, the problems with urban traffic, such as traffic congestion, environmental pollution, and energy shortage, are getting worse. Faced with these problems, many countries have focused on the application of Intelligent Transportation Systems (ITS) technologies. Floating car system, a new kind of ITS technology, has been proposed to collect traffic information. Compared with traditional traffic information acquisition systems, the lack of floating car data (FCD) may lead to an incomplete historical database when using fixed time granularities. Moreover, insufficient floating car data sample cannot guarantee data reliability. Too large or too small size of data storage particle cannot give the optimal data precision and storage costs at the same time. Therefore, proper data storage particle size plays a key role in establishing a precise and reliable FCD historical database with cost-effective storage space.

Although extensive research has been undertaken in developing ITS strategies and floating car systems, few articles have been devoted to study data storage particle size optimization of floating car historical database. Sun et al. (2011) proposes particle size monitoring for power consumption statistics of telecommunications industry based on fuzzy analytical hierarchy process (FAHP) and evaluation hierarchy. Yang et al. (2008) introduces a method of the dual granularity's degree, which can provide data of different granularity's degree with respect to dynamic requirements. Lv and Che (2009) analyzes the granularity model of data warehouse and puts forward a data granularity partition of banks considering acceptable minimum data storage particle size and the amount of stored data. Zhong (2004) investigates the granularity design from the following four aspects: requirements, data modeling, dimensions and time granularity. The paper concludes the principles of granularity design that the data granularity of the data warehouse must be altered with the changing of the requirements.

As demonstrated, most recent research focuses on physical storage of computer media. However, few articles have explored the data storage particle size optimization of historical link-based traffic information database for floating car systems when taking into consideration the FCD characteristics. Based on minimum description length (MDL) principle, this paper proposes a data storage particle size optimization model, which tries to balance data precision with data storage space for floating car systems. Meanwhile, a case study in Beijing is introduced and optimal particle size partition is given.

The remaining part of this paper is organized as follows: Sect. 44.2 introduces the concept and optimization of particle size for floating car traffic information. In Sect. 44.3, particle size optimization model based on the MDL principle is proposed. Based on the data collected in Beijing, the optimal particle sizes for different periods are recommended in Sect. 44.4, and the conclusions are given in Sect. 44.5.

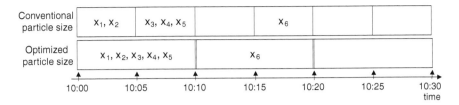

Fig. 44.1 Data storage particle size

44.2 Concept and Optimization of Particle Size for Floating Car Traffic Information

The storage particle size of floating car traffic information describes the length of time interval that information storage unit crosses. Figure 44.1 shows the concept of the conventional storage particle size and the optimized storage particle size of information. Usually, in floating car systems, the storage particle size for a road link is 5 min. After optimization, the storage particle size differentiates with the conventional particle size, e.g. 10 min shown in Fig. 44.1.

The storage space of floating car traffic information is the space needed for saving floating car traffic data in all time intervals during a day. For different particle sizes, any single storage unit occupies the same space of physical medium though crossing different length of time interval (Fig. 44.1 shows as an example). Thence, more storage units need for a road link by a smaller time particle size; in contrast, less storage units can meet the requirement for data storages in all time intervals by a larger time particle size, which saves storage space.

As a consideration of data precision, the storage particle size cannot be too large, while a large storage particle size leads to large differences in data samples and diminishes in reliability of traffic information in storage units for lacking enough data sample. In practical, even though smaller storage particle size improves the precision of data sample in storage unit, the storage space expands accordingly. Therefore, the storage particle size for data as a key factor related to the precision of history database and the optimization of storage space is of theoretical and practical significance.

44.3 Storage Particle Size Optimization Model of Floating Car Data

The FCD storage particle size optimization model is based on the Minimum Description Length (MDL) principle. The fundamental idea behind the MDL principle is that any regularity in a given set of data can be used to compress the data, i.e. to describe it using fewer symbols than the number of symbols needed to describe the data literally. This principle was put forward by (Jorma 1978) in1978 when he researched on generic coding. Its basic principle is that for a given set of instance data, in order to save the storage space, generally a certain model is adopted to conduct the compression coding and then the compressed data is saved. At the same time, the model is also kept to recover the instance data accurately in the future. So the total stored data length is the sum of compressed data length and model length, which is called total description length. MDL Principle aims at choosing the model that has the minimum total description length.

MDL follows the Occam's razor principle that unnecessarily complex models should not be preferred to simpler ones (Zhang and Mühlenbein 1993). Its essence is choosing the most appropriate particle size for describing objects, in order to achieve an optimal trade-off between model complexity and fitness of the data. It provides a quantitative measure to describe model conciseness: the total description length of data consists of two parts: the model coding length and the data coding length (Jiang et al. 2011; Vitanyi and Li 2000). The expression of coding cost is as follows:

$$Cost(Model, Data) = Cost(Data|Model) + Cost(Model). \quad (44.1)$$

Where,
Cost (*Model, Data*)   Total coding cost;
Cost (*Data |Model*)   Cost of the data coding;
Cost (*Model*)   Cost of the model coding

Back to the storage particle size optimization problem of FCD, if less data storage space (i.e., with larger particle size) is required, the price in data precision will be paid. Therefore, with reference of this idea mentioned above, the following model is adopted to optimize the data storage particle size.

$$\min\left(\sum_j \frac{1}{2} N_j \log S_j^2 + \frac{1}{2} J \log N\right) \quad (44.2)$$

where,
S_j^2 data sample variance of each storage unit under the condition of time interval number as j;
N_j data sample number of each storage unit under the condition of time interval number as j;
J time interval number;
N the total number of data samples

The term on the left side of plus stands for error term, and it reflects the ability of the compressed data to recover its original status, i.e., the ability of the storage numerical values in corresponding particle size can represent the original sample numerical values, or the price paid in the data precision under the corresponding particle size. The term on the right side of plus stands for the complexity term of the model, and it reflects the storage size to store information. The formula seeks

Fig. 44.2 Data source

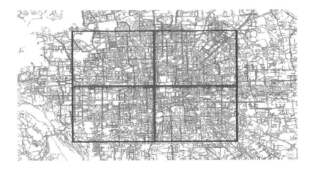

for an optimal particle size to balance the error term and the complexity term and achieve the minimum comprehensive cost.

44.4 Case Study

Experimental data is from Beijing floating system on 15th August 2010 with 14920063 pieces of floating car data included. The road network in Beijing contains 122848 road links. The research area is surrounded by the blue rectangle in Fig. 44.2, and 11440756 pieces of floating data, which is about 77.68 % of the whole data in the floating system, are collected in the area. The experiment is analyzed based on the data on road link No. 59566202171.

Figure 44.3 shows the variation of standard deviation and speed under different time periods of the experimental day. There are 288 time intervals of a day and number of time interval 0 stands for 00:00 while number of time interval 288 stands for 24:00. It can be observed that the standard deviation centralizes in the value of ten during the night, and it fluctuates around the value of five after 108 time intervals (i.e., 9:00 during the day). The values of the speed also indicate different tendencies of the whole day and the two turning points are shown as 9:00 and 23:00 in the Fig. 44.3. There are obvious difference between the nightly

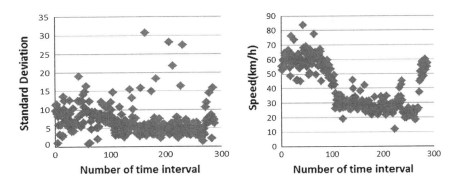

Fig. 44.3 Standard deviation and speed under different time periods of a day

Table 44.1 The MDL Principle based optimization results for daytime period

Particle size (min)	Number of time intervals	Total sample size	Error term	Complexity term	Error term + complexity term
5	168	1851	1324	274	1598
10	84	1851	1424	137	min★1561
15	56	1851	1483	91	1574
20	42	1851	1537	68	1605
25	34	1851	1586	55	1641
30	28	1851	1567	45	1612
35	24	1851	1602	39	1641
40	21	1851	1556	34	1590
45	18	1851	1566	29	1595
50	17	1851	1657	27	1684
55	15	1851	1639	24	1663
60	14	1851	1652	22	1674

periods and daytime periods. Therefore, for the sample link, the storage particle size is optimized based on different time periods, i.e., daytime (9:00–23:00) and night (0:00–9:00).

The calculation results with the FCD on road link No.59566202171 during daytime (9:00–23:00) and night (0:00–9:00) according to the Eq. (44.2) are as follows:

From Tables 44.1 and 44.2, it is obvious that: with the increasing of particle size, the number of time intervals decreases, meanwhile the complexity term also diminishes so the cost to save the model is lower. However, as complexity terms become smaller, error terms referring to the cost of precision for compressing data into a storage unit are larger. For a comprehensive consideration of complexity

Table 44.2 The MDL Principle based optimization results for night period

Particle size (min)	Number of time intervals	Total sample size	Error term	Complexity term	Error term + complexity term
5	120	606	576	166	742
10	60	606	577	83	660
15	40	606	585	55	640
20	30	606	595	41	636
25	24	606	597	33	630
30	20	606	602	27	629
35	17	606	581	23	min★604
40	15	606	662	20	682
45	13	606	669	18	687
50	12	606	601	16	617
55	11	606	616	15	631
60	10	606	612	13	625

terms and error terms, the minimum value of the sum of the two terms leads to the balance between the costs. According to the calculation results, the particle size is 10 min when the value of complexity term and error term is the minimum during the daytime, and during the night, particle size is 35 min. Therefore, it can be concluded that the total cost is minimum when particle size is 10 min during daytime and 35 min during night.

44.5 Conclusion

In this paper, a data storage particle size optimization model is proposed to optimize data precision, storage space, and comprehensive cost. Moreover, a dataset of Beijing is examined and analyzed, and the optimal data storage particle size (35 min at night while 10 min in the day) is recommended.

Acknowledgments This research is supported by National 973 Program of China (No. 2012CB725403) and National Key Technology R&D Program (No. 2011BAG01B01).

References

Jiang J, Xuan H, Hao S, Zhan X, Li H (2011) Statistical shape modeling based on minimum description length optimization in medical images. J Image Graph 16(5):879–885 (in Chinese)

Jorma R (1978) Modeling by the shortest data description. Automatica 14:465–471

Lv H, Che X (2009) Data granularity partition in warehouse. Comput Eng Des 9:2323–2325, 2328. (in Chinese)

Sun Y, Duan W, Yang S (2011) Optimization method of particle size of power consumption data for telecommunication industry. Mod Manag Sci 4:28–30 (in Chinese)

Vitanyi PMB, Li M (2000) Minimum description length induction, Bayesianism, and Kolmogorov complexity. IEEE Trans Inf Theor 46:446–464

Yang S, Zhang C, Li R (2008) Determination of granularity's degree in data warehouse. J Hebei Acad Sci 25(2):15–18 (in Chinese)

Zhang BT, Mühlenbein H (1993) Genetic programming of minimal neural nets using Occam's Razor. In: Forrest S (ed) Proceedings of the fifth international conference on genetic algorithms (ICGA-93), Morgan Kaufmann, p 342–349

Zhong Q A study on data warehouse's granularity and its implement. Master Degree Thesis, Beijing University of Posts and Telecommunications.2004. (in Chinese)

Chapter 45
Design of Double Green Waves Scheme for Arterial Coordination Control

Chengkun Liu, Qin Yong, Haijian Li, Yichao Liang, Yalong Zhao and Honghui Dong

Abstract Urban traffic congestion is increasing seriously, the arterial coordination control is an effective way to improve the smooth of arterial road. This paper designs an arterial double green waves coordination control method, which divides the arterial road into two coordination control sections bases on traffic flow, and redistributes the traffic flow in time and space according to phase offsets cooperative setting. This method can make traffic flow run smoothly on arterial road and improve the traffic condition of arterial road. Taking Ronghua Road in Beijing as an example, this paper simulates and analyzes the arterial double green waves coordination control scheme based on the microscopic traffic simulation software VISSIM. The results show that after optimizing, the traffic delay of the arterial road is decreased, and queue spillover phenomenon is improved obviously.

Keywords Traffic engineering · Arterial coordination control · Double green waves · VISSIM simulation

---

C. Liu (✉) · H. Li
School of Traffic and Transportation, Beijing Jiaotong University, Beijing, China
e-mail: liuchengkunyu@163.com

Y. Zhao (✉)
e-mail: ylzhaobjtu@163.com

Q. Yong · H. Dong (✉)
State Key Laboratory of Rail Traffic Control and Safety Name of Organization, Beijing, China
e-mail: hhdong@bjtu.edu.cn

Y. Liang
Beijing Traffic Management Bureau, Beijing, China

45.1 Introduction

Traffic smooth of arterial road will affect the service quality of the entire urban road transport system greatly. Arterial coordinated control is an effective method to improve the traffic operation conditions of arterial road at present. So it is significant to improve the traffic running condition of arterial roads (Zhen-wen 2003). Domestic and foreign scholars have done a vast amount of research (Bing 2005). These researches take intersections as an interconnected system to study with, but ignore the influence that different section traffic flow has on the whole arterial road, and the queue spillover phenomenon will occur in the section which has more traffic flow finally.

45.2 Arterial Double Green Waves Coordination Control

This paper designs an arterial double green waves coordination control method according to different sections has different traffic flow. The arterial traffic flow is divided into two sections which adopt arterial coordinated control. And the connected intersection of two sections is named intercepted intersection. Through phase offset setting, traffic flow in the section which has less traffic volume will enter the section which has more traffic volume after staying at the intercepted intersection for some time. Then the traffic volume distribution of arterial road is relative balance and the traffic pressure of the section which has more traffic volume is alleviated.

45.3 Design Process of Double Green Waves Scheme

This paper divides design and simulation process of double green waves coordination control scheme into five parts: traffic investigation and problem analysis, single intersection optimization, arterial road section division, double green waves coordination control scheme design, simulation analysis. The process of control scheme design is shown in Fig. 45.1.

45.4 Signal Cycle and Phase Offset Design of Double Waves Scheme

45.4.1 Signal Cycle

In this control system, in order to let the intersections of the arterial road cooperate with each other, signal cycle length of every intersection must be the same (Zhang 2009). So the longest cycle length of signal intersection should be chose as the common signal cycle of this system. The common signal cycle is defined as

45 Design of Double Green Waves Scheme for Arterial Coordination Control

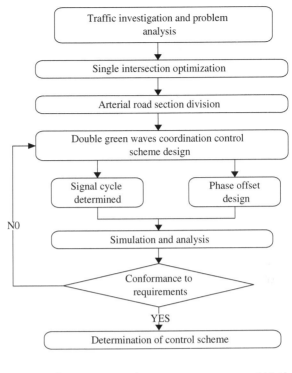

Fig. 45.1 Design and simulation process of double green waves control scheme

$$C = \max\{C_1, C_2, \cdots, C_n\} \qquad (45.1)$$

Where C denotes the common signal cycle of arterial double green waves coordination control scheme, C_n is the signal cycle of intersection n.

45.4.2 Phase Offset

Phase offset should be set based on the distance of intersections and vehicle running speed.. It denoted by ϕ, the calculation formula is given by

$$\phi = \frac{L_{a+i+1} - L_{a+i}}{V} - nC \qquad (45.2)$$

where L_{a+i} is the distance from the intersection i to the benchmark intersection a, V stands for is permissible speed, n is positive integer.

In this control system, the benchmark intersection of the green wave system before intercepted intersection is the first intersection. The benchmark intersection of the other green wave system is the intercepted intersection.

45.5 Example Analysis

This paper takes nine adjacent intersections of Ronghua Road in Beijing as example. To facilitate the narrative, the intersections from south to north are named J079–J087.

45.5.1 Traffic Investigation and Problem Analysis

The south and north entrances of J079–J086 intersections have four lanes, and the J087 have five lanes. But the west and east entrances of the nine intersections have few lanes. This paper analyzes the traffic flow on Monday, Tuesday, Friday and Saturday, and selects the evening peak hour 17:15–18:15 on Monday as the study hour. The traffic volume in every entrance of J079–J087 intersections is shown as Fig. 45.2.

All the nine adjacent intersections adopt three phases signal control scheme and use the same signal cycle (164 s). The amber time of all phases is 4 s. The all-red time of south and north straight phase and south and north left turn phase is 2 s, but this time of west and east phase is 4 s. The current signal timing schemes of nine adjacent intersections are shown as Table 45.1.

45.5.2 Scheme Design of Double Green Waves Scheme

Traffic flow from south to north is cumulative, and the distance between adjacent intersections is about 400 meters, so one-way green wave can be set up. The cycle length of arterial double green waves coordination control is 164 s.

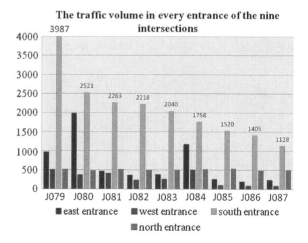

Fig. 45.2 Traffic volume of Ronghua Road adjacent intersections

45 Design of Double Green Waves Scheme for Arterial Coordination Control

Table 45.1 Current signal timing of Ronghua Road adjacent intersections

Intersection	South and north straight(s)	South and north left turn(s)	West and east(s)	Offset(s)	Adjacent intersections distance(m)
J079	100	14	30	117	0
J080	80	19	45	78	398
J081	74	14	56	41	402
J082	85	19	40	4	383
J083	80	14	45	131	431
J084	80	19	45	93	420
J085	90	14	40	62	418
J086	90	14	40	31	414
J087	90	14	40	0	445

The traffic data show that traffic flow of Ronghua Road is increasing from south to north, and traffic flow of south intersections are less, but the queue in south entrance of the J079, J080 two north intersections extend to the next intersection, there exists queue spillover phenomenon in these two intersections

The south to north traffic flow of J087, J086 and J085 intersections is less. In order to decrease the queue spillover phenomenon in J079 and J080 intersections, this paper chooses J084 as the intercepted intersection.

The first coordination control section chooses J087 intersection as the benchmark intersection. And the offset of J087, J086 and J085 intersections are set. Then the offset between J084 and J085 intersections is adjusted so that the traffic flow which runs smoothly meet red light at J084 intersection and the pile up effect is formed.

The second coordination control section chooses J084 as the benchmark intersection. And then set the offsets of J084, J083, J082, J081, J080 and J079 intersections. The signal timing of this scheme is shown as Table 45.2.

Table 45.2 Signal timing of arterial double green waves scheme

Intersection	South and north straight(s)	South straight and left turn(s)	South and north left turn(s)	West and east(s)	Offset(s)	Adjacent intersections distance(m)
J079	100	0	14	30	121	0
J080	80	4	15	45	91	398
J081	74	4	10	56	67	402
J082	85	0	19	40	40	383
J083	80	0	19	45	13	431
J084	80	4	15	45	150	420
J085	90	0	14	40	54	418
J086	90	0	14	40	27	414
J087	90	0	14	40	0	445

Fig. 45.3 Simulation testing process

Fig. 45.4 South entrance queue length optimization of queue spillover intersections

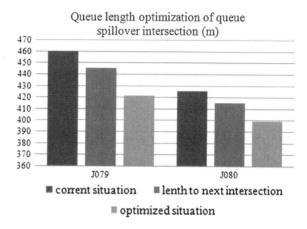

45.5.3 Simulation Evaluation

This paper adopts VISSIM to simulate the optimization scheme. VISSIM is capable to analysis the urban traffic operation under various traffic conditions. It is an effective tool for evaluating traffic engineering design and traffic organization optimization schemes (Chun-ying 2005; VISSIM 2006).

Establishing current model and double green waves scheme of the Ronghua Road based on fundamental data, input the traffic volume, signal timing scheme, traffic operation parameters, etc. The simulation testing process is shown in Fig. 45.3.

The simulation results are illustrated in Figs. 45.4 and 45.5.

The result shows that, the arterial double green waves coordination control scheme has some influence to improve the traffic quality. The south entrance queue length of J079, J080 intersections are decreased. The queue length of J079 intersection is improved 8.48 %, the queue length of J080 intersection is improved 6.12 %. The queue spillover phenomenon is improved. The delays of all the intersections in the arterial double green waves coordination control system are decreased except the J084 intersection.

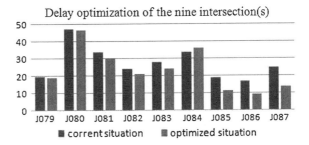

Fig. 45.5 The south to north traffic delay optimization of Ronghua Road adjacent intersections

45.6 Summary

Arterial road coordination control scheme is an effective method to improve traffic operation condition and alleviate traffic pressure. Aiming at the distribution of traffic flow is uneven and some intersections queue spillover. This paper designs a kind of arterial road double green waves coordination control scheme, and describes the design method and simulation process in detailed. Taking Ronghua road as an instance, a scheme is designed, adopting VISSIM microscopic simulation software to simulate and analyze current situation and double green waves scheme, the results show that the method has practical value. This work is supported by "the Fundamental Research Funds for the Central Universities" (2012YJS059), the National 863 Program (GrantNo.2006AA11Z231) and the National Natural Science Foundation of China (Grant No. 61104164).

References

Bing W, Ye L (2005) Traffic management and control [M]. People's Communication Press, Beijing
Chun-ying GAI (2005) Microscopic simulation system VISSIM and its application to road and transportation. J Highway 8:118–121
PTV VISSIM 4.20 User Manual [Z]. German: Planung Transport Verkehr AG, 2006
Zhang L(2009) Methods for arterial coordinated control of signal intersection under dissymmetry flow [D].Wuhan University of Technology, Wuhan
Zhen-wen H (2003) Ideas on the present situation and development of city ITS [J]. Traffic Eng Technol Natl Defense 1(2):10–13

Chapter 46
An Evaluation Indicator System of Low-Carbon Transport for Beijing

Siyuan Zhu and Xuemei Li

Abstract Establishing an indicator system is an effective way to find out the exited problems in development low-carbon transport. In this paper, we have analyzed the related principles and established the system of evaluation based on data dependency and principal components analysis (PCA). This system is composed by 23 indicators, taking the core of sustainable development. Based on the index system and weights of index, the paper gives an evaluated level of low-carbon transport in Beijing from 2006 to 2010. The conclusion is that Beijing's low-carbon transport development level is increasing continuously. Some positive results were achieved form public infrastructure construction, improving ecological environment and upgrading the industrial structure. But there are still some shortcomings, such as decline in bicycle riding, traffic jams and insufficient investment in assets, which reduce the overall level of the low-carbon transport.

Keywords Low-carbon transport · Principal components analysis (PCA) · Evaluation indicator system · Urban transportation

46.1 Introduction

As a major source of carbon emissions, transportation becomes an important field in international greenhouse gas emission reduction and climate change mitigation. Establishing an indicator system is an effective way to find out the exited problems in development low-carbon transport.

S. Zhu (✉) · X. Li
School of Economics and Management, Beijing Jiaotong University, Beijing 100044, China
e-mail: zhusyy@163.com

The existing research in low-carbon transport evaluation system, mainly concentrated in two parts. One is the index selection, and the other is the construction of the indicator system. Peris Mora et al. (2005) present when choose indicators need to meet the following two points: (1) Simple. The number of indicators and the method of calculation should be simplified as much as possible. (2) Goal-oriented. Correlation between indicators and the goal must be obvious, so that indicators can reflect the target's changes sensitively.

In the major application of statistical method, one is the Principal Component Analysis. From many observation variable original data to carry out comprehensive information and most independent several factors to explain the original data variables, make multidimensional variable dimensional reduction, thus simplifying the data structure. Another is Analytic Hierarchy Process (Saaty 1990). Comparing between every two factors have done and get the sort order by weightiness of all factors. With the development of artificial intelligence, scholars established new multi-criteria decision analysis method, such as Fuzzy AHP, Fuzzy Comprehensive Assessment (Lu et al. 1999). Anjali Awasthi (2011) uses Dempster-Shafer Theory, processed uncertain data and combined with AHP method to establish the evaluation indicator system of sustainable transportation development.

Together with a number of domestic and foreign literatures researches, foreigners have more empirical researches in different angles. But considering the data statistics and policies are different, foreign existing system can't completely suitable for urban traffic evaluation in china. In contrast, the domestic research is still in the starting period. Focus more on low-carbon traffic definition, system framework and policies. Given theoretical research and policy advice, be short of empirical research.

46.2 Methodology

46.2.1 Selection of Indicators

The selection of evaluation indicators is a repeated the experiment process, combing subjective experience and data analysis. The various steps are explained as follows.

(1) Initially draft of indicators

According to the existing research results and related system, we reference the form and principle of establish an evaluation system and find the widespread use of key indicator. Reference source mainly come from government agencies report and the periodical literature in both domestic and foreign, such as Herb Castillo (2010) and Nicolas Moussiopoulos (2010).

46 An Evaluation Indicator System of Low-Carbon Transport

(2) Selection of indicators

Data collect. Find the above reference data in "Beijing statistical yearbook", "China's environmental statistics yearbook", "Chinese urban statistical yearbook" and "the Beijing transportation development of the annual report" from 2001 to 2010. According to the properties divided into three categories: urban traffic, social economic and environmental energy.

Adjust the negative indicators. For criteria having negative impact, use reciprocal value to determine score. And then, make them available by data's standardization.

Remove uncorrelated and repeated indicators. In SPSS, we use Pearson correlation analysis to remove uncorrelated and repeated indicators in order to reduce index number, simplify the calculation amount. At the same time, select the indicators more objectively through the data analysis.

(3) The main indicators of the evaluation system

Through the selecting of relevance, establish a complete low-carbon transport evaluation indicator system, such as Table 46.1.

46.2.2 Quantification and Weighting

"The Principal component analysis" (PCA) basically eliminates the subjective judgment and be used in practice widely. So this paper uses it for quantification and weighting those indicators. PCA weights data by combining original variables into linear combinations that explain as much variation as possible. Mainly includes the following four steps:

(1) Calculating correlation coefficient matrix

Using standardized data calculate correlation matrix R, $R = \begin{bmatrix} r_{11} & \cdots & r_{1p} \\ \vdots & \ddots & \vdots \\ r_{p1} & \cdots & r_{pp} \end{bmatrix}$

Dim r_{ij} as Correlation Coefficient between X_i and X_j.

$$r_{ij} = \frac{\sum_{k=1}^{n}(X_{ki}-\overline{X}_i)(X_{kj}-\overline{X}_j)}{\sqrt{\sum_{k=1}^{n}(X_{ki}-\overline{X}_i)}\sqrt{\sum_{k=1}^{n}(X_{kj}-\overline{X}_j)}} \quad (i, j = 1, 2, \ldots, p)$$

(2) Calculating Eigenvalue and variance contribution

Solve equation, $|R - \lambda E| = 0$ and Calculate correlation matrix R's eigen value named λ, if $\lambda_1 \succeq \lambda_2 \succeq \ldots \lambda_n \succeq 0$ and then according to the factor accumulative total variance contribution over 85 to determine the number of main component.

Table 46.1 The main indicators of the evaluation system

Target layer	Rule layer	Indicator layer	Number	Property
Low carbon traffic development	Urban transportation	Private car growth	X1	−
		Number of the public transport vehicle operation in the end of year	X2	+
		Highway miles	X3	−
		Road area	X4	+
		Total length of public transport operating routes	X5	+
		Highway passenger quantity	X6	+
		Public transport passenger quantity	X7	+
		Taxi passenger quantity	X8	+
		Travel options-Bicycle	X9	+
		Travel options-Car	X10	−
		Travel options-Public transportation	X11	+
		Congestion conditions in workdays	X12	−
		The traffic accident happened several	X13	−
	Environment and Energy	Per capita green area	X14	+
		Air quality	X15	+
		Qualified rate of emission testing throughout the year	X16	+
		Ten thousand yuan in GDP energy consumption	X17	−
		Ten thousand yuan in GDP coal oil accounts for the proportion of total energy consumption	X18	−
	Social economy	Permanent population density	X19	−
		Per capita in GDP	X20	+
		Disposable income	X21	+
		Fixed assets investment in traffic	X22	+
		Financial expenditure in environmental protection	X23	+

(3) Determine the principal components number and the economic significance

To explain economic meaning of those components, key indicators attributes that had great impact weight are used. If the loads on 23 indicators on one of components are not far from each other, then rotate factors to make the principal components can be explained. This paper chooses the method of varimax orthogonal rotation to rotate factors.

(4) Calculation of main component scores and total points

We can calculation the score of main component by computing the Main component scores coefficient matrix. And then uses each of the principal component eigenvalue as root weight, to weighted summary for each main ingredient. Finally, we get the total score for evaluation.

Table 46.2 Eigenvalue and variance contribution

	Initial Eigenvalues			Extraction squares of the load			Rotating squares of the load		
	T	V	A	T	V	A	T	V	A
1	19.77	85.96	85.96	19.77	85.96	85.96	10.13	44.05	44.05
2	2.22	9.65	95.61	2.22	9.652	95.61	8.81	38.32	82.36
3	1.01	4.39	100.0	1.01	4.393	100.0	4.057	17.64	100.0

T total, V variance %, A accumulated %

46.3 An Application on Beijing

According to the establishment of the above evaluation system, we evaluate low-carbon development level in Beijing nearly 5 years by using the data from "Beijing statistical yearbook" and "the Beijing transportation development of the annual report" from 2006 to 2010.

46.3.1 Results

First of all, calculating correlation coefficient matrix, and the results show that there was significant correlation between 23 indicators. So the conditions of using PCA are satisfaction. Then, calculate the eigenvalue and variance contribution. From the Table 46.2, it is known that the first, second and third composition of the eigenvalue greater than 1, and three of main components of contribution rate is up to 100 %, which retain the original indicators all information, explain the three main ingredients can cover the whole indicators information.

Select the first three principal components, and calculates the corresponding main component of the correlation coefficient. Then rotate factor, the main components have explicable.

The first PC reflects the urban transportation condition, including city infrastructure, transportation passenger travel choice and road way, crowded degree, etc. The second PC reflects the city environment and economic development, including the environmental quality, and energy consumption, economic development level, etc. The third PC elements less, added that the urban traffic development, including: Qualified rate of Emission testing throughout the year and fixed assets investment in traffic.

Named three PC as f_1, f_2, f_3, scores as f, with each of the principal component eigenvalue for the root weight. Name three eigenvalues as λ_1, λ_2, λ_3. Use the equation is said for:

$$f = \frac{\lambda_1}{\lambda_1 + \lambda_2 + \lambda_3} f_1 + \frac{\lambda_2}{\lambda_1 + \lambda_2 + \lambda_3} f_2 + \frac{\lambda_3}{\lambda_1 + \lambda_2 + \lambda_3} f_3$$

Will the data of Beijing 2006–2010 are generation into the above formula, get all the main component scores and total score such as Table 46.3.

Table 46.3 Main component scores and total score

Years	f_1	f_2	f_3	f
2006	0.265	0.060	−0.779	−0.001
2007	0.398	0.087	−0.436	0.130
2008	0.491	0.939	−1.081	0.379
2009	−0.051	1.077	0.472	0.472
2010	0.317	1.727	−0.788	0.654

46.3.2 Analysis

(1) Total score analysis

Based on the above calculations of the Beijing low-carbon transportation development total score, it is easy to see in the 5 years, Beijing's low-carbon traffic comprehensive evaluation scores have been tend to growth, explained this 5 years Beijing has achieve the goal of low-carbon traffic benign development. From the absolute number of the growth, from 2006 to 2008, comprehensive score turned from negative to positive. The growth become moderated after 2008, but still maintained a good momentum.

(2) Prin1 analysis

First principal component scores is not very high, experienced three years growth until 2009. After the Olympic Games in 2008, directly investment in low-carbon traffic has been cut back. We analyze the reasons from both positive and negative aspects below.

First, the causes of positive factors promote the growth. One is increasing number of urban transportation infrastructure construction. From 2006 to 2010, Beijing's highway mileage and road area are continuing to increase, alleviate the pressure of too much motor vehicle, and improve travel road conditions. Another is the development of public transport. The number of public transport, operating routes, the passenger in the total length are increasing year by year, dense urban public transportation network, promote residents travel choice public transportation, and promote the development of the low-carbon traffic.

Negative reasons are following two points:

(a) Proportion of bicycle travel has dropped year after year.

Bicycle travel is a kind of convenient and health way to go out, which is no pollution and saving energy. The promotion of bicycle play an important role in develops low-carbon traffic. However, bicycle travel declined year by year in Beijing from 30 % in 2006 to 16.4 % in 2010. Cause this result both subjective and objective. Subjective reason is that low-carbon travel consciousness among residents is not strong enough. As the city economic development and living standards improving, vehicle travel becomes more popular. Even bicycling and walking are considered a symbol of the backward. The objective reason is the rapid expansion of car serious encroach the bicycle road space, making the bicycle travel environment worsening and security danger has also increased.

(b) Traffic congestion increased carbon emissions

Traffic congestion and carbon emissions are closely linked; the serious traffic congestion will greatly increase the energy consumption and exhaust emissions. However, from the data of "Congestion conditions in workdays", Beijing is in moderate congestion permanently. This situation become well in 2008, but the momentary relief was followed by a growing trend. This is a big obstacle in the process of low-carbon transport.

(3) Prin2 analysis

The second principal component is the only one maintains the upward trend of score among the three PC. For a number of reasons list below: (a) Economic development. Beijing has maintained a fast speed of economic growth. By adjusting industrial structure and developing the third industry, the per GDP and finance income increase year after year steadily, tremendously support the development of low-carbon transport. (b) Environmental improvement. In the environment, Beijing's air quality is continuous improvement and financial expenditure is increasing year by year. (c) Energy efficiency improving. Greenhouse gas main composition is carbon dioxide, carbon dioxide emissions mainly from fossil fuel burning and cement, lime, steel and other industrial production process. Beijing made a notable achievement in energy saving and emission reduction. Coal and oil accounts for the share of energy consumption decline every year and promotion in the new energy reduce gas emissions directly.

(4) Prin3 analysis

There is a large fluctuation in the third principal component score. Prin3 mainly explained qualified rate of emission testing throughout the year and fixed assets investment in traffic. Because of the two indexes ascension obviously, the score increase a lot in 2009. One of the reason of the promotion is a construction in several routes needs a mass of investment. Another is abolishment of consists car, vehicle exhaust emission condition has improved since 2009.

In addition, from the detail in Beijing's fixed assets investment we can find that investment is mainly used for highway construction more than hub stations. The results show that the public transport services don't get enough attention. Because the high demand of travel, the existing public travel density nets can't meet the demand. And become one reason to evoke the families to buy private cars.

46.4 Conclusions

In this paper, we have analyzed the related principles and established the system of evaluation based on data dependency and Principal components analysis (PCA). This system is composed by 23 indicators, taking the core of sustainable development. Based on the index system and weights of index, the paper gives an evaluated level of low-carbon transport in Beijing from 2006 to 2010. The evaluation process accord with theoretical requirements, the output of the results can be confirmed by specific data and used for further analysis. The data of indicators are from government institutions, available and powerful. This paper presented general methods and calculating, they have general application ranges.

The conclusion is that Beijing's low-carbon transport development level is increasing continuously. Some positive results were achieved form public infrastructure construction, improving ecological environment and upgrading the industrial structure However, there are still some shortcomings we found by this evaluation: proportion falls on bicycle travel; traffic congestion; incorrect structure of traffic assets investment.

The current work evaluates only consider the objective factors. Limit in available data from the government institutions which can't fully reflect the low-carbon development. Future work will involve assessment of people's subjective factors and the indicators have yet to be further improved.

Acknowledgments This paper was supported in part by "National Natural Science Foundation Projects (50978023)" and "Railway Ministry Projects (2010Z012)".

References

Awasthi A, Chauhan SS (2011) Using AHP and Dempster–Shafer theory for evaluating sustainable transport solutions. Environ Model Softw 26:787–796
Castillo H, Pitfield DE (2010) ELASTIC—a methodological framework for identifying and selecting sustainable transport indicators. Transp Res Part D: Transp Environ 7:179–188
Lu RS, Lo SL, Hu JY (1999) Analysis of reservoir water quality using fuzzy synthetic evaluation[J]. Stochastic Environmental Research and Risk Assessment 13(5):327–336.
Moussiopoulos N, Achillas C, Vlachokostas C, Spyridi D, Nikolaou K (2010) Environmental, social and economic information management for the evaluation of sustainability in urban areas: a system of indicators for Thessaloniki, Greece. Cities 10:377–384
Peris-Mora E, Diez Orejas JM, Subirats A, et al. (2005) Development of a system of indicators for sustainable port management. Marine Pollution Bulletin 50:1649–1660
Saaty TL (1990) How to make a decision: the analytic hierarchy process. Eur J Oper Res, North-Holland 48:9–26

Chapter 47
Design and Implementation of Regional Traffic Information Disseminating System Based on ZigBee and GPRS

Weiran Li, Wei Guan, Jun Bi and Dongfusheng Liu

Abstract In order to help drivers get traffic information more easily and accurately, a new regional traffic information disseminating system based on ZigBee and GPRS is introduced in this paper. The system consists of two main components: an Information Sending Device (ISD) fixed on the roadside which sends traffic information as well as an Information Receiving Device (IRD) equipped in vehicles which receives traffic information. Real-time traffic information is sent to ISDs through GPRS network by traffic information management center. Then ISD disseminates the information timely and accurately to IRDs through ZigBee network. At last the traffic information is broadcasted vocally to drivers using Text To Speech (TTS) technology by IRD. The hardware structure and software design are presented respectively. The system is tested in Chegongzhuang Street, Beijing. The test results show that the system works well and drivers can obtain traffic information accurately and timely.

Keywords Traffic information dissemination · GPRS · ZigBee · Text to speech (TTS)

W. Li (✉) · W. Guan · J. Bi · D. Liu
MOE Key Laboratory for Transportation Complex Systems Theory and Technology, Beijing Jiaotong University, Beijing 100044, China
e-mail: 11120886@bjtu.edu.cn

W. Guan
e-mail: wguan@bjtu.edu.cn

J. Bi
e-mail: jbi@bjtu.edu.cn

D. Liu
e-mail: 10120973@bjtu.edu.cn

47.1 Introduction

The critical point of a traffic information disseminating system is to transmit traffic information to drivers clearly and timely, which plays an important role in Intelligent Transportation System (ITS). Xiao et al. (2008) proposed a scheme which uses the existing radio equipment to disseminate traffic information. This method fails to consider the fact that the traffic information disseminated by FM doesn't work when the location of traffic information and the location of drivers are not the same; Qin et al. (2006) put forward a new traffic information disseminating system by the Internet. But it is hard for drivers to get the traffic information on the Internet; Shahjahan et al. (2008) introduced a new idea that disseminating traffic information via Short Message Service (SMS). In this scheme, drivers must send a SMS to SMS server, which is not convenient for drivers. Guan et al. (2008) proposed a novel method of disseminating traffic information via Variable Message Signs (VMS). This method does work well to some degree in guiding traffic flow. But when there are obstructs, for example a big bus, in front of the vehicle, the diver cannot see the information in VMS clearly and timely. What's more, the information displayed in VMS is a little simpler. Actually, there are various road conditions, the VMS, however can only show three road conditions, i.e. free, light congestion and severe congestion by three colors, i.e. green, yellow and red, respectively. So it is hard to judge road states that are between two colors.

In this paper, we design a new traffic information disseminating system based on ZigBee, (Bilgin and Gungor 2012) and GPRS (Ghribi and Logrippo 2000), (Ionel et al. 2012). The system is made up of two main components: an Information Sending Device (ISD) fixed on the roadside which sends traffic information as well as an Information Receiving Device (IRD) attached to vehicles which receives traffic information. Real-time traffic information is sent to ISDs through GPRS network by traffic information management center. Then ISD disseminates the information timely and accurately to IRDs through ZigBee network. At last the traffic information is broadcasted vocally to drivers using Text To Speech (TTS) technology by IRD.

47.2 Overall System Configuration

The overall configuration of the traffic information disseminating system is shown in Fig. 47.1. The traffic information management center is charged of traffic information collecting, processing and producing dynamic traffic guidance information. Then, according to the information's destination, the center transmits the information to the specific ISDs through GPRS network. Finally, the ISDs broadcast the traffic information to vehicles by ZigBee network. The ZigBee module in ISD is a coordinator node of ZigBee network and in IRD is an end node.

Fig. 47.1 Overall system configuration

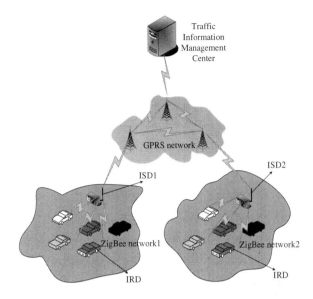

47.3 Hardware Design

47.3.1 Hardware Design of ISD

The hardware structure of ISD is shown in Fig. 47.2.The Cirrus Logic' EP9312 is chosen as the MCU, which has many interfaces that can be connected with peripheral devices. Serial port1 connects with GPRS module; serial port2 connects with ZigBee module; storage connects with MCU via SPI bus.

47.3.2 Hardware Design of IRD

The hardware structure of IRD is shown in Fig. 47.3. The IRD is attached to the vehicle, so it should be designed compactly. We choose winbond' W77E58 as MCU, which has two full duplex serial ports and is compatible with MCS-51 instruction set. Serial port0 connects with ZigBee module and serial port1 connects with TTS module.

Fig. 47.2 Hardware architecture of ISD

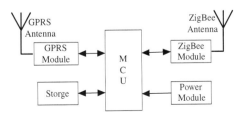

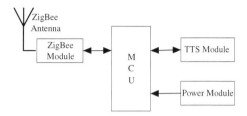

Fig. 47.3 Hardware architecture of IRD

47.4 Software Design

47.4.1 Software Design of ISD

The WinCE5.0 is selected as MCU's embedded operation system and use Embedded Visual C++4.0+SP4 as the developing environment. The multi-thread technology is used to realize the three modules' functions. The main program flow chart of the system is shown in Fig. 47.4.

GPRS module which connects ISD and upper server plays an important role in the system. At first, we need initialize the module, including setting the serial port parameters. By sending a serial of AT instructions and checking the responses, the MCU can control GPRS module to have a connection of TCP/IP with the upper server. We create two threads: m_hReadThread and m_hWriteThread. The m_hReadThread is used to read data from serial port and the m_hWriteThread is used to write data to serial port. Because GPRS wireless network is not absolutely stable, we propose a heartbeat mechanism to guarantee the connection between ISD and server is always reliable. The protocol format of heartbeat data is shown in Table 47.1. The ISD sends heartbeat data every 3 s and the server sends a packet of answer data once receiving heartbeat data. The protocol format of answer data is shown in Table 47.2. If the ISD doesn't receive answer data three times continuously, the ISD will rebuild a connection with the server. And if the ISD doesn't receive answer data after rebuilding two times continuously, the ISD will restart GPRS module. Furthermore, if the ISD doesn't receive answer data after restarting the module two times continuously, the ISD will restart the system.

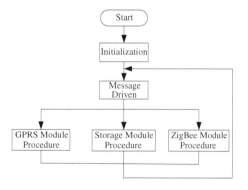

Fig. 47.4 The Main Program Flow Chart of System

Table 47.1 Heartbeat data protocol format

| $ | S | ID | yy-mm-dd | hh-mm-ss | "heb" | "0" | BCC |

Table 47.2 Answer data protocol format of heartbeat

| * | r | ID | yy-mm-dd | hh-mm-ss | "heb" | "0" | BCC |

Table 47.3 Request data protocol format

| & | S | "net" | "0" | BCC |

Table 47.4 Traffic information data protocol format

| $ | S | "guide" | Data length | ID number | Content | BCC |

The Block Check Character (BCC) is used as the data check. The function is CString MakeCheckSum (CString strData).

ZigBee module fixed to ISD is a center node of ZigBee network, and fixed in IRD is an end node of network. When the end node enters ZigBee network, it will send a packet of request data to the center node. The request data protocol format is shown in Table 47.3. Once the center node receives the request data, the ISD will broadcast the latest traffic information. The traffic information data protocol format is shown in Table 47.4. ID number is added in format, aiming to let ISDs decide whether to receive the traffic information or not. If a ISD has received a certain traffic information already, it should not receive it again.

To ensure every IRD in ZigBee network can receive a certain traffic information once, a timer is used to deal with this problem. When expire a timer, the ISD will broadcast the latest information again. The program flow chart of ZigBee module' receiving and sending information is shown in Fig. 47.5.

47.4.2 Software Design of IRD

The KEIL C51 is used as the developing environment of Winbond' MCU. The main program flow chart is shown in Fig. 47.6. When detecting ZigBee signal, the IRD will be driven by external interrupt. Then it will send a packet of request data to ISD. Having connected with the ISD, the IRD is just waiting for the interrupt from serial port0. When the interrupt is triggered, MCU will read out the data from buffer to TTS module. And TTS module synthesizes the data to speech. At last, the vocal traffic information is broadcasted to drivers. The function of speech synthesis is bool Send2TTS(char *buf, unsigned char len) of which buf is the pointer to character string to be synthesized and len is the length of character string.

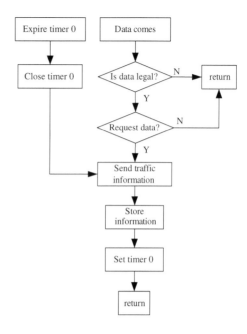

Fig. 47.5 The program flow chart of ZigBee module

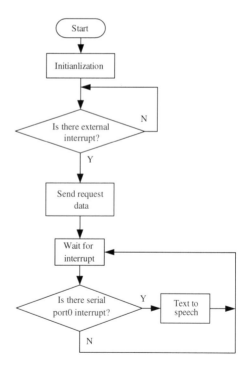

Fig. 47.6 The main program flow chart of IRD

Fig. 47.7 The result map of distance test

47.5 System Test

To verify the feasibility of the system, we carry out system test in Che-GongZhuang Street, Beijing:

Step 1: Open the server in traffic information management center, and set it to send a packet of data every 1 s.

Step 2: The ISD fixed on the roadside is shown in Fig. 47.7, which is the center node of ZigBee network. Once the ISD receives the data from upper server, it will transfer it to vehicles immediately through ZigBee network.

Step 3: The IRD attached to vehicles is used to receive the traffic information from an ISD.

When the ZigBee router, as is shown in Fig. 47.7, is not added, the IRD cannot receive the data by a distance of about 485 m to ORIGIN. When the ZigBee router is added, the distance lengthens to 856 m. At present, the mean speed of vehicle on urban road is 50 km/h, so the time from vehicle receiving the traffic information to reaching the center node is approximately 62 s. It is enough for drivers to choose their routes.

47.6 Conclusions

In this paper, we apply GPRS wireless communication technology and ZigBee wireless network technology to ITS and propose a novel traffic information disseminating system. We designed and implemented two kinds of hardware components, ISD and IRD. The hardware structure and software design of two devices are given respectively. At last, system test was carried out in Chegongzhuang Street, Beijing. The test results show that this scheme we proposed is feasible and practical. Drivers can obtain traffic information accurately and timely and can also have enough time to make the decision of road selection.

Acknowledgments This paper is sponsored by the National High Technology Research and Development Program of China (2011AA110303).

References

Bilgin BE, Gungor VC (2012) Performance evaluations of ZigBee in different smart grid environments. Comput Netw 56:2196–2205
EP9312 Data Sheet
Ghribi B, Logrippo L (2000) Understanding GPRS: the GSM packet radio service. Comput Netw 34:763–779
Guan JZ, Zheng CQ, Zhu XL, Liu JK, Wang YS, Qiao L, Ji YL (2008) VMS release of traffic guide information in Beijing olympics. J. Transp. Syst. Eng. Inf. Technol. 8
Ionel R, Vasiu G, Mischie S (2012) GPRS based data acquisition and analysis system with mobile phone control. Measurement 45:1462–1470
Qin MG, Wang YQ, Cui ZF, Zhu YY (2006) The design and realization of an advanced urban traffic surveillance and management system. In: Proceedings of the 6th world congress on intelligent control and automation
Shahjahan M, Nahin KM, Uddin MM, Ahsan MS, Murase K (2008) An implementation if on-line traffic information system via short message service (sms) for Bangladesh. International joint conference on neural network
W77E58 Data Sheet
Xiao ZH, Guan ZQ, Chen DJ (2008) The research and application of fm circuit based on CD4046 in the traffic information release system. International conference on computer science and information technology
ZigBee alliance. ZigBee specification. http://www.ZigBee.org

Chapter 48
Studying Electric Vehicle Batteries Consumption with Agent Based Modeling

Jinjin Fu and Xiaochun Lu

Abstract To develop electric vehicle is an approach to increasing the competition of electric vehicles industry, safeguarding energy security and boosting low carbon economy. It is well known that batteries can cause environment problem. Several problems must be solved before electric vehicle technology is applied in large scale. One of the problems is how many batteries should be equipped to meet the demand of certain electric vehicles. In this paper, we have built a batteries consumption model of electric vehicles based on agent. Simulation result shows that the optimum ratio of electric vehicles and lead-acid batteries is 1:2.93 and the optimum ratio of electric vehicles and lithium batteries is 1:1.36 as well. If electric vehicles are equipped with lithium batteries instead of lead-acid batteries, 53.4 % of batteries will be reduced.

Keywords Electric vehicles · Batteries · Environmental protection · Agent Based Modeling

48.1 Introduction

Environmental requirements and rising oil prices make electric vehicles become the new hotspot of vernicle development, because of its advantages of pollution-free, high energy efficiency and energy diversification. Increasingly more countries

J. Fu (✉) · X. Lu
School of Economics and Management, Beijing Jiaotong University, Haidian District, Beijing 100044, People's Republic of China
e-mail: 11120648@bjtu.edu.cn

X. Lu
e-mail: xclu@bjtu.edu.cn

make corresponding development plans for electric vehicles. "The development plan of energy-saving and new energy vehicles industry (2012–2020)" issued by the State Council and other standards make the development of electric vehicles irresistible. However, the current refurbished rate of battery in China is 85 %, which is far lower than the percent of 95 % in the developed countries (XuejiaoYan and Xiang Li 2007). Whether the development of electric vehicles will have the disastrous impact on the recycling of the used batteries in China and what the proper production of batteries is the most friendly to the environment are taken into account, which will receive detailed analysis below.

48.2 Literature Review

For the moment, there is very little research on used batteries of electric vehicles. Research on the batteries of electric vehicles mainly has following several aspects.

The author (Chau et al. 2011) described a new adaptive neuro-fuzzy inference system (ANFIS) model to estimate accurately the battery residual capacity (BRC) of the lithium-ion (Li-ion) battery.

Simulation of the performance of the PHEV throughout its battery lifetime (EricWooda 2011) shows that battery replacement will be neither economically incentivized nor necessary to maintain performance in PHEVs. The results have important implications for techno-economic evaluations of PHEVs which have treated battery replacement and its costs with inconsistency.

For all batteries, it remains a challenge to simultaneously meet requirements on specific energy, specific power, efficiency, cycle life, lifetime, safety and costs in the medium or even long term. Only lithium-ion batteries could possibly attain all conditions in the medium term.(Sarah J. Gerssen-Gondelach, André P.C. Faaij 2012).

48.3 Model Establishment

There are two aspects in the study, the batteries and the electric vehicles. Several batteries need to be replaced before the end of vehicle life. The uncertainty of battery life and electric vehicle life make the model more complex, it is suitable to use the agent-based model. Specific conceptual model is shown in Fig. 48.1.

48.3.1 Introduction of Any Logic Simulation Platform

The platform used in the experiments is AnyLogic 6.7.1 University. The programming language is Java. AnyLogic is the modeling and simulation tools which was introduced by the XJ Technologies. Its application areas include: control

48 Studying Electric Vehicle Batteries Consumption

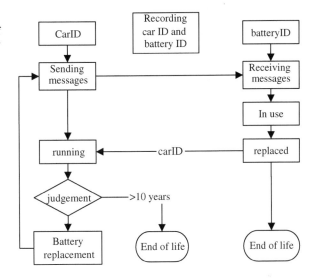

Fig. 48.1 A conceptual model of the usage process of an electric vehicle and battery

systems, traffic, dynamic systems, manufacturing, supply chain and logistics sector and so on.

State transition model of electric vehicles is shown in Fig. 48.2 and State transition model of batteries is shown in Fig. 48.3.

In Fig. 48.2, X_n stands for the state of the electric vehicle, $X_n \in \{1, 2, 3\}$. $X_n = 1$ means the electric vehicles are in the car factory, the electric vehicles are being produced. $X_n = 2$ means electric vehicles are in the state of being in use, the state of the electric vehicle is divided into the state of driving and the state of replacing the battery. $X_n = 3$ means electric vehicles are in the state of being scrapped, that is to say the electric vehicles come to the end of its life. xt_{12} stands for the time of transferring from X_1 to X_2. In the model, we set xt_{12} obeying the exponential distribution with $\lambda = 1/12, xt_{12} \sim$ Exponential (1/12). xt_{23} stand for the time of transferring from X_2 to X_3.

In Fig. 48.3, Y_n stand for the state of the batteries, $Y_n \in \{1, 2, 3\}$. $Y_n = 1$ means the batteries are in the state of waiting, it means it is an unused new battery.

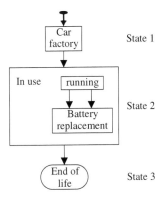

Fig. 48.2 State transition model of electric vehicles

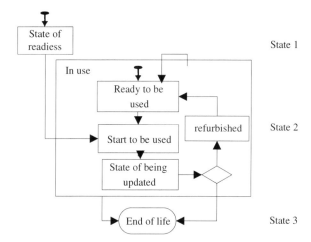

Fig. 48.3 State transition model of batteries

$Y_n = 2$ means the batteries are in use, it is a composite state, including all state changes in Fig. 48.2. $Y_n = 3$ means the batteries have been refurbished more than three times, or be discarded instead of being refurbished.

In the model, it must meet the following conditions before transferring from state Y_2 to state Y_3:

$$\begin{cases} \text{if } r|3 \quad r \text{ stand for refurbished times of the battery}, r = 0, 1, 2, 3; \\ \text{Transition probability } p_{23} = 0.15 \quad P_{23} = P(Y_n = 3|Y_n = 2) \sim \text{Binomial}(1, 0, 15) \end{cases}$$

48.3.2 Simulation Experiment

The most widely used secondary batteries with the most mature technology are the lead-acid batteries (LA), its production account for about 90 % of all the secondary batteries (Jianqiang Li 2011). Frequent "excessive blood lead" events occurred in recent years triggered a heavy hit to the lead-acid battery industry leaded by the Department of environmental protection. The lead-acid battery production has reduced, which make the lithium batteries have broad development prospect. In this paper, we set both lead-acid batteries and lithium batteries for example.

The existing literature data shows that the average life expectancy of lead-acid batteries is about 2 years (Hui Wang 2009). The average life of the electric vehicles is about 15 years. The life of the lithium battery is probably about 6 years (Mingjing Nie 2011). Domestic sales of electric vehicles in 2011 are 10,000. Table (48.1.)

48.3.3 Analysis of Simulation Results

For the life of Lithium batteries and Lead-acid batteries varies, the least batteries needed to keep electric vehicles running can be got. At the same time, the number of batteries used in electric vehicles and the number of scraped batteries can be used to study the regulation how electric vehicles consumed batteries.

48.3.3.1 Analysis of Simulation Experiment 1-Lead-Acid Battery

Through modifying the total number of lead-acid batteries, we found that the total number of lead-acid batteries is at least 29,300 to ensure the model running. Therefore, in this paper, there will be 29,300 lead-acid batteries as the basic population parameter of the model. The total number of batteries will increase by 20 and 50 % on this basis to comparatively analyze the battery consumption and scrapped batteries.

1. The number of the batteries is 29,300 in the environment

Under this condition, the curve of the number of total battery consumption, the ones in use and scrapped batteries can be shown in Fig. 48.4.

Battery life and car life are randomly distributed. In order to eliminate the uncertainty, this paper carries out 50 times simulations and make the analysis about them, the results obtained is shown in Table 48.2. M stand for the number of electric vehicles, N stand for the number of batteries production, N_1 stand for the number of battery consumption, N_2 stand for the number of battery used once, and N_3 stand for the number of batteries used twice, N_4 stand for number of scrapped batteries.

2. Comprehensive analysis

There is a comprehensive analysis of the three experiment data to study the impact of changes in battery production and get the data in Table 48.3.

The ratio of electric vehicles and the lead-acid batteries is 1:2.93.Take the change of N_3: N into account, when the 50 % increase in battery production N, and N_3: N will decrease from 0.81 to 0.54 with a change of 27 %, it is moderately sensitive with the sensitivity coefficient of 54 %.

Table 48.1 Parameter settings

Parameter	Value (lead-acid battery)	Value (lithiumbattery)
Number of electric vehicles	10,000	10,000
The life of the electric vehicle	T(13,15,17)	T(13,15,17)
Useful life of Lead-acid batteries for one time	T(1,2,3)	T(5,6,7)
The refurbished rate of used batteries	85 %	85 %

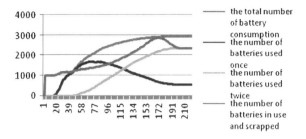

Fig. 48.4 The simulation curve of battery consumption

Table 48.2 Analysis of simulation results

Variable	Battery consumption (N_1)	Batteries used once (N_2)	Batteries used more than once (N_3)	The craped batteries (N_4)
Mean of the Samples	29,183	5,643	23,540	23,275
Sample standard deviation	34.30	141.10	142.18	339.01
Sample average error	4.85	19.96	20.11	47.94
t value	2.01	2.01	2.01	2.01
Confidence interval	9.75	40.10	40.41	96.35

Note: Simulation times is 50, confidence level is 95 %

Table 48.3 Comprehensive analysis of changes in battery production

Lead-acid battery production (N)	M:N	N1:N	N2:N	N3:N	N4:N
N = 29,300 (+0)	1:2.93	1.00	0.19	0.81	0.80
N = 35,160 (+20 %)	1:3.516	0.97	0.31	0.69	0.56
N = 43,950 (+50 %)	1:4.395	0.90	0.46	0.54	0.40

48.3.3.2 Analysis of Simulation Experiment 2-Lithium Battery

In the same way with the simulation experiment 1, there is a comprehensive analysis and we get the data in Table 48.4.

We can see that the most environmentally friendly ratio of electric cars and lithium battery is 1:1.36. Take the change of N3: N into account, when the 50 % increase in battery production N, and N3: N will decrease from 0.48 to 0.18 with a change of 30 %, it is moderately sensitive with the sensitivity coefficient of 60 %.

48.3.3.3 Realistic Effects

"The development plan of energy-saving and new energy automotive industry (2012–2020)" defines the development target of the new energy vehicles: To 2020, the cumulative production is over 5 million. The number of batteries 5 million electric vehicles consume is shown in Table 48.5.

Table 48.4 Comprehensive analysis of changes in battery production

Lead-acid battery production (N)	M:N	N_1:N	N_2:N	N_3:N	N_4:N
N = 13,500 (+0)	1:1.36	1.00	0.52	0.48	0.22
N = 16,320 (+20 %)	1:1.63	0.95	0.70	0.30	0.20
N = 20,400 (+50 %)	1:1.2.04	0.83	0.82	0.18	0.18

Table 48.5 The battery consumption of 5 million electric vehicles (unit:10,000)

	Battery consumption (N_1)	Batteries used once (N_2)	Batteries used more than twice (N_3)	The craped batteries (N_4)
Lead-acid battery	1459.64	284.15	1179.03	1168.57
Lithium battery	678.08	356.22	322.46	152.08

48.4 Conclusions

In this paper, we get the ratio of electric vehicles and lead-acid batteries is 1:2.93 and the ratio of electric vehicles and lithium batteries 1:1.36. The two scale factors have the practical significance for battery production planning. We get the following conclusions:

1. The battery production affects the total number of scrapped batteries

It can be seen from Tables 48.4 and 48.5, with the increase of battery production, the total number of battery consumption and the number of the scrapped batteries will increase. This is due to that some batteries were directly abandoned before they are fully refurbished, which is unfavorable to our environmental policy. Therefore, it is important to determine the production of the batteries according to the number of electric vehicles on the market and make plans for producing batteries.

2. The lithium battery is more environmentally friendly than lead-acid battery

If in China the cumulative production of electric vehicles in 2020 is five million, lead-acid batteries will reach 14.6 million, the lithium battery production is to reach 6.8 million, which apparently has 53.4 % reduction compared with lead-acid batteries. The ratio of the number of scrapped batteries and the number of lithium battery consumption is 0.22, which is much smaller compared with 0.80 of lead-acid batteries. However, the lithium battery technology is not mature and the price of lithium batteries is relatively higher than lead-acid batteries. Therefore, making more effort to the lithium battery research is very important to the significant popularity of electric vehicles and our energy-saving environmental protection.

3. In the present model, some data cannot be obtained temporarily were not considered due to the emerging area of electric vehicle, which needs more study.

(1) The distribution function of the life of electric vehicles and battery life need to be studied, which is assumed to obey the triangular distribution here.
(2) The annual production of electric vehicles is a dynamic process of change, which cannot be reflected in the agent-based model, how to overcome this problem of agent-based model in combination with other methods needs further study.
(3) In this paper, in order to simplify the calculation, battery in each electric vehicles is set to one, how to calculate the batteries in an electric vernicle in a more scientific and rational way needs further analysis.

Acknowledgments The article is supported by the two projects as follows:

1. Project supported by Beijing Education Commission: An optimization research on Beijing Railway Station and urban transit system by simulation.

2. Project supported by National Natural Science Foundation of China (Grant No. 71132008).

References

Chau KT, Wu KC, Chan CC (2011) A new battery capacity indicator for lithium-ion battery powered electric vehicles using adaptiveneuro-fuzzy inference system. Energy Convers Manag 45(11–12):1681–1692
Gerssen-Gondelach JS, André Faaij PC (2012) Performance of batteries for electric vehicles on short and longer term. J Power Sources 212:111–129
Li J (2011) Analysis of battery of electric vehicle. Larg Technol 9:20–21
Nie M (2011) How long is the lithium battery in the electric vehicle. Netw Glob Battery 12(1)
Wood E, Alexanderb M, Bradleya TH (2011) Investigation of battery end-of-life conditions for plug-in hybrid electric vehicles. J Power Sources 196(11):5147–5154
Wang H (2009) The causes and countermeasures of lead-acid battery's vulcanization. Power Technol 12(11):26–29
Yan X, Li X (2007) The significance of research on green energy recycled technology of used lead-acid battery and establishing its standard system. Brand and Standardization 2:15–16

Chapter 49
Low-Carbon Scenario Analysis on Urban Transport of a Metropolitan of China in 2020

Xiaofei Chen and Zijia Wang

Abstract This article discussed possible ways of implementing effective energy conservation and GHG emission reduction measures by providing: the forecasts of mid-to-long term city-wide carbon emission rate; and the analysis of potential low-carbon transport solutions. According to the characteristics of the transport system in a metropolitan in China, the comprehensive carbon emission calculation model established in this article includes road traffic and urban rail transit. Existing data were utilized with regression analysis to project the prospective traffic data in the baseline scenario at the target year of 2020 to calculate the emission amount. Four low-carbon scenarios were set in accordance with the goal of "low carbon transportation, green trip", and the effectiveness of each low-carbon scenario was evaluated by comparing them with the baseline scenario in terms of the respective GHG emission rate. The mode switching that increases the ridership of urban rail transit turned out to be the most effective outcome.

Keywords Low carbon transport · Carbon emission · Scenario analysis · Forecasting · Energy conservation and emission reduction

---

X. Chen
Department of Mechanical and Industrial Engineering, University of Toronto,
5 King's College Road, Toronto ON M5S 3G8, Canada
e-mail: xiaofei.chen@mail.utoronto.ca

Z. Wang (✉)
College of Civil Engineering, Beijing Jiaotong University, Beijing 100044, China
e-mail: hnzijia@gmail.com

49.1 Introduction

In China, the negative environmental consequences caused by GHG emissions from transport sector in metropolitan regions grow rapidly along with continuous urbanization process. At the stage of high-speed economic development, the household consumption level in a metropolitan in China rises quickly, resulting in dramatic increase in overall urban residential trips, which in return, increase the carbon emission rate in urban passenger traffic. In fact, the total GHG emissions volume stays at a very high level, despite the recent achievements from traffic and purchase restrictions imposed upon household automobile and actions that encourages the development of rail transportation. Setting up baseline and relevant policy-supported low-carbon scenarios to forecast total GHG emissions from transport industry in this metropolitan and evaluating the effectiveness of implementing certain energy conservation and emission reduction policies such as "public transit priority" are of great significance to this metropolitan and also China for coming up with a successful strategy to control the GHG emissions and to assume its commitment to international emission reduction obligations.

49.2 GHG Emissions Calculation Model of Transport Section of the Metropolitan

Firstly, the calculation model for each fuel category of urban transport was formulated to obtain the overall energy consumption. In accordance with the terminal consumption of each energy resource, the calculation method was established as follows.

For diesel fuel:

$$E_{diesel} = Q_f e_f + N_{d-bus} S_{d-bus} e_{d-bus} \tag{49.1}$$

where E_{diesel} refers to consumption of diesel fuel, Q_f for ton kilometre travelled per year (ton·km), e_f for fuel economy (L/(ton·km)) (Jia et al. 2009), N_{d-bus}, S_{d-bus}, e_{d-bus} for vehicle population, kilometres travelled per year (km), and fuel economy (L/km) of diesel buses, respectively.

For gasoline:

$$E_{gasoline} = \sum_{i=1}^{3} (N_i S_i e_i) \tag{49.2}$$

where $E_{gasoline}$ refers to gasoline consumption, i for relevant vehicle category, including taxis, private cars and cars owned by enterprises and public institutions, N_i for vehicle population of vehicle category i, S_i for km travelled per year of vehicle category i, e_i for fuel economy of vehicle category i (L/km).

For compressed natural gas:

$$E_{gas} = N_{g-bus} S_{g-bus} e_{g-bus} \tag{49.3}$$

where $N_{g-bus}, S_{g-bus}, e_{g-bus}$ refers to vehicle the population, kilometre travelled per year (km), and fuel economy (kg/km) of compressed natural gas driven buses, respectively.

While for electricity consumption:

$$E_e = N_{e-bus} S_{e-bus} e_{e-bus} + E_{illu} + \sum_{j=1}^{n} E_{j-vehicle} + \sum_{j=1}^{n} \sum_{k=1}^{n} E_{jk-station} \tag{49.4}$$

where $N_{e-bus}, S_{e-bus}, e_{e-bus}$ denotes the vehicle population, kilometre travelled per year (km), and fuel economy (kg/km) of electricity-based buses, respectively; E_{illu} for electricity consumed by road illumination system, which is recorded by power department, and $E_{j-vehicle}$, $E_{jk-station}$ refer to the vehicle energy consumption of rail line j and the station energy consumption of station k in rail line j, respectively.

Using the energy consumption obtained through the above model and the CO_2 emission factor of each fuel category in ICPP, in combination with the proportion of coal electricity out of overall electricity supply and CO_2 emission factor for electricity generation in the grid, the overall GHG emissions of urban transport of this metropolitan can be calculated with the following model:

$$C = \sum_{i=1}^{3} (E_i \times EF_i) + (E_e \times \zeta \times \psi) \tag{49.5}$$

where C denotes the overall GHG emissions of urban transport per year; E_i stands for consumption of fuel category i, including diesel fuel, gasoline, compressed natural gas (L or kg); EF_i is the emission factor for fuel category i (kgCO$_2$/L or kg CO$_2$/kg); E_e refers to electricity consumption of urban transport of this metropolitan, including consumption of electricity-based buses, road illumination system and urban rail transit system (k·Wh); ζ denotes the sharing of coal electricity out of overall electricity generated in the grid, which is 81.81 % (China electricity council 2011); ψ is CO_2 emission factor for coal electricity during generation stage in the grid (Ma 2002) (CO$_2$ kg/k·Wh).

49.3 Scenario Analyses on GHG Emissions of Urban Transport of the Metropolitan in 2020

49.3.1 Baseline Scenario

In terms of the projected calculation data obtained above, with the rail transit network planning details and passenger volume forecasting, the GHG emissions of baseline scenario for the transport of this metropolitan in target year 2020 was projected, which is 30.085 million tCO_2.

49.3.2 Low Carbon Scenarios

(1) Scenario 1: More strict control on vehicle population

The vehicle population of non-operating cars will reach 5.96 million in 2020 under the baseline scenario. It is necessary to hold more strictly control on vehicle population's growth so as to make sure the population of private cars and buses in is controlled at 5.5 million in 2020. In that case the GHG emissions of transport will be 28.203 million tCO_2, with emission reduction of 1.882 million tCO_2, or 6.26 % compared to the baseline scenario.

(2) Scenario 2: Decrease average trip distance

Optimization of urban layout will result in gradual decrease in average residential trip distance. In the baseline scenario, the average trip distance of non-operating cars is 10.78 km; and that of rail transit is 16.35 km. Relevant low-carbon scenario is set as: average trip distance of private car is 8 km, whereas the average trip distance of rail transit is 12 km. Under this situation, the GHG emissions of transport will be 25.648 million tCO_2, with emission reduction of 4.437 million tCO_2 or 14.75 compared to the baseline scenario.

(3) Scenario 3: Increase the ridership of public transport

In the baseline scenario, the structure of the traffic of this metropolitan is: the ridership of buses is 28.9 %, urban rail transit accounts for 10.0 %, taxis occupies 7.1 %, and the ridership of non-operating cars is up to 34.0 %.

Under the guidance of the "public transit priority" policy, with the rapid development of rail transit network in this metropolitan, the ridership of public transport modes, especially that of rail transit, will rise sharply. Accordingly, the low-carbon scenario is set as following: In 2020, the ridership of public transportation is up to 50 %. Specifically, the ridership of buses is 20 %, urban rail transit 30 %, taxi 5 %, and bicycles 25 %. In contrast, use of private cars decreases to 15 %, while the rest share of trip is assumed by walking. In terms of the scenario setting, GHG emissions in 2020 will be 23.426 million tCO_2, achieving a decrease of 6.659 million tCO_2 (Table 49.1).

Table 49.1 Comparison of scenarios 1, 2 and 3

Scenarios	Baseline scenario	Low-carbon scenario 1	Low-carbon scenario 2	Low-carbon scenario 3
GHG emissions (million tCO_2)	30.085	28.203	25.648	23.426
Emission reduction (million tCO_2)	/	1.882	4.437	6.659
Proportion (%)	/	6.26	14.75	22.13

49.4 Conclusion

The article completed the following researches and reached the corresponding conclusions:

(1) Under the current developing trend in policy environment and technical specifications, the total projected GHG (CO_2) emissions from transport sector at 2020 in the researched metropolitan of China will reach 30.085 million tCO_2; private-vehicle is the major contributor among all transport modes at 16.89 million tCO_2.
(2) As indicated by the analysis of low carbon scenarios, limiting the growth in private-vehicle ownership, reducing the frequency of mid-to-long range travel and the average trip distance, and prompting the public transit oriented policies are all possible solutions to reduce carbon emission. The most effective practice involves a shift in public travel behaviour.

References

Jia S, Peng H, Liu S (2009) Review of transportation and energy consumption related research. J Transp Syst Eng Inf Technol 9(3):6–16

China electricity council. The percentage of thermal power of total generating capacity. Jan 2011 http://tj.cec.org.cn/

Ma Z (2002) Evaluation studies of several major energy greenhouse gas emission factors. China Institute of Atomic Energy, Beijing

Chapter 50
Impact Study of Carbon Trading Market to Highway Freight Company in China

Li Chen, Boyu Zhang, Hanping Hou and Alfred Taudes

Abstract This paper studied the influence of the introduction of carbon trading system to small size highway fright transport companies in China based on cost-benefit analysis. Considering a carbon trading market consists of large number of identically small size companies, where each company is distributed the same amount of carbon credits for free and could sell or buy extra credits from the market. If the market is perfectly competitive, it will reduce the profit of the companies in the long run. The government could regulate the profit indirectly by free carbon credits and penalty for over emission.

Keywords Carbon trading · Carbon market · Highway freight transport

50.1 Introduction

Transport industry is one of the most important industries for developing Low Carbon Economy because of its high carbon intensity. Since 2009 China started to pay great attention on the development of low carbon transport. In February 2011

L. Chen · H. Hou
School of Economic and Management, Beijing Jiaotong University, Shang Yuan Cun 3, Beijing 100044, China

B. Zhang (✉)
School of Mathematical Sciences, Beijing Normal University, XinJieKouWai Street 19, Beijing 100875, China
e-mail: zhangboyu5507@gmail.com

A. Taudes
Institute for Production Management, Vienna University of Economics and Business, Augasse 2-6 1090 Vienna, Austria

Transport Ministry of China published "the guiding opinions of establishing low carbon transportation system" report, which stated the targets and assignment of low carbon development for transport industry. According to the report, the targets of CO_2 emission intensity reduction of road transport during 12th Five-Year are energy consumption per transport turnover of commercial vehicles decreased by 10 % by 2015 compared to 2005, of which passenger vehicles decreased by 6 % and cargo vehicles decreased by 12 %.

In order to achieve the goals of CO_2 emission reduction of 12th Five-Year's plan, several low carbon policies have been taken into effect. For example, fuel economy standards, motor vehicle license auction, motor vehicle limit line measures etc. However, there are no market-based policies in China.

In such context, this paper will conduct a detailed study in highway freight industry and demonstrates different degrees of carbon trading system impact to a single enterprise. The paper is organized as follows: Sect. 50.2 introduces the carbon trading schemes; Sect. 50.3 presents the effects of carbon trading scheme on road freight transport company, and demonstrates the company's optimal strategy under different situation; Sect. 50.4 analyzes the equilibrium price in the carbon trading market; Sect. 50.5 offers suggestion of regulation measurements to the government; and Sect. 50.6 concludes.

50.2 The Model

50.2.1 Carbon Trading Schemes

In accordance with many recent carbon market scheme proposals in China (Zhuang 2006), (Fu 2009), (Lewis 2010), this study follows the carbon trading regulations of EU ETS. According to EU ETS, carbon trading schemes can be taken out through two types. European Union Allowances (EUAs) are permits (called carbon credits) created and distributed under cap and trading (cap-and-trade) scheme which operates under the Kyoto Protocol. Companies receive carbon credits from their national allowance plans and the total quantity of credits is administered by EU state governments. Another type of carbon trading schemes is project based trading which is generated from projects that reduce greenhouse gas (GHG) emissions compared to a no-project scenario. Clean Development Mechanism (CDM) and Joint Implementation (JI) initiative are the two project based trading systems of EU ETS. CDM allows industrialized countries with a GHG commitment to invest in emission reducing projects in developing countries, such as China, India and Brazil. JI credits are from the projects which are cooperated within and implement in industrialized countries.

By 2009, Chinese road freight market comprised of more than 42,92,000 transport companies but lack of scale. Each company had only 1.7 trucks on average and 41 % of them had exactly one truck (Chen et al. 2011). In order to integrate with Chinese realities and keep the model simple, we only consider the cap-and-trade scheme where subjects are small size companies, but do not include project based trading schemes.

Assume that the road freight market consists of identically small size transport companies, where each company is distributed the same amount of carbon credits for free and could sell or buy extra credits from the carbon trading market. If a company emits CO_2 that exceeds what is permitted by its credits, then a penalty will be charged for the over emission.

50.2.2 Road Freight Market

Considering a highway freight market consists of large numbers of identical small size companies and assuming that these companies satisfy the following four assumptions:

(i) Perfect information: companies have perfect information for both the road freight market and the carbon trading market, which means freight rate and carbon credit price are known to all companies.
(ii) Perfect rationality: companies aim to maximize their profits, and they can make profit from road freight as well as carbon credit trading.
(iii) Identical service: each company has the truck with same load capacity T-ton and provides homogeneous service.
(iv) Identical cost: companies have the same operating cost C Yuan per ton-km.

From (i) and (ii), in a perfect information market which consists of many rational companies, homogeneous products have a unique equilibrium price (Nicholson and Snyder 2008). In the road freight market, both freight rate and operating cost are approximately linearly dependent on the distance travelled and the gross weight of the truck, as well as the CO_2 emissions (Forkenbrock 1999). Suppose that the equilibrium price of the road freight market is B Yuan per ton-km, and a truck emits E ton CO_2 per ton-km. Therefore, each company can gain net profit $p = T \times (B - C)$ Yuan per km, and produces $T \times E$ ton CO_2 emissions. Non-transportation emissions (e.g., from buildings or office work) are excluded in this paper since the percentage of non-transportation emission in total emissions of a company is very small (McKinnon and Piecyk 2009). In sum, a road freight market could be characterized by parameters p, T and E.

Following the four assumptions, in the case without carbon trading system, company strategy can be simply described by the monthly distance travelled d and net profit p. For instance, monthly profit of a company using strategy d is

$$P(d) = dp \qquad (50.1)$$

Clearly, $P(d)$ is monotonically increasing in distance. Suppose a truck can run at most D kilometers per month under the limits of driving hours and speed. Therefore, a rational company ought to run the maximum distance in order to achieve the optimal payoff $P(D) = Dp$. The CO_2 emission is DTE ton.

50.2.3 Carbon Trading System

Under a carbon trading system, each company must have carbon credits for CO_2 emission, where the amount of credits specifies the maximum weight of CO_2 emissions that the company is allowed to emit per month. Since the government intends to reduce the current emission level, supposing that each company is distributed N carbon credits for free, N is less than DTE. In addition to the credits given by the government, companies can buy or sell extra credits through the carbon trading market. z denotes the price of a unit carbon credit which allows company emits 1 ton CO_2 per month. If a company produces CO_2 emission more than permitted, a penalty will be charged at rate of y Yuan per ton for extra emission. Therefore, the state of a carbon trading system can be described by a vector of three variables (N, y, z), where N and y are set by the government and z is regulated by the market.

In order to calculate the incremental cost caused by the carbon trading system, we further make two assumptions (Hourcade et al. 2007).

(v) No transaction fees means that transaction fees for carbon credits are negligible.
(vi) Invariant road freight market means that parameters p, T and E are constant and do not change with the carbon trading system (N, y, z).

Assumption (vi) implies that small size companies are unable to reduce the CO_2 emissions or operating costs by improving their technologies or trucks, and can not affect the freight rate of the road freight market, and are not allowed to pass the incremental cost to consumers. Therefore, all carbon market related incremental cost is the burden to companies and reflected by changes of their profits.

With the carbon trading system, the strategy of a company could be defined by (d, n), where d represents the monthly distance travelled and n is the amount of carbon credit transactions, where $n > 0$ means that the company buys n carbon credits from the market and $n < 0$ means selling the carbon credit. Therefore, a company using strategy (d, n) is allowed to emits $N + n$ ton CO_2 per month, and actually produces dTE ton CO_2 emission. A rational company will not keep the extra carbon credits since selling the rest credits can increase profits. Hence, variables satisfy $0 \leq N + n \leq dTE \leq DTE$.

For given road freight market (p, T, E), the impacts of carbon trading system to a small size transport company could be analyzed through a payoff function of the company strategy (d, n), the government policy (N, y) and the carbon credit price z

$$P((d,n), (N,y), z) = dp - nz - (dTE - N - n)y \quad (50.2)$$

Where dp is the profit from road freight transport, nz is the cost (income) of carbon credit transaction and $(dTE-N-n)y$ is the penalty for over emission.

50.3 Strategies of Companies

In a perfectly competitive carbon credits market which includes large numbers of companies and carbon credits, the price z will not be affected by the decision of a single company. Therefore, we rewrite Eq. (50.2) as

$$P_{(N,y,z)}(d,n) = d(p - TEy) + n(y - z) + Ny \qquad (50.3)$$

Where $E_{(N, y, z)}(d, n)$ denotes the payoff to a company using strategy (d, n) in a carbon trading system (N, y, z).

Denote the strategy that leads to the optimal payoff by (d^*, n^*), we call it the best response to state (N, y, z). Notice that $P_{(N, y, z)}(d, n)$ is a linear function of d and n, for any given (N, y, z), the best response strategy is unique and the values (d^*, n^*) are decided by the sign of $(p - TEy)$ and $(y - z)$. According to the order of p/TE, y and z, there are three different cases.

(a). If $y > p/TE$ and $z > p/TE$, the best response strategy is $(d^*, n^*) = (0, -N)$, which means the company stops transport operating and sells all its credits. In this case, the income is entirely from the carbon transaction, and the company gets profit $P_{(N, y, z)}(0, -N) = Nz$.

(b). If $y > z$ and $p/TE > z$, the best response strategy is $(d^*, n^*) = (D, DTE - N)$, which means the company maximizes the distance and buys carbon credits for its extra emission. In this case, the payoff is $P_{(N, y, z)}(D, DTE - N) = Dp - (DTE - N)z$, which is always less than the optimal profit without the trading system, Dp.

(c). If $p/TE > y$ and $z > y$, the best response strategy is $(d^*, n^*) = (D, -N)$, which means the company maximizes the distance but sells all the credits. In this case, the company pays the penalty for its emission but earns money from both the freight transport and carbon trading. The profit is $P_{(N, y, z)}(D, -N) = Dp - DTEy + Nz$.

50.4 Equilibrium Credit Price

In the carbon trading market, transport companies are both buyers and sellers and the price of a carbon credit is decided by demand and supply. Since these companies are identical, they have the same best response strategy. In cases (a) and (c) where the price z is larger than p/TE or y, the companies tend to sell their credits which result in the decrease of price. On the other hand, in case (b) where z is less than both p/TE and y, the companies are willing to buy credits which will increase the price. Equilibrium price of the carbon trading market is the price when carbon credit demand equals to the supply. Therefore, the unique equilibrium is $z^* = \min\{p/C, y\}$, which means the price of unit carbon credit equals to the smaller one of the transport profit and the penalty.

At this equilibrium, all companies have the same payoff

$$P((d*, n*), (N, y), z*) = Dp - DTEy + Nz* \quad (50.4)$$

and no company can get higher by changing its strategy while the others keep theirs unchanged. Notice that parameters in Eq. (50.4) satisfy $N < DTE$ and $z* \leq y$, we have $P((d*, n*), (N, y), z*) < Dp$. As a result, although a company may gain higher payoff from carbon trading in some situations, such as in cases (a) and (c) in Sect. 50.3, the long run profit is always reduced by the system.

50.5 Government Regulation

Equation (50.4) implies that the benefit of the transport companies at the equilibrium price is entirely decided by the government policy (N, y). For convenience, denote $P((d*, n*), (N, y), z*)$ by $P*(N, y)$. If $y < p/TE$, $z* = y$ and the payoff function is written as $P*(N, y) = Dp - (DTE - N)y$. Notice that $DTE > N$, $P*(N, y)$ is linearly decreasing in the penalty y and linearly increasing in the amount of free carbon credits N. On the other hand, if $y > p/TE$, $z* = p/TE$ and the payoff function is written as $P*(N, y) = Np/TE$. In this case, the payoff is independent of y but linearly increasing in N.

Therefore, the government can regulate the carbon credit price by changing the penalty y in interval $[0, p/TE]$. However, the regulation carbon credit price is not effective when $y > p/TE$. Because then the penalty is higher, and the price will stabilize at the transport profit per unit CO_2 emission. In both cases, the company's average profit can be improved by increasing the free carbon credits.

50.6 Conclusion

Based on the analysis in Sects. 50.3–50.5, this paper found that when the transport profit per unit carbon emission is less than the penalty of extra emission per unit and the price of unit carbon credit, the optimal strategy of the transport company will be to stop road transport operations and sell all the carbon credits. In contrast, when the transport profit per unit carbon emission is higher, the optimal strategy is to maximize the transport distance, and sell the distributed carbon emission credits if the penalty of extra emission per unit is less than the price of unit carbon credit,

If the carbon trading market is a perfectly competitive market, the carbon trading system will always reduce profit of small size transport company in the long-term. It's because when the supply meets the demand in the carbon market, the market price of carbon credit $z*$ will tend to stabilize to the equilibrium $z* = \min\{p/TE, y\}$. The profit of highway freight company in the market is equal to the previous profit minus the cost of carbon trading, therefore, the carbon trading system will always reduce long-term profit of small transport companies.

The government can regulate the long-term profit of the small highway freight companies through free carbon emission credit and carbon emission fines. On the one hand, the government can improve the company's average profit by the increase of free carbon emission credits. On the other hand, when the extra emission penalty is less than the profit of the transportation per unit carbon emissions, reducing the punishment can also increase company's total profit. However, when the penalty is higher than the net transport profit, the second regulation method will no longer be effective.

Acknowledgments This study was supported by the Planning Fund Program for Social Science in China, Ministry of Education of China (Grant No.10YJA630059), and the Nature Science Fun project "Integration of Logistics Resources and Scheduling Optimization" (Grant No. 71132008).

References

Chen L, Taudes A, Wang C, Hou H (2011) In : Proceedings of Forum on Integrated andSustainable Transportation System (FISTS), IEEE pp. 344–350
Forkenbrock DJ (1999) External costs of intercity truck freight transportation. Transp Res Part A: Policy Pract 33:505–526
Fu L (2009) On legislation of EU emissions trading scheme. Areal Res Dev 1:124–128
Hourcade JC, DeMailly D Neuhoff K, Sato M (2007) Differentiation and dynamics of EU ETS industrial competitiveness impacts. Clim Strateg Rep Climate Strategies, London
Lewis JI (2010) The evolving role of carbon finance in promoting renewable energy development in China. Energy Policy 38:2875–2886
McKinnon A, Piecyk M (2009) Measurement of CO_2 emissions from road freight transport: a review of UK experience. Energy Policy 37:3733–3742
Nicholson W, Snyder C (2008) Microeconomic theory: basic principles and extensions, South-Western Publishing, Nashville
Zhuang G (2006) The emissions trading scheme of the EU and its implication to China. Chin J Eur Stud 3:66–68

Chapter 51
The Importance and Construction Measures of Chinese Low-Carbon Transportation System

Xinyu Wang and Yurong Gong

Abstract With the rapid development of transportation industry and China's economy, the construction and development of low-carbon transportation system is an important support to the development of low-carbon economy and sustainable economy in China, and is the only way to realize the low-carbon and sustainable development of the transportation of the transportation industry. Based on the existing problems in the transportation system, the article analyses the importance of developing the low-carbon transportation system, and propose measures of establishing the low-carbon transportation system.

Keywords Low-carbon transportation system · Energy · Development

51.1 The Importance of Chinese Low-Carbon Transportation System

Transportation industry is the second large oil consumption industry, and China's low carbon energy saving key industries. In 2009, China had become the world's largest car producer. The automobile purchasing and consumption quantity is ranked first in the world. Rapid growth in automobile production and consumption stimulate and promote the huge domestic market demand potential, increase the

X. Wang (✉) · Y. Gong
School of Economic and Management, Beijing Jiaotong University,
Beijing 100044, P.R.China
e-mail: 11120627@bjtu.edu.cn

Y. Gong
e-mail: yrgong@bjtu.edu.cn

fossil resources consumption and utilization. With the development of automobile industry, transportation energy consumption increases. At present it has already accounted thirty percent of the total energy demand Jun (2010). Therefore, construction and development of low-carbon transportation system can make the transportation industry to minimize the energy consumption and energy saved, emission reduced and scientific development, also it is the objective demand and inevitable choice for promoting the transportation industry's sustainable development.

51.1.1 The Development and Construction of Low-Carbon Transportation System can Save Energy and Reduce Energy Consumption

The transportation industry's energy consumption is high, energy consumption in national energy consumption accounted for a high proportion. With the development of economy, the traffic transportation industry energy consumption ability appears an upward trend. During the period of the 11th Five-Year Plan, the transportation industry energy consumption growth rate is higher than that of the whole society energy consumption growth rate all the time, except the year of 2006 and 2007 Zhou (2010). Since 2010, the growth rate started to get larger. In this context, development and construction of low carbon transportation system, has a positive significance, no matter for the transportation industry or for the macro economy situation.

51.1.2 The Development of Low-Carbon Transportation System is Our Country's Active Response to Global Climate Change

Vehicle exhaust is an important factor, leading to air pollution and climate change. At present, as clean energy is still not been large-scale promoted, diesel oil and gasoline is still the main fuel for transportation. For example, automobile exhaust fumes will contain much harmful substances, its composition is very complicated, more than 100 species, including carbon monoxide, hydrocarbons, nitrogen oxides, suspended solid particles and so on. According to statistics, the harmful gas discharged by a car a year is 3 times greater than the car's weight. According to the department of transportation prediction, by 2020 the number of China's private cars will exceed 100 million and exhaust quantity will be very great. Vehicle exhaust emission has become one of the main sources of pollution of China's major city. Exhaust pollution caused by vehicle increasingly affects the life of people.

Now, China's various types of vehicle average fuel consumption per 100 km is larger than the developed country over 20 %. Besides, China's per capita energy amount is very low. However, traffic transport industry's development needs energy support. Effective conservation and rational using of energy is not only related to the sustainable development of transportation industry, but also related to the safety of energy resource in China. Therefore, to reduce transportation energy consumption, reduce environmental pollution, build and develop low carbon-transportation system has become the primary task faced by China's transportation industry's development.

51.1.3 Development and Construction of Low-Carbon Transportation System is Transportation Industry's Inevitable Choice to Make Sustainable Development Come True

Since 2000, China's transportation industry gets quick development. Electric railway's length grows from 14,900 km in 2,000 to 36,200 km in 2010 and freeway's length develops from 16,300 to 74,100 km. Waterway and the pipeline transportation have also been developed rapidly. But, the sustainable development ability of the transportation industry is faced with grim challenge. According to the statistics, China's highway, waterway transportation energy consumption accounts for over 1/3 of the total national petroleum consumption, it has already become the main source of green gases and air pollutions. Transportation energy utilization rate is obviously lower compared with developed countries in the world. With the investment that the state gives to the transportation industry increases and a variety of modes of transportation develops, energy consumption will increase gradually. To accelerate the construction of low-carbon transportation system is in urgent need.

51.2 The Construction Measures of Chinese Low-Carbon Transportation System

51.2.1 To Adjustment the Transportation Tools Combination and Energy Consumption Structure and to Develop the Low-Carbon or Carbon Free Transportation Tools

Transportation tools are the protagonist to realize the construction of low carbon transportation mode, we should construct the inland waterway infrastructure, including water, groundwater, rivers and lakes, improve the proportion waterway transport accounted in an integrated transport, produce waterborne energy-saving emission reduction effect Dai (2009). We should pay attention to the seamless joint between water, land, air transportation. On the combination of transportation, we should develop the low carbon or carbon free transportation tools, including city subway, light rail, bicycle, electric railways and other low carbon of carbon free public transportation tools. At the same time, we also should adjust the energy consumption structure in the transportation system, change the petrochemical energy utilization structure gradually to realize low carbon clean energy driving of the transportation system Table 51.1.

51.2.2 To Strengthen Energy-Saving Emission Reduction Research and Develop Traffic Low Carbon Technology

The low carbon technology in the transportation area refers to the new technology that can efficiently control greenhouse gas emissions existing in various links of transportation. The low carbon technology can be divided into two types: one is the carbon reduction technology, refers to energy saving and emission reduction technology, such as the clean and efficient utilization of coal, oil and gas resources and the coalbed gas exploration and development, focusing on improving energy efficiency. The second class is carbon removing technology, typical is carbon capturing and storage, carbon dioxide recovery and utilization of polymerization.

Table 51.1 Transportation line odometer of five transportation ways

	Railway		Road		Waterway	Airport		Pipeline
	Total mileage	Electric mileage	Total mileage	Highway	Inland waterway	Total mileage	International line	Oil and gas mileage
2000	6.87	1.49	140.27	1.63	11.93	150.29	50.84	2.47
2001	7.01	1.69	169.80	1.94	12.15	155.36	51.69	2.76
2002	7.19	1.74	176.52	2.51	12.16	163.77	57.45	2.98
2003	7.30	1.81	180.98	2.97	12.40	174.95	71.53	3.26
2004	7.44	1.86	187.07	3.43	12.33	204.94	89.42	3.82
2005	7.54	1.94	334.52	4.10	12.33	199.85	85.59	4.40
2006	7.71	2.34	345.70	4.53	12.34	211.35	96.62	4.81
2007	7.80	2.40	358.37	5.39	12.35	234.30	104.7	5.45
2008	7.97	2.50	373.02	6.03	12.28	246.18	112.0	5.83
2009	8.55	3.02	386.08	6.51	12.37	234.51	91.99	6.91
2010	9.10	3.62	400.82	7.41	12.42	276.5	107.0	8.5

Unit ten thousand kilometers

51.2.3 To Develop Bicycle and Bicycle Pedestrian System Actively, and Improve the Bicycle Service Industry's Level

Firstly, we should establish dedicated bicycle road system, improve the pedestrian system, making the motor vehicle and non-motor vehicle shunting to provide a safe, comfortable, efficient bicycle traffic environment.

Secondly, we should establish and improve bicycle rental and service network. For example, in Denmark Odense bicycle service industry is very developed, cyclists enjoy numerous "convenience", Bremen 42 bicycle club sites are densely covered high streets and back lanes, replacing 1,000 private cars. Improving bicycle rental and service network is the guarantee of low carbon (zero carbon) traffic or the foundation of the realization of public transportation modes Xiong (2010).

51.3 Conclusion

The development of transportation industry needs the support of energy, effective conservation and rational using of energy, is not only related to the sustainable development of transportation industry, but also related to the safety of energy resource in China. Transformation the transport development mode and development of low power consumption mode of transport has already become the key and main content of controlling the deterioration of the environment, alleviating zoology pressure, building a resource-saving, environment-friendly society, stimulating person and natural's harmonious development.

References

Dai Y (2009) The necessity and governance model of the development of low carbon city in China [J]. China population. Resources and environment 19(3):12–17

Jun Q (2010) Low carbon traffic: The new route of economic development [N]. First financial daily (10)

Xiong Y (2010) The development of green traffic and promotation low carbon transition[J]. The construction and management of traffic (4):30–31

Zhou X (2010) The analysis of the current situation of transportation energy consumption and future trend[J]. Chinese and foreign energy (15):9–17

Chapter 52
Planning Model of Optimal Modal-Mix in Intercity Passenger Transportation

Makoto Okumura, Huseyin Tirtom and Hiromichi Yamaguchi

Abstract Environmentally sustainable transportation becomes an important issue as well for intercity passenger transportation, where modal shifting from energy consuming airline and bus service to energy efficient high speed railway is the most feasible measure. But due to the less flexibility and fixed to locations of railway improvements, strategic redistribution of network-wide demand onto the improving rail lines is required (Okumura and Tsukal 2007). This paper presents an optimal modal-mix planning model in intercity passenger transportation, which aims to design a modal mix network of least CO_2 emissions and less total travel time, as well as less intermediate transfer cost, considering feasibility and economical sustainability of the service frequency. The proposed model is formulated as a mixed integer linear programming model, which can be numerically solved by general solver programs.

Keywords Modal-mix · Intercity transportation · Network design · Sustainability

---

M. Okumura (✉)
International Research Institute of Disaster Science, Tohoku University, Room 152, RIEC No.2 Building, 2-1-1 Katahira, Aoba-ku, Sendai 980-8577, Japan
e-mail: mokmr@m.tohoku.ac.jp

H. Tirtom · H. Yamaguchi
Department of Civil Engineering, Graduate School of Engineering, Tohoku University, Room 152, RIEC No.2 Building, 2-1-1 Katahira, Aoba-ku, Sendai 980-8577, Japan
e-mail: tirtom@cneas.tohoku.ac.jp

H. Yamaguchi
e-mail: h-ymgc@cneas.tohoku.ac.jp

52.1 Introduction

Environmentally sustainable transportation concept was first proposed by OECD as urban transportation and urban planning context, but recently discussion were going on intercity passenger transportation field as well, for example EU's idea of high speed railway service substitution of shorter feeder airline service. Considering large difference of energy intensity and unit CO_2 emissions, we can say that modal shifting from energy consuming airline and bus service to energy efficient high speed railway is the most feasible measure (Meng and Wang 2011).

We want to check such possibility of strategic redistribution of network-wide demand onto the improved rail lines, and resulted reduction of energy-use and CO_2 emissions. This paper presents an optimal modal-mix planning model in intercity passenger transportation, which aims to design a modal mix network of least CO_2 emissions and less total travel time, as well as less intermediate transfer cost, considering feasibility and economical sustainability of the service frequency.

52.2 Outline of Planning Problem

As the objective of network design problem, total travel time, total generalized cost, asset cost, as well as total CO_2 emissions can be considered (Andersen et al. 2009). In order to formulate a multi-criteria optimization problem, a single objective function is often synthesized by weight parameters for each single component objective (Balakrishnan et al. 1997). Recently, minimax problem which minimize the largest unsatisfactory level in several objective component was formulated into linear minimization function (Chang et al. 2000).

This paper considers the problem of finding multi-modal route traffic flow shares for given OD traffics and necessary rail, bus and air frequencies on each link which minimize total CO_2 emissions. In order to avoid too long detouring of passengers and too many transfers between different modes, we consider total travel time and total transfer cost of passengers as other objective components to be minimized. In order to secure the feasibility and economical sustainability of the service frequency on each link, we set the frequency providing enough capacity for the assigned traffic flow on the link. Furthermore, the assigned passenger numbers on that link must be larger than the required passenger number to sustain the frequency level.

52.3 Model Formulation

52.3.1 Variables and Parameters

In our network design, each city is represented by a node n ($n \in N$) and connecting arcs between nodes i, j through different modes m ($m \in M$) are indicated by

$(i,j) \times m \in A$. In order to express transit connection between modes explicitly, each node is divided into arrival node by mode m as $n_- \times m$ and departure node by mode m' as $n_+ \times m'$, and transit arc between them is indicated by $(m,m') \times n$. Also, amount of OD traffic between zones $(k,l) \in K \times K$ is given by T_{kl}. Endogenous and exogenous variables are explained in Table 52.1.

52.3.2 Objective Function and Minimax Problem

In this model, we pick up total passenger travel time P, total passenger transit cost Q and total CO_2 emissions associated with transport operations V as the objective components to be minimized:

$$P = \sum_i \sum_j \sum_m t_{ij}^m \sum_k X_{ij}^{km} \qquad (52.1)$$

$$Q = \sum_n \sum_m \sum_{m'} \tau_n^{mm'} \sum_k Y_n^{kmm'} \qquad (52.2)$$

$$V = \sum_i \sum_j \sum_m c_{ij}^m F_{ij}^m. \qquad (52.3)$$

These 3 objective components are not suitable for integration because their scales are different. Therefore, we set new p, q, v values below to be scaled down between 0 and 1 by using ideal values P^*, Q^*, V^* and evaluation values P_0, Q_0, V_0.

$$p = \frac{P - P^*}{P_0 - P^*}, q = \frac{Q - Q^*}{Q_0 - Q^*}, v = \frac{V - V^*}{V_0 - V^*} \qquad (52.4)$$

Here, in order to optimize all of three objectives, we consider minimax problem by the introduction of λ representing the most inferior objective, and sufficiently small positive constant ε, as follows:

Table 52.1 Endogenous and exogenous variables

Variable	Explanation
X_{ij}^{km}	Traffic amount on an arc originated from node k by mode m
$Y_n^{kmm'}$	Amount of transit passengers from mode m to m' at node n
B_k^m	OD trips originated from node k using mode m
A_n^{km}	OD trips between k and n using mode m
Z_{ij}^m, F_{ij}^m	Existence of service and frequency on an arc for mode m
t_{ij}^m	Travel time on an arc for mode m
$\tau_n^{mm'}$	Transit cost between modes m and m'
h^m, g^m	Seat capacity and max. operable frequency of mode m
c_{ij}^m	CO_2 emissions per one service/flight on an arc
d_{ij}^m and e_{ij}^m	Fixed and variable cost of maintaining service on an arc

$$\min_{X,Y,B,A,Z,F} \lambda + \varepsilon(p+q+v),\ p \leq \lambda,\ q \leq \lambda,\ v \leq \lambda \qquad (52.5)$$

52.3.3 Constraints

First, we describe the conditions for the preservation of traffic amount. Regarding to the arriving traffic at each node n, the following two equations are satisfied:

$$\sum_{i \in N^-(n)} X_{in}^{km} = A_n^{km} + \sum_{m' \in M} Y_n^{kmm'} \ \forall n \in N, \forall k \in K, \forall m \in M \qquad (52.6)$$

$$\sum_m A_n^{km} = T_{kn} \ \forall n \in N, \forall k \in K \qquad (52.7)$$

Similarly, regarding the passengers departing from each node n, following two equations are satisfied:

$$B_n^m + \sum_{m' \in M} Y_n^{km'm} = \sum_{j \in N^+(n)} X_{nj}^{km} \ \forall n \in N, \forall k \in K, \forall m \in M \qquad (52.8)$$

$$\sum_{l \in K} T_{nl} = \sum_{m \in M} B_n^m \ \forall n \in K \qquad (52.9)$$

Next, the constraints about the frequency set up will be described by Eq. (52.10)–(52.13).

$$F_{ij}^m \leq g^m Z_{ij}^m \ \forall (i,j) \times m \in A \qquad (52.10)$$

$$\sum_{i \in N^-(n)} F_{in}^m = \sum_{j \in N^+(n)} F_{nj}^m \ \forall n \in N, \forall m \in M \qquad (52.11)$$

$$\sum_k X_{ij}^{km} \leq h^m F_{ij}^m \ \forall (i,j) \times m \in A \qquad (52.12)$$

$$\sum_k X_{ij}^{km} \geq d_{ij}^m Z_{ij}^m + e_{ij}^m F_{ij}^m \ \forall (i,j) \times m \in A \qquad (52.13)$$

Finally, followings are added as the domain of variables.

$$X_{ij}^{km} \geq 0,\ Y_n^{kmm'} \geq 0,\ B_k^m \geq 0,\ A_n^{km} \geq 0 \qquad (52.14)$$

$$Z_{ij}^m = \{0,1\},\ F_{ij}^m \geq 0 \qquad (52.15)$$

As mentioned above, the problem which takes Eq. (52.5) as objective function and takes Eqs. (52.1)–(52.4) and (52.6)–(52.15) as constraints turns into a mixed linear programming problem containing a small number of 0–1 variable (Z_{ij}^m). Therefore, proposed mixed integer linear programming model can be numerically solved by general mathematical software packages.

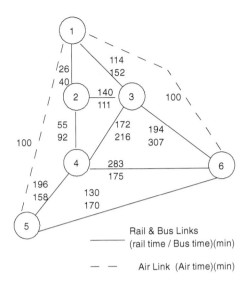

Fig. 52.1 Sample network

52.4 Numerical Example and Conclusion

The proposed model has been applied to a small network consists of 6 nodes and 20 links of 3 different modes (railway, bus and airline) as shown in Fig. 52.1.

We used LP Solve package for solving the equations by the methodology explained in Sect. 52.2 using above data. Resulting P, Q, V values are shown in Table 52.2. Figure 52.2 illustrates optimal network shape with link frequencies and passenger numbers on links. Considering the low unit emission of CO_2 by rail comparing to the air and bus, the result network is mainly consisted by rail links. However, for between city 1 and 6, where trip time by rail via city 3 is 308 min, too larger, direct air service of 100 min is provided. Similarly, bus service is provided on link 4–6, where trip time of rail is much longer than bus. For link 4–5, where bus service is slightly faster than rail, co-existence of bus and rail is observed. As described above, the proposed model successfully give a best-mix design of inter-city modal-mix network.

Table 52.2 Resulting objective values

Objective	Total travel time	Total transfer cost	Total CO_2 emiss.
Min. travel time (P)	3.890.636	18.240	113.590
Min. transfer cost (Q)	7.765.335	0	30.333
Min. CO2 emission (V)	6.004.489	0	22.558
Optimal solution (PQV)	5.413.639	0	35.781

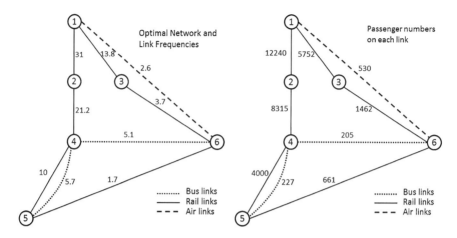

Fig. 52.2 Resulting network with link frequencies and passenger numbers on links

In conclusion, we have presented an optimal modal-mix planning model to design a modal mix network of least CO_2 emissions from the viewpoint of transport operators. We applied the model on a sample network successfully using mixed linear integer programing tools. Resulting optimal network shape and required frequencies for sustainable operation for given OD demand were also presented. This paper provides an upper-level model to consider operators behavior of the network design problem. For the future study, passengers' route choice behavior should be modeled as the lower-level of the network design problem.

Acknowledgments This study is supported by the JSPS KAKENHI Grant Number 21360239.

References

Andersen J, Crainic TG, Christiansen M (2009) Service network design with asset management: formulations and comparative analyses. Transp Res Part C 17(2):197–207

Balakrishnan A, Magnanti TL, Mirchandani P (1997) Network design. In: Dell'Amico M, Maffioli F, Martello S (eds) Annotated bibliographies in combinatorial optimization. Wiley, New York, pp 311–334

Chang Y-H, Yeh C-H, Shen C-C (2000) A multiobjective model for passenger train services planning: application to Taiwan's high-speed rail line. Transp Res Part B 34:91–106

Meng Q, Wang X (2011) Intermodal hub-and-spoke network design Incorporating multiple stakeholders and multi-type containers. Transp Res Part B 45:724–742

Okumura M, Tsukai M (2007) Air-rail inter-modal network design under hub capacity constraint. J East Asia Soc Transp Stud 7(CD-ROM)

Chapter 53
Research on the Optimization Scheme of Beijing Public Bicycle Rental System Life Cycle

Kaiyan Jiang and Hao Wu

Abstract Public bike-rental system is a kind of green, low carbon, economical and convenient city public slow traffic system. In the management mode of government leading and enterprises participating, public bike-rental system's normal operation depends on close link of capital and liability. It is a large engineering system which needs strict control. Based on the theory of life cycle, this article analyzes and divides the life cycle of public bike-rental system into five stages. Taking example by the experience of the other countries and cities, we put forward the corresponding optimization scheme with the problems in different stages.

Keywords Public bicycle · Rental system · Life cycle · Rental mode · Beijing

53.1 Introduction

With superiority of being suitable for the short distance travel and providing "peer-to-peer" service, in Beijing, the optimization of public bicycle rental plan could greatly enhance the competitiveness of the public transportation, pull the coordination of the comprehensive transportation and promote the sustainable development of urban traffic.

K. Jiang (✉) · H. Wu
School of Economics and Management, Beijing Jiaotong University,
Haidian District, BeiJing 100044, China
e-mail: jiangkaiyan2011@163.com

53.2 Public Bike-Rental System's Life Cycle

In the mode of government leading and enterprise managing, the public bike-rental system should be the all process including concept stage, building stage, promotion stage, using stage, and recovery stage. The recovery of two kinds of bicycles makes recovery and promotion stage and using stage change into each other. One is recycling discarded public bicycle, by material processing to use again; Another is recycling private vehicles and modifying them. The process would reduce the volume of private bicycle, may guide the needs to the public bicycles, and so the publicity function of public bicycle. As shown in Fig. 53.1.

53.3 Beijing's Development of Public Bike-Rental System of Necessity and Feasibility

53.3.1 Necessity

Traffic pressure. From 2003 to 2012, the proportion of green travel in Beijing fell to 18.1 % from 34.7 %. Beijing and Mexico City were listed as the city with the worst traffic by the worldwide appraisal of "IBM Commuter pain index".

According to the 2010 Beijing annual traffic report, compared to the double difficulties in transferring buses and car congestion in peak period, bicycle's movement speed is very competitive. From Table 53.1 we can see that the average speed of bicycle in the morning rush hour was only 0.3 km/h slower than the bus travel's speed.

Environmental pollution. France once banned motor vehicle travel for 13 h, reducing noise by 3/4, air environment also improved significantly. Promoting the use of bicycle with zero pollution benefits the improvement of the noise and environmental pollution.

53.3.2 Feasibility

Policy support. Building up the good bicycle traffic environment conforms to the "Beijing transportation development outline" (2004–2020) (Lu and Li 2010). According to the Beijing Traffic Committee's plan, by 2015, the travel proportion of bicycle will reach 20 % in Beijing (Dan et al. 2011).

Fig. 53.1 Each stage of public bike-rental system's life cycle

Table 53.1 Comparison of 2010 Beijing vehicle's average travel speed during peak hours

Transportation	7:00 a.m–8:00 a.m (km/h)	17:00 p.m–18:00 p.m (km/h)
Bicycle	7.7	7.1
Bus	8.0	7.9
Car	17.8	18.1
Subway	13.0	14.2
Walk	4.0	4.2

Economic value. In view of wear, steal, preferences and other reasons, when the average car can be used for 2 years, the average market price is 300 yuan and bike maintenance costs 50–100 Yuan, then 2 years of private bicycle use cost about 350–400 yuan. Nowadays, the pilot public bicycles in Beijing cost only 100 yuan every year without maintenance cost (Wang and Li 2007). From the social perspective, a new bus costs about 0.4–1 million Yuan, though more buses of every line cannot cover all around the area. However, as 500 bikes cost only 150 thousand Yuan, shuttling by bicycles is far cheaper.

53.4 Beijing Bike-Rental System Development Course

Beijing bicycle rental service experienced the beginning stage, rapid spread stage, to be saturated tendency, gradually decline period, the government's rebuilding stage (Song and Crann 2008). At present, Beijing public bike-rental pilot operation is started up by government's capital investment, whereas the operators were in charge of specific business and bearing the expenses (Wang and Li 2008). Advantages of enterprises' operation and government sectors' functions can be embodied in the whole life cycle of the bike-rental service, making public bike-rental system long run down.

53.5 Lease Optimization Scheme of Public Bicycle Based on the Life Cycle

53.5.1 Concept Stage

The plan of lease point and bicycles. Summarizing the domestic and foreign experience, initial construction of lease points can be within the scope of 500–3,000 m away from rail transit sites, with a density of a point in per 700 m (Dai et al 2007). The number of bicycles in each point should be various according to the 4 kinds demand as transit point, settlement point, recreational point and campus point. Overmuch bikes are unfavorable. The plan can be adjusted according to demand again in the future.

Planning of bicycle lanes. Nowadays, in Beijing bike lanes are occupied by motor vehicles along the 70 % of subways of line 2 and the southern section of line 4. Independent bicycle right, high quality, safe and orderly bike lanes are the foundation to promote the bicycle transport. In addition to financial support, the government should provide rights to use the land, bike lanes and other aspects of security.

53.5.2 Construction Stage

High-tech security technology. In the operations of public bicycle-rental system, security is the guarantee of quality of service. Like main parts of public bicycles in France are not the general to ones in markets, Shanghai current public bicycles are installed GPS chips, these security measures reduced the possibility of stealing. However, Beijing's public bicycle is only equipped with a halo lock.

Global positioning technology to allocate bicycle. The location technology could help public bicycle-rental system administrators find the position of bicycles in time to implement the deployment or emergency aid program.

53.5.3 Promotion Stage

Social consciousness and publicity mechanism. In 2007,Paris's"VELIB" propaganda led the bicycle boom called "green revolution" (Hailei 2012); The government ministers, mayors, officials in Dutch cycle to work, and 70 % out workload of the civil servants is completed in B + R way, making an good exemplary role.

Formulation of deposit. Reasonable formulation of the deposit and price is the key to promote the use of public bicycles. The high cost of renting will make part of the potential customers leave, and finally turned to other means of transportation.

Formulation of use price. For the low price of bus and rail traffic in Beijing, as public welfare facilities, public bicycle rental prices should not be too high. Currently, Beijing adopts a "free & low cost" timing charging mode, namely: 1 h's ride for free, further $1 for per hour a day. Accumulated charge is no more than 10 yuan.

53.5.4 Using Stage

Perfect the management system. It is not only service hotlines are few for bicycle-rental points in Beijing, but also the repair and maintenance is not in time, causing bicycle-rental concerns. In the Netherlands, all the cities and rural areas have special bicycle thrusters, free repair tools and medicines were placed every bar and bicycle parking lot, avoiding the overtime debits for bicycles' failures.

The balance of cost and benefit. According to the figures, 2,000 bicycles of pilots in Beijing got rent 2,000 yuan for the first month. This means that the government needs to invest a great amount of capital for operation and maintenance. At present, however, 60,000 vehicles get Hangzhou about 5 million RMB a year by public bicycle rental and more than 30 million RMB by the hull advertising. In the earlier stage, government subsidies can be the main finance source. When the system developed mature, the urban traffic structure has been improved, it is an advice to make up for public welfare by the commercial development, and invest the part of original human and funds that were used to solve the traffic congestion into the day-to-day running and expanding of the construction.

53.5.5 Recovery Stage

Recycling link can extend the public bicycle-rental system life cycle, avoid recession. The recovery of private and public bicycles can present "cycle to cycle" of the product life cycle form.

The prevalence rate of Beijing bicycle is high while utilization rate is low. More than 90 % of residents have private bicycle, about 75 % of the residents use bicycle a day not more than 1.5 h. Bicycles are mainly used for leisure, shopping, commuting, etc., idle in the rest of the time.

After residents' idle bicycles having been modified as standardization and put into public bicycle transport system operation, the cost is lower than the purchase of a new bicycle. It can remove the citizens' troubles about the loss of bicycle parking places and steal. In order to recall massive private bicycles, operators can provide preferential policy about rental bicycle to sellers.

53.6 Endnotes

As more and more people pay attention to the importance of saving energy and environmental protection, bicycle rental will be important role of short-distance travel of Beijing residents, also an important tool for rule blocking. Beijing public bicycle-rental system is still in the exploration, but already showed the forehead of vigorous development prospects.

References

Dai Y, Xu W, Xing LB, Li J (2007) Based on the lifecycle of the car green recovery system planning and management. Journal of Southwest University, pp 157–160

Dan W, Qianwen L , Shuquan T (2011). Half of public rental bicycle idle. Beijing Daily. http://news.xinhuanet.com/fortune/2011-12/14/c_111241357_2.html

Hailei H (2012) Public bike rentals in Beijing. Beijing Times, 2012-7-17. http://www.yangtse.com/system/2012/07/17/013796009.s

Lu QH, Li J (2010) Analysis of application status of information technology in distribution center. Science and technology innovation Herald, p 228

Song RL, Crann D (2008) Based on the product life cycle cost management of environment. Master thesis, Wu Han University of Technology

Wang QH, Li J (2007) Planning and management of modern Urban. Architecture & Building Press, China, p 86

Wang GP, Li J (2008) Speed up construction of public bicycle transportation system to solve the bus's problems of last kilometer. Hangzhou daily, pp 03–21

Chapter 54
The Primary Condition of Bicycle Microcirculation System Benign Operation in Urban: Taking Hangzhou and Beijing for Example

Hao Wu and Xiao You

Abstract With increasingly fast urbanization process, motorization makes urban traffic crowded and air pollution. Considering about environmental protection and trip, the application of bicycle microcirculation system may these problems. This paper analyses the primary condition of bicycle microcirculation system demographically runs in urban by contrasting Hangzhou and Beijing. Government can promote the supply rationalization of bicycle microcirculation system and stimulate demand by funds, policy and technology, thus promotes bicycle microcirculation system demographically dunning in urban.

Keywords Bicycle rental · Low carbon transport · Microcirculation system · Benign operation

54.1 Introduction

Along with the rapid development of China's economy and the acceleration of urbanization, Chinese urban especially big cities traffic problems and air pollution is serious. Car growth too fast, to ensure the smooth operation of car, bicycle lanes are occupied in different degree, especially in the big city fewer and fewer people trip by bikes. However, congestion is still serious. We need reasonable allocation

H. Wu (✉)
School of Economy and Management, Beijing Jiaotong University, Beijing 100044, People's Republic of China
e-mail: hwu1@bjtu.edu.cn

X. You
Shanghai Industry Head Company, Shanghai 200003, People's Republic of China
e-mail: bkhtan@bjtu.edu.cn

between urban road space and different transportation. Public management model of bicycle has become the consensus of the public and the government.

Today, congestion is the common fault of the major cities in China, facing the busy period and the busy road traffic load, many local shifted to bicycle microcirculation system construction. In foreign country, riding bike instead of walking has become a trend of solving the traffic problem, Paris, Copenhagen, London and other cities has become a successful model of bicycle microcirculation system construction (Liu 2009).

In May 2008, Hangzhou took the lead in putting out and implementing bicycle micro circulation system in domestic. According to principle of the government guidance, company operations, policy guarantee, social participation, relying on the public transport system, such a system has been created. Now, the bicycle microcirculation system operates in an orderly way in Hangzhou. It may relieve road traffic pressure of Hangzhou (Shi 2008). In 2011, the survey of satisfaction degree on ten people's livelihood projects, satisfaction degree of this project is far ahead by 99 %.

By the end of 2008, public bicycle microcirculation system rental model takes the lead in the Maizidian community of Chaoyang district in Beijing. Demand is very big, but ultimately failed because income can not cover outcome. By the end of 2010, Beijing advocated green environment protection trip, bicycle microcirculation system was brought up again to ease traffic congestion, But now, only a small portion of the area still is as pilot operation such as Chaoyang park.

Benign operation of bicycle microcirculation system requires a certain conditions, not all of cities the can smoothly establish and implement the system (Zhou 2010). This paper analyses the primary condition of bicycle microcirculation system benign operation through contrasting Hangzhou to Beijing.

54.2 Analysis of Bicycle Microcirculation System in Hangzhou

54.2.1 Analysis of the Status

1 May 2008, public bicycle service system officially began to operate in Hangzhou. Since 2007, Hangzhou has put forward to learn French Paris in the building of public bicycle service system. Therefore, the public bike system has been taken into the public traffic system planning and city public transport system of Hangzhou, namely the subways, buses, taxis, water bus, public bicycle, and put forward the public bicycle is an important component of urban public transport. Hangzhou municipal government thinks that public bicycle as public products must be led by government, so it is operated and managed by Hangzhou city public transport group. In 2008, the government invested 150 million yuan, other funds is borrowed through the bank financing by enterprise. In 3 years, Hangzhou public

bicycle service net quickly expanded whole city from the west lake scenic area. In the center area of Hangzhou, public bicycle service point is able to be found about every 300 m. 40 service points can realize the self-service 24 h. According to the planning of the Hangzhou, to 2015, the quantity of public bikes is about 100,000.

54.2.2 Government's Status in Public Bicycle Service System

Government plays a leading role in the public bicycle service system. Although cover area of Hangzhou is incomparable to Beijing or Shanghai, traffic pressure of Hangzhou is no less than Beijing's. Development of the tourism industry led to the traffic problems, lake traffic pressure near the west is big, parking influence the visitors and citizens to travel. Based on the development of the public demand, the public transport group-only a bus company, pure state-owned enterprise, is responsible for the development of public bicycle service system. The public transport group has set up wholly owned holding of the Hangzhou Public Bicycle Transport Service Development Co., LTD, which will be in charge of operations.

By the end of 2011, Hangzhou has invested to public bicycle service system nearly 400 million yuan. In view of the business development process, Hangzhou municipal government think that if commercial operation funds will not be sufficient to make up for the whole operation, the government will subsidy and ensure public bicycle normal operation by public finance. Since then, bicycle microcirculation system forms by the government supporting resources and enterprise developing resources through benefits, and normally operates. So government has no doubt such effect on the bicycle microcirculation system in Hangzhou.

54.3 Analysis of Bicycle Microcirculation System in Beijing

54.3.1 Analysis of the Status

Traffic congestion has become a serious problem that obstructs citizens' work and life and the city's rapid development in Beijing. Traffic congestion has caused air pollution. In 2005, public bicycle rental service network first was set in Maizidian community. Network operation bills for business after less than three months. In 2008, during of Beijing Olympic Games, public bicycle rental market reached the peak. However, after Olympic Games, public bicycle rental industry operation is difficult. Leasing company business is difficult, the bicycles were damaged and lost, rental outlets were eroded, there was no attention. By December 2010, Beijing municipal government issued the comprehensive measures to alleviate traffic jam, limited motor vehicles, and advocated public transport and bicycle trip, bicycle rental system was taken into transport planning again. So far, bicycle rental outlets are trial operating in Xicheng district and Chaoyang district.

54.3.2 Analysis of Government Promoting Mechanism

As an obviously difference from foreign cities, Beijing's bicycle rental service is completely dominated by the enterprise, lacks government regulation and corresponding policy support. And a single enterprise financial and human resource is limited, so the process of network construction is slow. The Beijing municipal government invests billions every year for public transportation construction, but as public welfare undertakings, bicycle rental service was incorporated into the policy of the government consider category until 2010 year. Beijing has no unified promoting and management mechanism, because of multiple management of commercial bureau, the public security bureau, transport committee, industrial and commercial bureau, the application of bicycle rental project often need pass many hurdles that let the leasing company expend setbacks.

In addition, road planning problem is also important reasons of influence the bicycle rental according to the consumer reflect. The paths are narrow. In addition to government gives policy mechanism support, also should pave roads for benign operation of bicycle microcirculation system.

For enterprise, the government's unified planning support is very important in order to solve difficulty of different operation mode. In order to protect the environment, low carbon traffic is advocated, the government should take the bicycle rental industry into the whole public service system, and give capital and policy support, in order to promote the development of the bicycle rental service.

54.4 Comparative Analyses of Operation Conditions of Bicycle Microcirculation System in Beijing and Hangzhou

First, the government vigorously support bicycle microcirculation system will promote bicycle microcirculation market initially established. Compared to Beijing and Hangzhou, Hangzhou bike microcirculation market is established by the government leading, got $400 million government finance support capital; In contrast, the government supported only in policy, no successive funds and technology support, the bicycle microcirculation system can't benign operation in Beijing.

Second, in Hangzhou, government encourage the bicycle rental network operators to rent to each merchants around and the enterprise, the enterprise can be attracted to this through advertising of the inhabitants of the bicycle rental, and at the same time, the operations of the enterprise lease rental outlets around the store to get large profits, subsidies bike rental and brings due to maintenance cost; but in Beijing, bicycle microcirculation leasing points don't provide such services, the government does not provide the platform for each enterprise, so operators has no incentive to this activity. We found that the policy of the government will strongly affect the functioning of the bicycle microcirculation system.

In addition, the government limits user of bicycle microcirculation system will affect the benign operation. Bicycle microcirculation system and urban public transport system is closely related in Hangzhou, the public transport card can use freely in bike rental outlets; However, bicycle rental market is more controlled in Beijing, only residents living in Beijing for a full years can get leasing qualification. Such a control limits the bicycle microcirculation system benign operation. From the point of view of demand, the bicycle rental demand mainly comes from the floating population, and Beijing's policies limited the demand. Not only that, bicycle rental outlets lack reasonable layout, and failed to take into account the actual demand of residents.

54.5 Summary

Through analysis of Beijing and Hangzhou their respective bicycle microcirculation system, this paper thinks that the primary condition of bicycle microcirculation system benign operation is support from government. First, support from the government can promote to increase supply of bicycle service. Second, support from the government to bicycle rental industry operators can increase the demand of support. Third, support from the government can promote the network layout and construction of bicycle micro circulation system.

To sum up, the support from government is decisive condition for bicycle microcirculation benign operation. The government's support of capital, policy and technology can promote matching supply with demand of the bicycle rental service, at the same time, the capital and technology support from government can promote the bicycle rental outlets rationalization and effective, make the bicycle rental outlets real network. Eventually promote the bicycle microcirculation system benign operation.

References

Liu W (2009) Research of traffic microcirculation system and bypass network in city. Plan Forum 6(25):21–24

Shi F (2008) Function characteristic of city road microcirculation system. Res City Dev 15(3):23–26

Zhou J (2010) Microcirculation theory and bypass traffic. City Transportation 8(3):41–48

Chapter 55
The New Energy Buses in China: Policy and Development

Jingyu Wang, Yingqi Liu and Ari Kokko

Abstract With the advent of "low carbon" economy, new energy vehicles are increasingly favored by the Chinese government and manufacturers. New energy buses have become an important channel for the promotion of new energy utilizations. Based on the summary of policies, this paper conducts a thorough research on the technology and promotion achievements on new energy buses. We have found that the promotion achievements have difference with plans and gaps exist in different cities. In the paper we discuss the policy efficiency, the correlation between achievements, policies and the influence from oil price. We draw the conclusions that clear direction and detailed plans will enhance the new energy bus promotion and rising oil prices will promote new energy buses as well.

Keywords New energy bus · Promotion achievements · Government policies

55.1 Introduction

New energy vehicles– defined as vehicles with fuel or power systems that are not based on traditional internal combustion engines—include fuel cell vehicles (FCEV), hybrid power automobile (HEV) etc. In 2010, China's annual oil

---

J. Wang · Y. Liu (✉)
School of Economics and Management, Beijing Jiaotong University,
Beijing 100044, People's Republic of China
e-mail: liuyq@bjtu.edu.cn

J. Wang
e-mail: 11120813@bjtu.edu.cn

A. Kokko
Department of International Economics and Management,
Copenhagen Business School, 2000 Frederiksberg, Denmark
e-mail: ako.int@cbs.dk

consumption reached 455 million tons, 55 % of which depending on foreign oil. Energy has undoubtedly become the constraints of China's automobile industry development. The government and a large number of automobile manufacturers have therefore focused on new energy vehicles. In January 2009, the Ministry of Science and Technology, Ministry of Finance, National Development and Reform Commission and Ministry of Industry and Information Technology jointly started a project called "ten cities one thousand new energy buses" which plans to launch 1,000 new energy vehicles per city in 10 cities annually in a period of 3 years to enable the new energy vehicles to occupy 10 % of the automobile market. Buses, taxies, service cars and etc. are all involved in the project. At present, the number of the cities that are involved has reached to 25. The first group of cities includes Beijing, Shanghai, Chongqing, Changchun, Dalian, Hangzhou, Jinan, Wuhan, Shenzhen, Hefei, Changsha and Kunming. The second group of cities includes Tianjin, Haikou, Zhengzhou, Xiamen, Suzhou, Tangshan, Guangzhou. The third group of cities includes Shenyang, Chengdu, Huhehaote, Nantong, Xiangfan. Subsequently, the government put forth a series of policies and measures to promote the development of new energy vehicles.

As the government lays emphasis on public transportation, the paper discusses the policy efficiency based on the research of the current status of the new energy buses and related policies. Section 55.2 summarizes the central government policies and the comparison between local government policies. Section 55.3 describes the current status of new energy buses from technology and quantity perspectives. Section 55.4 discusses the policy efficiency, relevance, achievements. Section 55.5 shows our conclusions: First, the government should give a clear definition of the development form to avoid wasting resources. Second, having detailed plans is the key point of promoting new energy buses.

55.2 Policies

55.2.1 Central Government

Facing high technology competition, China launched the National High Technology Research and Development (the so-called 863 program) in November 1986 to improve technological innovative capabilities. Through this program, the government attracts talent, personnel, colleges and enterprises to high-tech research by providing funding and preferential policy. The new energy industry is included in this program.

In January 2009, the four ministries launched a promotion program called "ten cities one thousand new energy buses" followed by a series of government documents to accelerate the new energy vehicle development in public transportation. Table 55.1 (National Development and Reform Commission 2009) presents the subsidies for new energy bus over 10 m from the central government.

Table 55.1 Subsidies for new energy bus over 10 m (10,000 Yuan/per vehicle)

Type	Efficiency of saving fuel (%)	Hybrid power system (VRLA)	Hybrid power system (Ni-MH,LIB,LIP,EC) Max electric power (20–50 %)	Hybrid power system (Ni-MH,LIB,LIP,EC) Max electric power (over 50 %)
HEV	10–20	5	20	–
	20–30	7	25	30
	30–40	8	30	36
	40 以上	–	35	42
BEV	100	–	–	50
FCEV	100	–	–	60

We find that the central government provides strong financial support to the development of new energy buses. For example, the subsidies for electric buses reaches ¥ 500,000, which is 50 % of the average price of electric buses (¥ 100,0000), suggesting the consumer can buy the electric bus at half price even without local government subsides.

Based on the other new energy vehicle industry alliance, State-owned Enterprise Electric Vehicle Industry Alliance including Changan, Dongfeng, Yiqi, Shangqi Putian and etc. was established by State-owned Assets Supervision and Administration Commission of the State Council in August 2010. It indicates that substantive steps have been made to the new energy vehicle industry. The alliance was divided into three segments covering automobile group, battery group and service group and was expected to be the model of industry alliance and to provide the alliance specifications.

In summary, the Chinese central government has committed to providing lasting support to the new vehicle industry with incentives for both R&D and consumers.

55.2.2 Local Government

As the subsides of "863" program are mainly from central government, we will focus on the promotion policy of new energy buses from the local government

Table 55.2 Number of public buses to promote in Shenzhen, Shanghai and Beijing

	Type	2009	2010	2011	2012	Total
Shenzhen	HEV and BEV (bus)	380	470	930	2220	4000
Shanghai	HEV (bus)	150		550		
	BEV (bus)	120		883		1709
	FCEV (bus)	6		–		
Beijing	BEV (bus and car)	–	1000	4000	18000	
	HEV (bus and car)	–	–	1000	6000	30000

Table 55.3 Number of backup service plans in Shenzhen and Beijing

	Type	2009	2010	2011	2012
Shenzhen	Bus charging station (slow)	5	5	7	8
	Bus charging station (fast)	5	5	7	8
	Official car charging pile (slow)	500	600	700	700
	Public charging station (slow)	1625	2500	2750	3125
	Public charging station (fast)	36	46	54	64
Beijing	Charging pile	1200	6000	28800	36000
	Charging station	5	20	75	100
	Battery replacement station	0	1	0	1
	Battery recovery station	0	0	2	2
	Service station	1	3	6	10
	Information station	1	-	1	2

point of view. The selected cities have issued official documents since the pilot project was launched. Based on the economy situation, vehicle manufacturers and how open the city is, we selected three cities, Shenzhen, Beijing and Shanghai for deeper research and summarized their local promotion policies on new energy bus in Tables 55.2 and 55.3 (Shenzhen Government 2009; Shanghai Government 2009; Beijing Government 2009).

We can infer that Shenzhen has the largest number of new energy vehicles to promote and it has put emphasis on the hybrid power and electric buses, while Shanghai just plans to promote 1,709 new energy buses with FCEV also involved.

There is conspicuous difference between the two cities' policies. Shenzhen only includes charging station in its plans but with clear classification and numbers. Beijing does not classify the service station accurately, but with more battery replace stations and recovery stations involved in its service system (We cannot find the complete plan of Shanghai).

Shenzhen is the first city to provide subsidies to new energy vehicles from local government. Table 55.4 (Shenzhen Government 2009) shows its subsidies system. We find that Shenzhen gives clear subsidies plans about new energy bus and charging station while such official and clear plans are not found in Shanghai and Beijing.

Table 55.4 The subsidies for buses and service system from Shenzhen local government (10,000 Yuan)

	2009	2010	2011	2012	Total
HEV and BEV	8697	9870	9300	26328	54195
Bus charging station (slow)	3180	3180	4452	5088	15900
Bus charging station (fast)	1500	1500	2100	2400	7500
Official car charging pile	500	600	700	700	2500
Public charging station (slow)	1625	2500	2750	3125	10000
Public charging station (fast)	7200	9200	10800	12800	40000

55.3 The Development of New Energy Vehicles in China

55.3.1 Technology Development

New energy vehicles are defined as the vehicles using alternative fuel technologies including HEV, BEV and FCEV. Because of its challenge in material sourcing, high maintenance cost and imperfect service system, FCEV is out of favor with most manufacturers. So far, large and medium-sized manufactures are focusing on HEV and BEV.

Both Electric and hybrid power buses have higher energy efficiency and lower emitted carbon dioxide against regular bus. A hybrid power bus saves 6,300 liter of diesel per year and a plug-in hybrid power bus saves 8,400 liter of diesel per year.

55.3.2 Achievement of Promotion

China New Energy Bus Development Forum in 2010 indicates that the new energy bus development in public transportation is now in full swing. By the end of 2010, the number of new energy buses has reached 4,000. It is estimated that the number will reach 15,000 by the end of 2012. Figure 55.1 (Beijing Bus 2011; Shanghai Bus 2011; Shenzhen 2011) resents the number of new energy buses in Shenzhen, Beijing and Shanghai.

There is a gap between the actual and plan numbers, but the project did achieve good results. By 2011, the share of new energy bus has reached 6 % in Beijing and Shenzhen public transportation and 3 % in Shanghai.

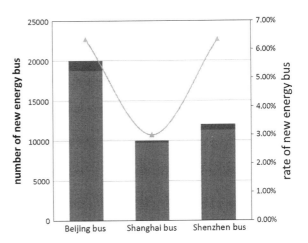

Fig. 55.1 Number and rate of new energy bus in Beijing, Shanghai and Shenzhen

55.4 Discussion

55.4.1 Policy Efficiency

According to the available data, the new energy bus basic technology indicators have met the public transportation requirement, which means the input and output of new energy buses are able, while the actual promotion number is less optimistic than the planned number.

First, most local policies ignored the exploration of marketing model, which led to the unclear future of new energy bus. Exploring marketing model on their own will no doubt add to the enterprises' burden and various different business models in parallel may disrupt vehicle markets.

Second, we can not see the focus of the government from its policies because the subsidies mentioned in the policies are all based on the bus cost and the fuel it saves.

Consequently, the policies are good in technology, but unclear direction and dimming marketing model have bad effects on their efficiencies.

55.4.2 Correlation Between Promotion Achievements and Policies

While what the selected cities get are the same from central subsidies perspective, each city's project involved in "863" program are not same, which results in different science funding and technology level. State-owned Enterprise Electric Vehicle Industry Alliance can not cover all the selected cities either. Viewed this way, the support the cities get from central government is still different.

Although the support that Shenzhen, Shanghai and Beijing get is not much different, there is a big gap among their achievements.

Shenzhen is the first city to provide subsidies to new energy vehicles from local government. Additionally, Shenzhen has its detailed plans for new energy buses and service system to help the consumers with the subsidies that they are entitled to.

Beijing has been actively exploring a business mode to look for better solutions to the battery purchase and its follow-up service. Therefore promotion plan of Beijing covers battery-recycle and information station. Furthermore, Beijing established new energy vehicle design and manufacture base in 2008 and the first new energy bus alliance in China in March 2009 called Beijing New Energy Vehicle Industry Alliance included Baic Group, Beijing Bus Group, Beijing Institute of Technology etc.

Shanghai has defined plans for only new energy vehicles involved in "863" program, but its plans for new energy bus seem not so clear.

So we infer that promotion achievements and local policies are closely related. The detailed plans of Shenzhen and the exploration of Beijing have encouraged the development of local new energy buses.

55.5 Conclusions

The paper analyzes the policies on new energy bus from central and local government and describes the promotion achievements of Shenzhen, Shanghai and Beijing. Then the paper discusses the policy efficiency, the correlation between achievements and policies.

Clear direction

Government should define the clear direction of new energy buses based on relevant research and experience. Moreover, government can promote certain type of new energy buses in certain cities based on its resources not only to make the direction clear but also to help to collect and analyze data and summarize experience.

Detailed and comprehensive plans

We have observed that the more detailed and comprehensive the plans are, the better achievements. Such plans should be provided to the consumers to make sound decisions.

Acknowledgments The research reported here was supported by the Chinese Ministry of Education Youth Fund Project of Humanities and Social Sciences (Project No.11YJCZH114), Beijing Planning office of Philosophy and Social Science (Project No. 11JGB033), and the Sino-Danish Center for Education and Research, SDC.

References

Shenzhen Government (2009) The new energy vehicle promotion plan of shenzhen. Available at: http://www.sz.gov.cn/cn/. Accessed 1 April 2012

Shanghai Government (2009) The new energy vehicle promotion plan of shanghai. Available at: http://www.shanghai.gov.cn/shanghai/node2314/index.html http://www.shanghai.gov.cn/shanghai/node2314/index.html. Accessed 1 April 2012

Beijing Government (2009) The new energy vehicle promotion plan of beijing. Available at: http://zhengwu.beijing.gov.cn/. Accessed 1 April 2012

National Development and Reform Commission (2009) The new energy vehicle promotion plan. Available at:http://www.sdpc.gov.cn/. Accessed 1 April 2012

Beijing Bus (2011) Available at:http://www.bjbus.com/home/index.php. Accessed 1 April 2012

Shanghai Bus (2011) Available at:http://www.84000.com.cn/. Accessed 1 April 2012

Shenzhen Bus (2011) Available at:http://www.szbus.com.cn/. Accessed 1 April 2012

Chapter 56
The Determinants of Public Acceptance of Electric Vehicles in Macau

Ivan Ka-Wai Lai, Donny Chi-Fai Lai and Weiwei Xu

Abstract This study aims to examine the factors that influence individual intention towards the adoption of electric vehicles. Questionnaire survey was conducted in public area in Macau. Data collected were analyzed by confirmatory factor analysis and structural equation modeling. The results of data analysis show that environmental concern and environmental policy are the antecedent factors of perception of electric vehicles, and that influences customer's behavioral intension to purchase electric vehicles. This study also finds that perception of benefit is one of the factors that influence the adoption of electric vehicles. Vehicle operators are more sensitive to fuel economy. They are seeking for future long-term fuel savings. Thus, government striving to promote low-carbon transportation needs to scale up its efforts to enhance citizens' environmental concern, establish proper environmental policy, and provide long-term financial and strategic support for electric vehicles.

Keywords Public acceptance · Environmental concern · Environmental policy · Electric vehicles

---

I. K.-W. Lai (✉) · W. Xu
Faculty of Hosiptiality and Toursim Mangement, Macau Unviersity of Science and Technology, Taipa, Macau
e-mail: kwlai@must.edu.mo

W. Xu
e-mail: weal_xu@163.com

D. C.-F. Lai
Department of Computer Science, City Unviersity of Hong Kong, Kowloon Tong, Hong Kong
e-mail: donnylai@cityu.edu.hk

56.1 Introduction

"Building a low carbon Macau, creating green living together" is the vision of the first environmental planning of Macau (Macaunews 2011). As a world tourism and leisure center, air pollution has become a key problem affecting tourism in Macau. One of major sources caused air pollution is carbon emissions from vehicles. Over the past few decades, research has been conducted to investigate various aspects about the developments of low-carbon transportation technologies. As a result there are already a number of potential alternatives to the conventional diesel/petrol combustion engines (Schulte et al. 2004), such as electric vehicle engine. Although electric vehicles have been introduced for a long period of time, they are not popular. Thus, there is a need to examine the factors on influencing public acceptance of electric vehicles.

Macau is a small city (total area = 29.9 km^2) with narrow roads and streets (total length = 413.4 km). There were 182,765 vehicles with a population of 563 thousands in 31 March 2012 (DESC 2012). One of every three citizens has a vehicle. The typical driving range of vehicles is less than 40 km (Tse 2011). Therefore, electric vehicles are very suitable for Macau environment. Macau government is developing environment policies include introducing tax preference for environmental light vehicles.

Many factors influence the car-purchasing behavior. These include situational factors such as regulatory environments (Collins and Chambers 2005). In addition to situational factors, psychological factors are equally important, for example, personal attitudes (Choo and Mokhtarian 2004). Although some empirical studies of public acceptance of hybrid vehicles have been conducted, there is little research that considers psychological factors and situational factors together. Also, there is a lack of research on the public acceptance of electric vehicles.

This study focuses on the public acceptance of full electric vehicles in Macau. This study addresses the need for an empirical study that estimates the situational and psychological factors that impact public acceptance of electric vehicles and tests the relationships among these factors. This research will contribute to help academic and government understand the factors that influence customer's purchase intentions of electric vehicles.

56.2 Factors Influencing Car-Purchasing Behavior

Twenty years ago, Ellen et al. (1991) identified important factors that motivate environmentally conscious behaviors. In this study, environmental concern and perception of environmental policy are combined with perception of benefit, perception of electric vehicles, and behavioral intension to purchase electric vehicles to form a research model as shown in Fig. 56.1.

56 The Determinants of Public Acceptance of Electric Vehicles

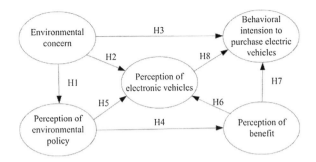

Fig. 56.1 Research model

Environmental concern is a general attitude against environmental deterioration (Fransson et al. 1994). Kahn's (2007) study indicated that environmentalists are more likely to purchase hybrid electric vehicles than non-environmentalists.

H1: A customer's environmental concern has a positive impact on the customer's perception of environmental policy.

H2: A customer's environmental concern has a positive impact on the customer's perception of electric vehicles.

H3: A customer's environmental concern has a positive impact on the customer's behavioral intension to purchase electric vehicles.

Irwin and Wynne (1996) stated that political context affects the validation of new technology. Environmental policy is a situational factor that should affect customer's perception of electric vehicles.

H4: A customer's perception toward environmental policy has a positive impact on the customer's perception of benefit.

H5: A customer's perception toward environmental policy has a positive impact on the customer's perception of electric vehicles.

The acceptance of a product is often affected by personal perception of benefit. Consumers may consider these benefits when they are making decision on purchasing a new vehicle.

H6: A customer's perception toward benefit has a positive impact on the customer's perception of electric vehicles.

H7: A customer's perception toward benefit has a positive impact on the customer's behavioral intension to purchase electric vehicles.

Positive perception of a product can make a customer more likely to purchase the product (Viardot 1998). Thus, the perception toward electronic vehicles should influence customer's purchasing behavior.

H8: A customer's perception toward electric vehicles has a positive impact on a customer's behavioral intension to purchase electric vehicles.

56.3 Research Method

The research question of this study is: what are the factors that affect public acceptance of electric vehicles in Macau? A questionnaire survey is used. The interviewer-administered survey was conducted on the street in March 2012. A filter question "do you drive your own car?" was asked. A total number of 310 completed questionnaires were collected in a month. However, two questionnaires were eliminated (e.g., for giving the same rating for all items), leaving 308 questionnaires as valid for analysis.

56.4 Findings

Confirmatory factor analysis (CFA) was performed to evaluate construct validity regarding convergent and discriminant validity. Table 56.1 shows the means, standard deviation, reliability, and standardized factor loadings of the constructs. The Cronbach's alpha for all components are higher than 0.6 and all factor loadings are above 0.5. Based on the guidelines of Hair et al. (2010), the reliability and construct validity of the study are accepted. Table 56.2 shows the correlation matrix of the five constructs. All of the correlation values among the constructs of the model are significant (p value < 0.01).

Table 56.1 Reliability and construct validity

	Mean	Standard deviation	Cronbach's alpha	Factor loadings	EVA	CR
EC	5.553		0.905		0.764	0.906
EC1	5.507	0.776		0.854		
EC2	5.490	0.793		0.894		
EC3	5.662	0.852		0.873		
PEP	4.320		0.933		0.830	0.936
PEP1	4.497	1.093		0.864		
PEP2	4.234	1.045		0.923		
PEP3	4.231	0.966		0.944		
PB	5.128		0.916		0.787	0.917
PB1	5.166	0.681		0.911		
PB2	5.198	0.678		0.875		
PB3	5.020	0.744		0.874		
PEV	4.398		0.929		0.809	0.927
PEV1	4.419	0.897		0.903		
PEV2	4.351	0.892		0.887		
PEV3	4.425	0.967		0.908		
BI	4.927		0.912		0.758	0.904
BI1	4.990	0.678		0.853		
BI2	4.860	0.678		0.836		
BI3	4.932	0.634		0.921		

Table 56.2 Construct correlation matrix (standardized)

	EC	PEP	PB	PEV
PEP	0.155			
PB	0.068	0.440		
PEV	0.230	0.554	0.621	
BI	0.362	0.335	0.588	0.506

Correlation is significant at the 0.01 level (2-tailed)

Structural equation analysis (SEM) was performed to test the research hypotheses empirically. Figure 56.2 shows the results of SEM analysis. The results of SEM provide that the model fit is acceptable and all hypotheses are valid.

56.5 Discussions

The results of this study prove that environmental concern is an initial factor that finally leads customer's behavior intention to purchase an electric vehicle. Environmental concern is a psychological factor that directly and indirectly influences three kinds of perceptions that mediate the link between environmental concern and the acceptance of electric vehicles.

Unsurprisingly, the results of this study indicate that perception toward environmental policy is positively correlated with perception of benefit and perception

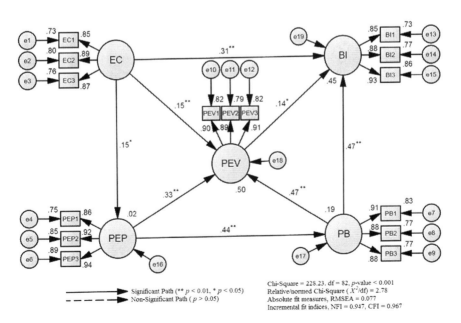

Fig. 56.2 Structural equation modeling results

of electric vehicles, and the perception of benefit and perception of electric vehicles directly affect the acceptance of electric vehicles. These findings are consistent with Kang and Park's (2011) study. Public becomes aware the environmental policy and believes the policy will be maintained continuously in order to build a low carbon society.

This study investigates an important factor—perception of benefit that economically affects consumers buying behavior toward electric vehicles. Vehicle operators are more sensitive to fuel economy. The cost saved from tax incentives is little and short-term compared with the cost of continually fuel savings. In case the price of gasoline is raised, the future fuel savings by the adoption of hybrid vehicles is less than the cost saving by the adoption of electric vehicles. Therefore, perception of benefit is the major determinant of public acceptance of electric vehicles in Macau.

This research model consists of psychological factors (environmental concern and perception of electric vehicles) and situational factors (perception of environmental policy and perception of benefit). The results of data analysis indicate that both factors play important roles in the adoption of electric vehicles as discussed above. This study contributes a research model that can be further investigated in order to explain the causal effects of these four factors for the adoption of environmental technologies.

For the practical implications, this study indicates that environmental concern is antecedent factor that stimulates the interest of electric vehicles. Macau government should educate public the importance of environmental protection and the environmental advantages of driving electric vehicles.

The introduction of electric vehicles requires forceful government's environmental policy. Macau is a small city. Limits on driving distance and lack of power should not be great issues for the adoption of electric vehicles in Macau. However, electric vehicles are expensive to own. Macau government can offer an electric vehicles program to subsidize vehicle owners to replace their exiting gasoline vehicle with new electric vehicle. Also, the batteries need to be recharged. Macau government should establish the supporting infrastructure for electric vehicles such as provide charging facilities in public car parks.

This study indicates that the acceptance of electric vehicle as a common transportation equipment will be major determined by the perception of benefit. That is the cost advantage of electric vehicles compared with vehicles that use gasoline. Consumers care long-term running costs. Macau government should provide long-term financial support such as free public parking for electric vehicles.

This study only focuses on the citizens on their acceptance of electric vehicles in Macau. This study may not be generalized to apply to other countries with basically different cultures. Future research is suggested to focus on other countries. Also, this study is only concerned about the public acceptance on electric vehicles. It may not be generalized for other environmental technologies. Future study is suggested to investigate whether the similar concept can be employed to other environmental products like LED lighting.

References

Choo S, Mokhtarian PL (2004) What type of vehicle do people drive? the role of attitude and lifestyle in influencing vehicle type choice. Transp Res A 38:201–222

Collins C, Chambers SM (2005) Psychological and situational influences on commuter-transport-mode choice. Environ Behav 37:640–661

DESC (2012) Macau Census and Statistics Department. http://www.dsec.gov.mo/CMSPages/c_mn_indicator.aspx

Ellen PS, Wiener JL, Cobb-Walgren C (1991) The role of perceived consumer effectiveness in motivating environmentally conscious behaviors. J Pub Pol'y Mark 10:102–117

Fransson N, Davidsson P, Marell A, Garling T (1994) Environmental concern: conceptual definitions, measurement methods and research findings. Department of Psychology, University of Goteborg, Goteborg

Hair JF Jr, Black WC, Babin BJ, Anderson RE (2010) Multivariate data analysis, 7th edn. Prentice-Hall Inc, Englewood Cliffs

Irwin A, Wynne B (1996) Misunderstanding science? The public reconstruction of science and technology. Cambridge University Press, Cambridge

Kahn ME (2007) Do greens drive hummers or hybrids? Environmental ideology as a determinant of consumer choice. J Env Econ Manag 54:129–145

Kang MJ, Park H (2011) Impact of experience on government policy toward acceptance of hydrogen fuel cell vehicles in Korea. Energy Pol'y 39:3465–3475

MacauNews (2011) Macau to have first environmental planning until 2020, 1 Apr 2011. http://www.macaunews.com.mo/index.php?option=com_content&task=view&id=1248&Itemid=4

Schulte I, Hart D, van der Vorst R (2004) Issues affecting the acceptance of hydrogen fuel. Int J Hydro Energy 29:677–685

Tse WC (2011) Road testing of electric vehicles in Macau. J Asian Electr Veh 9:1491–1495

Viardot E (1998) Successful marketing strategy for high-tech firms, 2nd edn. Artech House, Boston

Chapter 57
Synthetical Benefit Evaluation of High-Speed Rail, Take Beijing-Shanghai High-Speed Rail for Example

Han-bo Jin, Hua Feng and Fu-guang Cui

Abstract Through research and analysis on the status of domestic and international high-speed rail, this paper established the Theoretical Framework of the economic benefits, Social benefits and Synthetical benefits of high-speed railway. And this paper summarized and summed up the factors affecting the efficiency and evaluation methods, and finally got a more standardized evaluation criteria. Beijing-Shanghai high-speed rail, for instance, has a strong comprehensive benefits. The corresponding conclusions and policy recommendations in this article is below; firstly high-speed rail line that has been put into operation, planning or under construction, should be made the necessary adjustments and evaluation; Secondly, the Government should give different types of subsidies to high-speed rail for different comprehensive benefits; and last, for high-speed rail line, we should give full play to the advantage of high-speed rail such as obtain the largest passenger flow, and increase the transport density.

Keywords High-speed rail · Synthetical benefits · Economic benefits · Social benefits · Low carbon transportation

57.1 Introduction

Since the reform and opening up, China's economic development momentum is very rapid over the years. At the same time, Chinese society has also experienced various problems brought about by the economic boom, which for the

H. Jin (✉) · H. Feng
School of Economics and Management, Beijing Jiaotong University, Beijing, China
e-mail: hamburger81@163.com

F. Cui
Air Force, Beijing 95880, China

development of China, both opportunities and challenges. With the improvement of people's living standards, basic living expenses other than living expenses are also increasingly diversified and enlarged, which, traffic travel expenses, and the degree of diversification of transportation travel demand go hand in hand. In today's society, life and work pace is accelerating, and high-speed railway has become the new ways for people's travel choice.

Background based on the above issues, as well as its practical and theoretical value, this paper set mainly the two research objectives. First, trying to find the synthetical benefits of the high-color rail to be assessed from two aspects of economic and social benefits; second, combined with the Beijing-Shanghai high-speed rail specific data to evaluate the synthetical benefits of the Beijing-Shanghai high-speed rail, and give a specific feasible optimization recommendations.

Clear the above research objectives and research methods, the technology roadmap of this paper is organized as follows (Fig. 57.1).

57.2 Literature Review

Theoretically, Kiyoshi Kobayashi (1997) Emphasized the high-speed rail system played an important role in affecting the regional economy by constructing a system model composed of a number of cities connected by high-speed rail system; U. Blum K (1997) proposed high-speed rail to a certain extent connected with the city with a transition to an expansion of the functional areas or the economy as a whole corridor, and from the near, medium and long term analysis of the economic zone's economic integrity.

Empirically, Feng and huang (2010) take Tianjin Intercity Railway for example, the high-speed rail will be able to achieve good economic and social benefits,

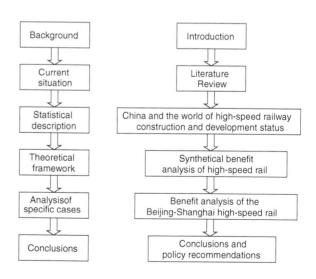

Fig. 57.1 This article line and content arrangements flowchart

and maximize the benefits of high-speed rail to make reasonable suggestions; Feng and Xue (2011) studied on synthetical benefits, with government support policy for high-speed railway in China, the article points out that in the long run high-speed rail can achieve not only the direct economic benefits, but also the great social good. Therefore, the high-speed railway construction and operation of effective government policy support.

57.3 China and the World of High-Speed Railway Construction and Development

Japan is the world's first country built for high speed railway line. In the late 1950s, Japan began to build a dedicated high-speed rail line. France following the Japanese Tokaido Shinkansen high-speed railway to proceed with economic and technical studies. 1971 the French government approved the construction of the Paris-Lyon TGV Southeast Line; Germany to build high-speed rail is the same in order to ease the growing tension capacity due to increased demand; America had little to build a high-speed rail. The U.S. Federal Railroad Administration in 2001 proposed the development of a high-speed rail vision to form a 13,600 km high-speed rail network.

Chinese-owned high-speed rail Beijing-Tianjin Intercity, Chang Ninetowns, Harbin-Dalian line, Passenger Line, Zhengzhou-Xi'an high-speed railway, Wen-Fu line, Beijing-Shijiazhuang line, Han-Yi Line, Hong Kong and broad and deep, the Beijing-Shanghai line. As of the end of 2010, China's new high-speed railway operating mileage of 5,149 km, separate 17,000 km mileage under construction. Operating mileage of 7,531 km is the longest of the world's high-speed rail operation mileage in the construction of the largest countries. Also the most technologically, the strongest integration capabilities, the highest speed.

57.4 Synthetical Benefits and Methods of Analysis of High-Speed Rail

57.4.1 The Economic Benefits of High-Speed Rail

$$\text{The economic benefits of high} - \text{speed rail} \\ = \text{operating income} - \text{operating costs} + \text{profit on sales}$$

Operating income is the income of high-speed rail in the operation process, basically equal to the face value of the sum of high-speed rail operators sold tickets. Operating costs is the cost spent by the high-speed rail in the operation process, including the following factors: the train station depreciation expenses, administrative expenses, interest on loans and other. Sales profits, high-speed rail operations, vehicle sales profits, advertising revenue, station sales, profits, and so the sum of profit.

57.4.2 The Social Benefits of High-Speed Railway

The vigorous development of high-speed rail can bring what social benefits and social assessment of how to become a very meaningful subject. Ultimately, the social benefits of the following six high-speed rail.

(1) To promote regional economic development. (2) Economical use of land. (3) Low carbon transportation. (4) Time saving. (5) Transport capacity replacement. (6) Increase opportunities for employment. In summary, the social benefits of high-speed rail is more than the sum of the effectiveness of the six areas.

57.4.3 The Synthetical Benefits of High-Speed Railway

The synthetical benefits of high-speed railway, high-speed rail in the operation process of the economic benefits and social benefits collectively.

$$\begin{aligned}\text{The synthetical benefits of high} - \text{speed rail} \\ = \text{high} - \text{speed rail economic efficiency} + \text{high} \\ - \text{speed railway social benefits}\end{aligned}$$

Depending on the effectiveness, high-speed rail can be divided into three categories, as shown in Fig. 57.2.

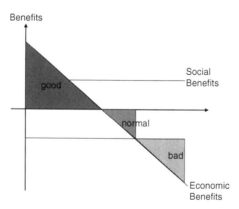

Fig. 57.2 According to the classification schematic of the overall efficiency of high-speed railway

Table 57.1 Beijing-Shanghai high-speed rail Beijing part of the statistical results

Date	Statistical number of days (day)	Daily average on the car number (people)	Average daily number of trains (train)
2011.12	18	38586	64.7
2012.1	29	45484	75.2
2012.2	29	42105	73.2
2012.3	29	48747	65.4
2012.4	14	51793	70.4
Total	119	45155	70.2

Source Beijing Railway Bureau

57.5 Benefit Analysis of the Beijing-Shanghai High-Speed Rail

57.5.1 The Economic Benefits of High-Speed Rail

Analysis of operating income

Beijing-Shanghai high-speed rail operating income is the amount of ticket revenue to calculate the known conditions of the ticket revenue for the number of passengers and fares.

According to my field research and study, I have 126 days from December 13, 2011 to April 2012 16 a total of 119 days in the Beijing bureau of the number of passengers, the statistical results shown in Table 57.1.

Ticket side, from the Beijing South Railway Station to Shanghai Hongqiao Station, first-class seat 960 yuan, second seat 555 yuan, the proportion of second-class seat is about 80 %. By this calculation, ticket revenue is up to the daily average of more than 48 million yuan, 17.5 billion yuan of annual ticket revenue. Estimated average annual ticket revenue in 2020 is stable at around 22.5 billion yuan.

Analysis of operating costs

The total investment of 220.9 billion for the Beijing-Shanghai high-speed rail project company registered capital ratio of 50 %, the rest of the approximately 110 billion yuan of investment through bank loans and issue bonds to raise. Preferential policies in accordance with the Bank to the Ministry of Railways "loan interest rates generally fall 10 %", the loan interest rate of around 6 %, interest payable annually at 6.6 billion yuan, which is a day to pay interest of 18 million yuan. Repayment of principal in accordance with the 20-year period, each year, 5.5 billion, or about 15.07 million yuan a day. According high-speed rail has opened the law, the annual depreciation rate of about 3–4 %, or 220.9 billion investment, the annual depreciation of fixed assets cost about 6.6–8.8 billion.

In summary, expected in 2012 as a whole, the economic benefits of the Beijing-Shanghai high-speed railway operating income of 17.5 billion yuan, 21.7 billion, operating costs, profit on sales of 350 million yuan, and the economic benefits of the sum to −38.5 billion; at the same time in 2020 economic benefits can be stabilized at about 1.25 billion yuan.

57.5.2 Social Analysis of the Beijing-Shanghai High-Speed Rail

The Beijing-Shanghai high speed railway line is line bridge 244, a total length of 1059.7 km, accounting for 80.4 % of the full range of the total mileage. Calculation, the value of the Beijing-Shanghai high-speed rail to save land for more than 20 billion yuan to 1,000 yuan per square meter of land.

Beijing-Shanghai high-speed rail operational vehicles CRH380A is energy saving and environmental protection of high-speed trains, mainly reflected in the low-power, lightweight, dirt collection. Kilowatt calculations, about $ 500 million worth of energy saving and environmental protection.

The experience of Japan and France Rail conservative forecast, the Beijing-Shanghai high-speed rail operators can create 200,000 labor positions, in accordance with the per-capita wages of 2,000 yuan, 400 million yuan of revenue.

Beijing-Shanghai line average transport density passenger of more than 30 million passengers, cargo of more than 8,000 tons, respectively 5 times and 3.5 times the average of the national railway, has reached the limit of passenger and freight mix of double track railway transport capacity. The opening of the Beijing-Shanghai high-speed rail can liberate part of the transport capacity to displace. Replacement-effective transport capacity is expected to about 80 billion.

Various parts of social benefits to the sum of processing available to the social benefits of the sum of the Beijing-Shanghai high-speed rail to 31.25 billion yuan.

57.5.3 Conclusion

Managed to support the situation compared to the national high-speed railway, Beijing-Shanghai high-speed rail is still relatively optimistic about the prospects. The one hand, the direct economic benefit of the Beijing-Shanghai high-speed rail is −38.5 billion can be expected after eight years into a 1.3 billion yuan, the Beijing-Shanghai high-speed rail without a government subsidy self-financing projections, but recovery of the funds of the project and also the pressure of this interest payment is expected to take 30 years or more to complete. On the other hand, the social benefits of the Beijing-Shanghai high-speed rail is enormous, and I believe over time scale will be greater. So for such a railway, the Government should build and give part of the subsidies.

57.6 Conclusions and Policy Recommendations

57.6.1 Re-Evaluation of High-Speed Rail Line has been Put into Operation

Of more than 350 km per hour and 250 km Passenger Dedicated Line has been in operation for more than one year from the current situation, the passenger line passenger seriously lower than expected, there is a serious loss in varying degrees, such as the Wuhan-Guangzhou Railway, Zhengzhou-Xi'an high-speed rail. Consider the current Ministry of Railways debt, thus re-evaluate the need for these high-speed passenger dedicated line item, the evaluation focused on the comprehensive benefits of the high-speed railway line. If the line is within the acceptable time to recover the cost or break-even, the line can continue to operate; serious line loss but enormous social benefits, the government can consider giving subsidies; if the comprehensive benefits of the line is negative, you can consider a moratorium on operation or the adjustment and optimization.

57.6.2 Consolidated Income and Government Subsidies

Author believe that can be high-speed rail line operations are divided into three categories: first, the economic benefits and the synthetical benefits are positive, such lines can be profitable without subsidies by the Government; the second is the economic benefits for the negative but the synthetical benefits are positive, the government should be the appropriate subsidies to ensure the normal operation of the line; the last one is the economic benefits and the synthetical benefits are negative, this line should stop operating.

57.6.3 Give Full Play to the Advantage of High-speed Rail Access to the Largest Passenger Flow, Increase the Transport Density

High-speed rail compared to other modes of transportation, some of its own characteristics: First, rail transport, economic and technological properties determine the rail with low tariffs, and a large volume competitive advantage, as the popularization of transport should be the main service most low-and middle-income travelers, so the Ministry of Railways need to correct market positioning, rather than blindly at the request of the high-end crowd quickly seek comfort. Second, high-speed passenger dedicated line could attract a larger passenger, depending on the value of high-speed rail passengers, which is determined by the economic value of travel time savings. China's per capita income is still very low, the economic value of travel time savings is very low. Rail passenger market in China, cheap and basic level of comfort is more important than saving a few hours of travel time.

References

Feng H, Huang L (2010) To play a modern railway role of high-speed rail to maximize the benefits. J Chinese Railw 10:4–7
Feng H, Peng X (2011) The synthetical benefits of the high-speed railway in China and support policies of the. J Soc Sci Guangdong 3:12–19
Kiyoshi K, Okumura M (1997) The growth of city systems with high-speed railway systems. J Ann Reg Sci 31(1):39–56
U Blum K (1997) Introduction to the Special Issue: the regional and urban effects of high—speed trains [J]. Ann Reg Sci 31(1):1–20

Chapter 58
Strategy Research on Planning and Construction of Low-Carbon Transport in Satellite Towns: The Case of Shanghai

Luwei Wang and Xinsheng Ke

Abstract China is in an important period of the urbanization and the rapid development of urban construction stage, in the backdrop of global climate change, the coordinated development of economy and the environment between the central city and satellite towns facing enormous challenges. In this paper, the case study of Shanghai, the main problems of the current satellite towns transportation planning and construction, from the angle of low-carbon, analysis methods to solve problems, to adapt to the trend of the development of low-carbon transport.

Keywords Low-carbon transport · The satellite towns · Urban railway

In recent years, an important period of our country in urbanization and the rapid development of urban construction, the big city as center gradually form the metropolitan area and many satellite towns appear. With the increased distance, the increase in passenger traffic between the city's areas and satellite towns, the enormous pressure brought to public transport system. With increasing scale of city, it brought new problems, such as overcrowding, traffic congestion. In recent years, the increasing demands of building the low-carbon city, Shanghai as represented of many cities are gradually exploring low-carbon transport of satellite towns.

L. Wang (✉) · X. Ke
Beijing Jiaotong University, Haidian, China
e-mail: 11120710@bjtu.edu.cn

58.1 The Connotation of Low-Carbon Transport

Low-carbon transport is a fundamental characteristic of low energy consumption, low pollution, and low emission urban transport development mode. The core of building a low-carbon transport system is to promote urban transport systems, improve energy efficiency, and improve the structure of energy consumption (Huapu 2009). The low-carbon transport is not a new mode of transportation, but a new concept. Its core is to improve the energy efficiency, and optimize the development of transportation, guide people reasonable travel. Its purpose is to reduce energy consumption and carbon emissions, as passenger and cargo flow to provide a safe, convenient, comfortable and fair service, to meet demands of people's transportation.

58.2 The Main Problems of Transportation Planning and Construction in Current Satellite Towns

The development of satellite towns in Shanghai has been more than ten years, however, the integrated transport supporting is not perfect, resulting in these satellite towns in and out of the city public transport is inconvenient. Overall, in current satellite town transportation planning and construction has the following problems:

58.2.1 Transportation and Land-Use Planning is Not Enough Interactive

Traffic between the central city and satellite towns belong to a composite travel. Compound travel refers to use two or more modes in the whole travel process. Therefore, the configuration of them should not be limited to the analysis of characteristics of the various modes, need to consider end-to-end distributed patterns, reflecting the traffic and the intensification of land-use. Planning in satellite towns is still mainly traditional road grid-based land-use, especially rail transportation and the transportation hub of urban land-use don't layout embodies enough (Wenzhong 2003).

58.2.2 The Planning of Rail Transit Network is Not Enough in Satellite Towns

Comprehensive transportation planning only as a subsidiary of the content in satellite towns, depth and breadth are not enough. Between satellite towns and the center city, the traffic model is not clear enough, and rail transit network planning inadequate, construction of rail transit facilities have not done enough. The rail transit of Qingpu, Nanqiao, Nanhui in Shanghai don't connected to center city. Existing rail transit of Songjiang, Jiading, as Line 9 and Line 11 not only assumed the central city, but services for satellite towns, the travel speed is only 40–45 km/h.

58.3 Strategies on Planning and Construction of Low-Carbon Transport in Satellite Towns

The purpose of satellite towns closure the flow of the central city, control the size of population; transfer some economic functions of the central city, especially industrial, such as pollution industries to new satellite towns. Current satellite towns in China are not perfectly achieve, which mainly due to the construction of transportation facilities between the central city and satellite towns backward, influencing economic exchanges.

58.3.1 Build Public Transport-Oriented Systems

Various satellite towns are impossible to have the independent community function, therefore, the satellite towns to establish close contact with the central area is inevitable. For this reason, it is necessary to further build and improve the traffic system which large capacity, safe, fast, and cheap. The public transport-oriented system is a global planning, provides a new model of transportation construction and land-use (Zheng Ping and Cheng Na 2010).

Its model emphasizes comprehensive land-planning, at first set the urban railway station, the traffic in a within a radius of it via buses and the bus station link up with walking trails or bike lanes. The urban railway and loop of rapid transit, bike or walk lines superimposed together, they can seamlessly transfer. Traffic between the central city and satellite towns belong composite travel, the grading select model of composite transport trips shown in Fig. 58.1, and as a basis for the central cities and satellite towns, composite transport trips pattern as shown in Fig. 58.2.

Curitiba, in Brazil, is a world-recognized public transport model city, the city's transport axis along 5 high-density linear, transformation of the inner city; give priority to public transport instead of private cars. Currently, 75 % people in city

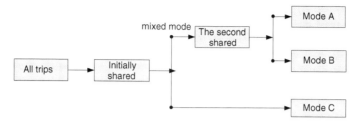

Fig. 58.1 The grading select model of composite transport trips

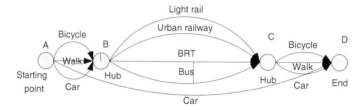

Fig. 58.2 The composite transport trips pattern

use public transport, 17,300 times one day, transport 1.9 million passengers, travel 230,000 miles, saving 700 million gallons fuel a year. We can see that public transport-oriented systems can not only ease the traffic pressure of satellite towns, as the land has been comprehensive planning and utilization, but also use energy more efficiently, reducing carbon emissions, to achieve low-carbon traffic.

58.3.2 Develop Low-Carbon Rail Transport Network

Compared with conventional ground public transport, rail traffic reflects the low-carbon characteristics: low energy consumption, low pollution, low emissions, low noise, and optimize the layout, contributing to industrial development (Dai Yixin 2009). The proportion of rail transit between satellite towns and central city should be higher than 60 % (Tokyo is 80 %).Shanghai as an energy-saving and new energy demonstration pilot city, naturally develop low-carbon rail transit network. Build a well-developed transportation network system and pay attention to develop urban railway, subway and light rail. Table 58.1 shows the characteristics of three rail transit system. Table 58.2 compared the energy consumption and emissions of different transportation modes.

Building rail transit (especially urban railway) as the backbone of the transportation system will be an important part of low-carbon traffic in satellite towns.

Table 58.1 The characteristics of three rail transit system

Item	Index	Light rail	Subway	Urban railway
The indicators of operational characteristics	Max speed (km/h)	70–80	80–100	80–130
	Operating speed (km/h)	20–40	25–60	40–70
	Max density (row/h)	40–90	20–40	10–30
	One-way capacity (row/h)	10000–30000	40000–60000	20000–50000
	Reliability	High	Higher	Higher
Overall system performance	Station spacing	300–800	500–2000	2000
	Average travel distance	Short	Long	Longer

Table 58.2 The energy consumption and emissions of different transportation modes

Transportation	Bus	Car	Subway	Urban railway
Energy consumption/[kJ·(per·km)$^{-1}$]	714	2795.1	322.4	302.5
Average emissions of CO_2/[g·(per·km)$^{-1}$]	19.4	133.9	4.7	3.6

58.4 Conclusion

In recent years, Shanghai has really done much work and achieves certain results in satellite towns of low-carbon transportation system. Overall, however, the development of low carbon transport is still in its infancy, is still a long way away from the comprehensive requirements for low-carbon transport system. The planning of satellite towns is an important issue in urbanization. Its development has an irreplaceable role to solve the urban problems of excessive expansion. In the new century and situation, we must adhere to the principle of low-carbon development and build low-carbon satellite towns and develop low-carbon transportation, to keep the coordinated development between economy and environment.

References

Dai Y (2009) The necessity for China's low-carbon urban development and governance model. J Resour Environ Chinese Popul (3)
Huang W (2003) Development of satellite towns in Shanghai. J Shanghai Adm Inst Newsp 3:33–35
Lu H (2009) Realization pathway of urban green transport. J Urban transp 6(7):23–27
Zheng P, Cheng N (2010) Urban planning based on the concept of low carbon transport strategy. J Huazhong Archit 28 (8)

Chapter 59
Study on Intensive Design of Urban Rail Transport Hub from the Perspective of Low-Carbon

Haishan Xia and Xiaobei Li

Abstract As an important node of three-dimensional development of urban space, rail transportation hub plays a positive role in guiding low-carbon urban construction and intensive development. The construction of rail transit has promoted integration of urban functions of surrounding lots, which can be seen as a catalyst for urban development. Targeted to low-carbon city construction, the paper will analyze main problems of urban rail transport hubs through investigations on urban rail transportation hubs in Beijing, analyze the integration role of rail transit according to catalyst theory, explore and summarize the practices of foreign successful cases, and finally propose solving strategies, hoping to promote the low-carbon development of mega-cities.

Keywords Low-carbon city · Rail transport · Intensification · City catalyst

59.1 Introduction

The intensive development of urban rail transport hub space has a positive effect on low-carbon urban construction and brought a range of linkage effects to sustainable development of urban space. As rail transit for the carrier, the rail transit complex can become a catalyst for update and development of urban space and promote large-scale urban redevelopment of surrounding areas. This paper will

H. Xia (✉) · X. Li
School of Architecture and Design, Beijing Jiaotong University, Beijing 100044, China
e-mail: haishanxia@163.com

X. Li
e-mail: 357806385@qq.com

take urban rail transportation hubs in Beijing for example to analyze main problems of rail transportation hubs in China, apply urban catalyst theory to explore catalytic roles of rail transport as well as intensive design strategies.

59.2 Problem Analysis

59.2.1 Investigation and Analysis of Rail Transportation Hubs in Beijing

In recent years, combined with rail transit construction, Beijing has built some large-scale transportation hub complexes. As two important rail transits in Beijing, Xizhimen and Xidan Joy City have gathered a lot of people and traffic. However, there are still some problems in intensive utilization of urban space.

(1) **Xizhimen rail transit complex**

In past three years, the questionnaires of Xizhimen rail transit complex have been disseminated thousands. The transfer space design has following problems: the crowd disorganized within the traffic transfer space; the transfer route is too long and the space-oriented unclear; the setting of pedestrian system imperfect and the continuity between indoors and outdoors is not strong.

(2) **Underground Commercial Street of Xidan Joy City**

Underground Commercial Street of Xidan Joy City is an extension of the commercial shopping center on the ground, large-scale and more types. However, in the research process it is found that the scene in Underground Commercial Street seems deserted. Through statistical analysis of the questionnaires, poor physical environment and poor sense of direction are two main reasons why shoppers do not want to go to the underground shopping (Fig. 59.1).

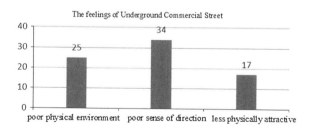

Fig. 59.1 The feelings of underground commercial street

59.2.2 Existing Problems of Rail Transportation Hub in China

Through above analysis and surveys about rail transport hubs in Beijing, there are following problems in construction of rail transportation hub in China:

(1) **The line network planning of rail transport is not closely linked with urban planning or urban design.** Urban design and analysis for the comprehensive development of the rail transport are not enough, causing functional convergence between rail transportation hub and urban space not close enough.
(2) **The development capacity of public space in rail transit station areas is small and single function.** Rail transit station has the important advantage of gathering crowd, which relates to a wide range of public activities in the city, but the development scale of commercial function is not enough, insufficiency, and fails to take full advantage of its economic and commercial value.
(3) **The underground pedestrian system in rail transit station areas is underdeveloped.** The underground pedestrian space is of fragmentation, the connection points with the ground is less and attracting scope small. Besides, space for walking is narrow and single form, lacking support of other city functions compatible with the walkathon, lacking of systematic and sustainability.
(4) **The design of transfer space is insufficient.** Traffic flow lines disorganized, poor reachability, space-oriented weak and other problems cause the contract between rail transportation hub complex and urban public transport weak, thus giving a lot of inconvenience to the transfer Shen (2009).

59.3 Urban Catalyst Theory: The Integration Role of Rail Transit Complexes

59.3.1 Urban Catalyst Theory

American architects Wayne Attoe and Donn Logan have proposed urban catalyst theory: there should be a series of limited effects in the city and each other plays a coordinating role by mutually stimulating. The introduction of urban catalyst as new elements can quickly stimulate related elements react and immediately cause a chain development of other projects. As a concept and method of urban design, how to play this promoting effect and guide low-carbon urban construction is worthy of further exploration.

59.3.2 The Catalytic Role of Rail Transit Complexes

During the construction of low-carbon city, a complex combining many functions into one can reduce land and transportation energy consumption and make pooling of resources and energy Xia (2006). This idea of intensification will be theoretical basis of design of urban rail transit.

Rail transportation complex should become a catalyst for the development of urban areas, catalyzing the development of surrounding areas through integrated functions of cluster effect and walking system of diffusion effect. According to this theory, the catalytic reaction of rail transit should have five basic elements: a catalyst—urban complex; excellent value—integrated functions of cluster effect and walking system of diffusion effect; catalytic mechanism—with the development of urban complex catalytic surrounding development to promote their excellent value; guide policy—put forward a series of controllability requirements to the development around rail transit stations; joint development—rail transportation and real estate, government and non-governmental jointly developed the urban complex.

According to urban catalyst theory, the construction of rail transit complex has injected new elements to urban development and is bound to drive the overall development of surrounding areas. The city catalysis of rail transit complex shows in the following three aspects:

(1) **The creation of positive space and gathering space of low-carbon city.** Combined with the construction of rail transit sites, the influenced area will have an effective urban design, increase parks, squares and other open space, improve the pedestrian system, set flag nodes, enhance the vitality of the region and the public's sense of belonging and improve the quality of urban space.
(2) **Rail transportation has promoted low-carbon space mode of metropolitan group-style.** The construction of rail transit can adjust the structure of urban space, guide urban land use to develop towards reasonable directions, drive the land of surrounding areas appreciation and construction density grow and promote urban spatial structure to transform to low-carbon space mode of multi-center group-style.
(3) **Three-dimensional urban space transformation and revival.** Rail transport can lead to large-scale urban development of the surrounding areas, drive the dynamism of the entire region, have a strong role in promoting the revival of old central area and improve land use efficiency of the new district. The three-dimensional development model of "Underground—ground—high level" has gradually become a standard trend Du (2006).

59.3.3 Intensive Means of Foreign Rail Transit Hubs

(1) **"Wing" integrated transport hub in Nagoya, Japan**

"Wing" integrated transport hub in Nagoya, Japan is located in the city center with land very tense, which is an ideal underground rail hub. In order to achieve intensive and efficient use of urban space, "Wing" integrated transport hub combines functional, spatial and transport organizations together, creating a huge space and value in a limited space and providing people with a perfect mode of multifunction integrated transport hub station.

(2) **Transbay hub in San Francisco**

Transbay hub in San Francisco is the gateway to New York City, an integrated transport hub of setting rail transit, coach passenger and urban road traffic as one. In order to adapt to the needs of urban new development and people travel, the project has conducted integrated arrangement. Different modes of transportation establish each transport function area and flow line system of relatively independent and full contact with surrounding areas. At the same time different levels, multi-directional channels have been set to facilitate passengers passing in and out and transferring, promoting the development and intensive use of the surrounding land.

59.4 Strategies Under New Perspective

The pursuit of a rational and efficient high-density urban space should be the premise of low-carbon urban construction. In order to achieve the intensive use of rail transit hub space, the following resolution strategies have been proposed:

(1) **Emphasis on urban design at rail transit hubs**

In order to solve the problem that the urban design of rail transit station area is weak, there should be an urban design on walking reasonable area as rail transit station for the core of spatial organization and radius of 500 m. Determine the function of comprehensive development and morphological mode of rail transport hub with the surrounding architecture, the scale of the space above and below ground and the underground pedestrian system, flat profile layout and so on Lu and Jing (2007).

(2) **Construct comprehensive walking system in the surrounding region of rail transit hub**

Rail transit complexes should vigorously develop underground pedestrian system and make full use of catalyst effect to stimulate development and construction of the underground walkway of surrounding areas. Concourse level of the station can be directly linked to the underlying of surrounding buildings by ways of overpass, underground sidewalks or commercial pedestrian streets and so on. Entrances should be set cross-blocks, maximize to suture the ground urban space severed by motorized transport, reduce the inconvenience of the ground vehicle flow, increase reachability of stations.

(3) **Focus on oriented design inside rail transit complex**

In response to the problem of weak oriented inside the rail hub, a simple and comfortable transfer space can be created by building means and unified, effective identification system. For example, the design of the entrance should be more visually significant and reflected in the morphology and identification; by strengthening certain spatial form (such as increase the main channels through space as well as atrium in the space center, etc.), elements (light, color, material, etc.), make passengers easier to determine a starting point, destination point and position in space, and guide the flow of people through building means as much as possible.

(4) **Enhance the landscape and environmental quality of rail transit station areas**

For low-carbon urban construction, centralization is not only to obtain space to have buildings clustered together, but focus on the protection of landscape environment. The construction of rail transit site provides an opportunity to the shaping of regional environmental quality. Urban public green spaces and squares should be set in conjunction with exits. Increase the rate of urban green space and public recreation space to effectively enhance the quality of local environment. At the same time integrate the architectural landscape of surrounding areas to form well landscape style of modern city.

59.5 Conclusions

During the construction of low-carbon city, compactness is an inevitable trend of urban architectural space. Through above research on intensive design and countermeasures of rail transport hub, the conclusions are as follows:

(1) According to urban catalyst theory, rail transit complex should play an active role in the construction of low-carbon city, promote intensive regions, functional efficiency and dimensional space.
(2) From the practical point, the rail transportation hub has great potential for integration of urban functions, promoting urban functional integration of surrounding areas and bringing vitality for the city with intensive spatial form.
(3) Successful rail transportation hubs have a variety of patterns to develop. Learn from intensive means of foreign rail transit hubs to direct the future construction for our country.

Acknowledgments This research was supported by: National Natural Science Foundation of China (51078022); The Fundamental Research Funds for the Central Universities (2009JBZ023); Doctoral Fund of Ministry of Education of China (20090009120013); The science and technology project of Ministry of Housing and Urban–Rural Development: Performance evaluation studies of low carbon building technologies and emission reduction (2012-R1-11).

References

Du H (2006) Planning and design study of urban traffic hub. City planning, 07:85–88
Lu J, Jing H (2007) The systematic and urban design of rail station area. Urban Plan, 2:32–36
Shen Z (2009) The core issues of the design of rail transportation hub complex. Time Archit, 5:27–29
Xia H (2006) Ecological transformation and overall design of urban architecture. Southeast University Press, Nanjing

Chapter 60
The Roles of Railway Freight Transport in Developing Low-Carbon Society and Relevant Issues

Guoquan Li

Abstract In this study, the situations of surface freights and potential demands to be suitable for railway container are analyzed, and the possible predominance ranges of railway in transport cost are estimated by the comparative analysis between railway and truck freight rates in transport distance. Moreover, by a case study, the effects of railway freight transport in the reductions of logistics costs and CO_2 emissions, and the savings of energy are derived. Finally, the relevant issues concerning the actual conditions are discussed.

Keywords Railway freights · Predominance range · Logistics costs · CO_2 emissions · Savings of energy

60.1 Introduction

When our world is just facing on the global warming problem, the great numbers of passengers and freights are increasingly flowing among different regions, different countries with the globalization of economy and industry. Although the flows among different countries can use ship, railway, motor, air or pipe, so many products in the Origin or Destination area or between OD are transported with surface means, especially trucks. Therefore, in order to construct a sustainable development and low-carbon society, Railway freight transport is expected to play

G. Li (✉)
Transport Planning and Marketing Laboratory, Railway Technical Research Institute,
2-8-38, Hikari-cho, Kokubunji-shi, Tokyo 185-8540, Japan
e-mail: ligq@rtri.or.jp

more important roles. In the meantime, because of the circumstances of the severe competitiveness in the freight transport market, it is not easy to reinstate its former status in freight market, especially as the island country of Japan, only performing in the relevant policies.

Therefore, as the first step, it is essential to analyze and grasp the effects of railway freight transport in the actual socio-economic activities, and discuss relevant issues of railway transport based on the actual situations of surface freights.

There are some relevant studies with many academic values and actual meanings for railway freight transport, according to the different situations in freight transport market (Li 2003, 2009a, b; Hino et al. 2000; Bontekoning and Priemus 2004; Bärthel and Woxenius 2004; Li et al. 2012). These researches showed that we must improve relevant services and technology of railway freight transport.

This study, at first, investigates the possible freights to be suitable for railway container in Japan's domestic freights, according to the actual freight situations. Then, the potential railway freights passing the analyzed target corridor distributed on the concerning regions are elucidated through building the database of freight flows. Then we estimate the possible predominance range of railway in costs through comparing the actual rate of railway freight to that of truck. With a case study, we derive the effects of railway freight transport in the reductions of logistics costs and CO_2 emissions and the savings of energy. And the relevant issues are discussed.

60.2 Domestic Freight Situations

According to the relevant inter-regional freight flow survey of Japan's government (Ministry of Land, Infrastructure, Transport and Tourism 2003), we can describe the changing tendency of domestic or inter-regional freights. The domestic freight tonnages decreased from 6.96 billion tons in the early 1990 s to 5.32 billion tons in 2008. The decreased freight tonnages were about 1.5 billion tons. On the other hand, the inter-regional freights grew from 1.62 billion tons in 1991 to 1.83 billion tons in 2008, during the same period.

Furthermore, The investigations of actual surface transport situations revealed that road haulage accounts for an overwhelming transport proportion in short-distance, and even in long distance of over 1,000 km, road haulage's share was still more than three times larger than that of railway transport.

Although the modal shift policy from excessive road haulage to railway transport or coastal shipping in Japan has been proposed from the 1980 s, the freight share of road haulage in domestic transport market is increasing, and the relevant share of railway or coastal shipping freights is decreasing, continuously. As for the road haulage, the shares in tonnage have gone up from 90.2 % in 1985 to 91.2 % in 2005, and that in ton kilometer from 47.4–58.6 %, in the same period.

Contrastively, the railway freights' shares have gone down from 1.8 % to about 1.0 % in tonnage, and from 5.1 % to about 4 % in ton kilometer, respectively. Additionally, the shares of coastal shipping changed from 8.1–7.8 % in tonnage, and from 47.4–37.2 % in ton kilometer, respectively. It is really the reverse phenomenon of the modal shift to the relevant proposed policy.

All as mentioned-above can be depicted two extremes. One is meant that there are really no suitable freights for railway. The other is that the roles or significances of railway freight transport can not still be understood by the society.

60.3 Potential Railway Freights Based on the Analyzing Corridor

In this study, we focus on the surface inter-regional freights in manufacture, the freights among different transport means such as chartered truck, trailer, rail container and other, are comparatively analyzed, in production scales of users, the shipments' time of freight, transport lot. And then the concentration ratio of freight's transporting destinations, and freight items sent from different regions are investigated. It is found that there are so many similarities between the freights of current railway transport and road haulage by chartered truck and trailer. Therefore, the freights of the chartered truck and trailer can be also seen as the possible freights to be suitable for railway container. They are considered as potential demands for railway transport.

With the use of the actual transport routine database of railway freights built in this study, and goods survey of the government (Ministry of Land, Infrastructure, Transport and Tourism 2003), the corresponding freight flow routines in the influence range of each terminal related to relevant analyzed target corridor are investigated, which be called as the analyzing corridor, according to the actual situations of surface freights. Based on these, the possible freights to use the analyzing corridor when they are transported by railway, following the routines' database of the actual railway freights, are extracted.

In this study, we use the corridor in railway network that links Kyushu area to other areas of Japan as target analysis. The analyzing corridor is concerning 138 terminals almost covering all regions of the country. The detailed contents of the potential railway freights indicate that although the actual result of railway freight in transport share of surface freight is higher than the average level of entire country, the transport share of road haulage in the potential railway freights is about 84.8 %, and the railway container share is only 15.2 %. Therefore, the actual share of railway freights is very low in its potentials. Therefore, in the actual inter-regional surface freights, there are the freights suitable for railway transport.

60.4 Possible Predominance Range of Railway in Transport Costs

It is needed to discuss the possible predominance ranges of railway transport according to the actual situations in multiple respects. But there are many difficulties to grasp all of the real transport conditions, concerning the business practices of users and operators, many factors in the freight market. Therefore, this study focuses on the truck freight rate in the mileage system and investigates the truckload transport rates using actual data from the users' survey (Cargo News 2002, 2006). And then, through the comparison with the transport rate of railway container, we can find the possible predominance range of railway in transport cost.

The main factors involved in the setting of truckload freight rates based on the mileage system are truck type (i.e., vehicle tonnage) and transport distance. In the conventional tariff based on ton-kilometer, the freight rate generally tends to decrease with increased transport distance and/or vehicle tonnage. In reality, many factors influence truck freight rate. When a user chooses a carrier to transport freight, the whole range of carriers available is generally considered. Charges involved in using expressways or ferries if necessary, as well as the term of the contract with the carrier, have an influence on freight rates. In addition, when multiple means of transport are available, users have the advantage in transport contract negotiations with carriers.

About the freight rate of railway container, because it is based on the 12ft container in Japan, only if we find the relationships between the railway transport costs and relevant distance, based on the actual transport cost of users' survey, the freight rate of railway transport can be derived.

In the potential railway freights, it is clear that all of benefits are obtainable not by railway, but the extent to which railway transport is beneficial should be determined. Generally, one of criteria to understand the obtainable benefits by railway is transport costs. On the premise that users choose the cheapest transport means, the possible predominance range of railway in distance can described through following condition.

Railway freight rate ≤ truck freight rate

It can also be seen as one of the fundamental standards against which the cost advantages of railway freight transport are judged. According to the comparative analyses of truck and railway freight rate, using the current transport data from the users' survey, the predominance range of railway container transport can be estimated as shown in Fig. 60.1.

For example, the freight rate of railway container will be cheaper than that of 10-tonnage truck when the transport distance is greater than 350 km. Naturally, the advantages of each transport means cannot be judged exclusively in term of the rates, but there do are freight convoys that may potentially be transported by railway with the cheaper freight rate.

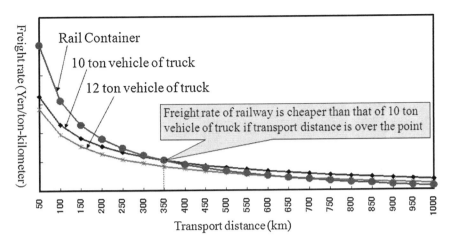

Fig. 60.1 Possible predominance range of railway freight in transport cost

60.5 Effects of Railway Freight Transport on Developing Low-Carbon Society

As stated previously, railway freight transport, in fact, can be described as current and potential freights, based on the current situations and restrictions on freight transport along the analyzing corridor. Social and economic benefits by railway include the current effects of the carriage of freight by railway and the potential effects that may be induced when the potential freights of railway could be shifted from road to railway.

The current effect denotes the level by which logistics costs, CO_2 emissions and energy consumptions are reduced due to goods currently being transported by railway.

First, the freight transported on the analyzing corridor is ascertained from the current inter-regional freight data, and logistics costs, CO_2 emissions and energy consumptions are estimated based on the routing of railway freight. Next, relevant amounts are estimated assuming that said freight is transported by road. The comparison of the two gives us the evaluated current effects of railway freight.

The potential effects are assumed that there will be the reductions in logistics costs, CO_2 emissions and energy consumptions if freight currently transported by road in the predominance range is shifted to railway.

Because the influential area of the analyzing corridors is concerning the entire country, the surface freight transported by road must match the relevant railway route along the analyzing corridor. As the results of a case study, the social and economic effects of railway freight transport can be derived as shown in Table 60.1.

The current effects in cost reductions are 7.8 billion yen per year, 390 thousand tons per year in CO_2 emissions reductions, and 150 million litre per year in savings

Table 60.1 Effects of railway freight transport based on the analyzing corridor

Effects (an year)	Cost reductions (Billion Yen)	CO_2 emission reductions (thousand ton-CO_2)	Savings of energy (million litre)
Current	7.8	390	150
Potential	16.3	810	310

of diesel. When the modal shift of potential freight transport from road to railway is really implemented, the railway freight transport would save 16.3 billion yen in costs, reduce 810 thousand tons of CO_2 emissions, and save 310 million litre of diesel.

60.6 Discussions on the Issues of Railway Freight Transport

In order to promote the railway freight transport for low-carbon society, some important issues must be sufficiently understood and discussed.

In freight transport market, the railway transport's merchandise is originally the train diagram with exact schedule to be offered to the users according to their needs. This is the basic and essential conditions for promoting railway freights. As for the actual situations of railway freight transport, it is difficult to make the corresponding freight diagram exactly matching with the needs of users, because the railway freight operator has to use the railway networks of plural passenger railway companies to run the freight trains in the gaps among the passenger trains. In the meantime, there are overcrowding schedule of passenger trains in many main corridors, where are called as bottlenecks. These bottlenecks are restricting the development of railway freights. How to arrange the freight trains, as affecting the whole railway network, is one of the most important issues to be soberly discussed further in the future.

And then, the operator of railway transport shall make greater efforts to provide the relevant services corresponding to the needs of users. Moreover, Corresponding to the change of socio-economy and industries, the radical issue in railway is to reconstruct the traditional system to be suitable for the new needs.

Government shall have more effectual and useful measures and policies in improving transport condition of the network, especially the bottleneck corridors, and in guiding users to use railway such as introducing the taxation of CO_2 emissions, environment-friendly grant, and so on. To the users, it is necessary to recognize the merits of railway transport not only in the environment-friendliness, but also in the efficiency of logistics and the savings of the energy and labor, and so on.

Therefore, the current issues of railway freight transport are not only concerning the relevant operators, but also concerning the entire society.

60.7 Conclusions

Through the comparative analysis of current railway transport and road haulage, based on a special case as Japan railway, this study can mainly be concluded as follows.

In the actual situations of domestic surface freights, there are still a large volume of potential freights to be suitable for railway. It is estimated that the possible predominance range of railway in transport cost is greater than 350 km of transport distance. With the case study of an analyzing corridor, using the criterion in transport cost, the effects of railway freight transport in the reductions of logistics costs and CO_2 emissions, and the savings of energy are derived. The results showed that it is possible that in developing low-carbon society, railway freight transport will play more important roles. Conclusively, the relevant issues concerning the actual transport conditions are discussed, in the improvement of railway transport system, government roles and users' recognitions. Only if the reasonable transport policies and measures to improve the relevant railway infrastructure and transport services are implemented, the railway freight transport will have furthermore effects.

References

Bärthel F, Woxenius J (2004) Developing intermodal transport for small flows over short distance. Transp Plan Technol 27(5):403–424
Bontekoning YM, Priemus H (2004) Breakthrough innovations in intermodal freight transport. Transp Plan Technol 27(5):335–345
Cargo News (2002) The current situation of transport tariffs and warehouse rates of main users. In: 24th revision version of survey, Tokyo (in Japanese)
Cargo News (2006) The current situation of transport tariffs and warehouse rates of main users. In: 28th revision version of survey, Tokyo (in Japanese)
Hino S, Kishi K, Satoh K, Chiba H (2000) Roles and persistence measures of railway freight transportation between hokkaido and honshu. Infrastruct Plan Rev 17:827–834 (in Japanese)
Li G (2003) Railway system for intermodal freight transport: its concept and relevant approach of improvement. Transp Policy Stud Rev 13(3):14–23 (in Japanese)
Li G (2009a) Intermodal freight transport. Railway Res Rev 66(3):30–33 (in Japanese)
Li G (2009b) Modal shifts in freight transport. Railway Res Rev 66(12):6–9 (in Japanese)
Li G, Muto M, Suzuki T, Okuda D (2012) Fundamental study on the possibilities of freight modal shift based on the actual situations of domestic surface transport. J Railway Eng JSCE 16:189–194 (in Japanese)
Ministry of Land, Infrastructure, Transport and Tourism, Japan: Regional Freight Flow Survey. http://www.mlit.go.jp/

Part III
SS-Industrial Security Under Low Carbon Development

Chapter 61
Preliminary Study on Coal Industrial Safety Evaluation Index System Under Low-Carbon Economy

Lei Zhang and Cheng Chen

Abstract Coal industry occupies a dominant position in our energy supply and consumption, which is directly related to national security, social stability and sustainable development. However, under the environment of low-carbon economy, with resources and environmental problems becoming increasingly prominent, the constraints on economic development are increasingly influencing the stable and healthy development of the industry. Therefore, it requires taking resources and environmental factors into the evaluation index system of the industrial security. Based on the coal industry security factors and principles of indicators system designing, the policy that states to promote energy conservation to achieve the strategic objectives of the low-carbon development, and characteristics of coal industrial development, establish a more systematic, comprehensive evaluation index system of low-carbon development on the coal industry.

Keywords Coal industry · Low-carbon development · Industrial safety · Evaluation index

L. Zhang (✉)
School of Economics and Management, Beijing Jiaotong University, Beijing, China
e-mail: emmachang1989@sina.com

C. Chen
Graduate School, The Chinese Academy of Social Sciences, Beijing, China
e-mail: jerrychan1988@sina.com

61.1 Introduction

Energy is the basic factor constraining economic growth and the basis of the existence and development of human society, we must insist on energy as the strategic focus of economic development, make sure sustainable development and efficient use of energy to support sustainable economic and social development. Chinese energy structure-"Lack of oil, less gas, rich in coal" decides that coals occupy a dominant position in our energy supply and consumption. As the earliest development and use of primary energy in human society, coal occupies the irreplaceable foundation position in national development, which is directly related to national security, social stability and sustainable development, therefore it has been very important in the world's energy consumption structure. At the same time energy is also an important guarantee to the thriving of a country. Especially in the last few years, in the context of the world's strong economic growth and rising global oil prices, with coal-clean, the mature and use of efficient technology, the value of coal itself has been greatly improved, while it also stimulated a substantial increase in global coal production, consumption and trade.

Data published by BP World Energy Statistics (2012) shows that in 2011, coal now accounts for 30.3 % of global energy consumption, the highest share since 1969. BP data shows that in 2011, coal consumption grew by 5.4 % in 2011, the only fossil fuel to record above average growth and the fastest-growing form of energy outside renewable, reaching 3.72 million tonnes of oil equivalent and accounting for about 68.6 % of the total global consumption in the Asia-Pacific region.

In 2011, China's coal imports continue to remain high, the total imports build another new height in recent years, and the scale of exports is still small, with imports disparities. China's coal import is 182 million tons, an increase of 10.8 %; export is 14.66 million tons, down 23 %; net import is 168 million tons, an increase of 15.2 %. At the end of December 2011, the whole society in China stocks 253 million tons coal, an increase of 16.6 % compared with the beginning. In recent years, as a number of new constructions, renovation, expansion and integration of resources coal technological transformation completed in succession and put into production, the release of coal production capacity gradually speed up, in 2011 the production capacity of China's coal industry is 95 million tons.

In the next 30–50 years, although the new energy and renewable energy development and promotion, will make the consumption proportion of coal in primary energy decline, but the dominant position of coal still will not change. Thus, the coal in the rapid development of China's national economy will continue to play an important role. With the worldwide resource depletion, and advances in technology, the development of China's coal industry has begun to gradually enter the stage of the "qualitative change". But in the process of change on the coal industry, China's coal industry is running a lot of problems: the world's major coal-producing countries have increased the intensity of coal production and export in the stimulation of international coal prices rising. China's energy structure-"Lack of oil, less gas, rich in coal"—decides the only coal-based primary energy structure in the historical period towards the sustainable development, which requires promoting the progress of the coal industry to meet the national growing energy demand. The key to the development of China's coal industry is to strive to narrow the gap of the efficiency, effectiveness, safety and environmental protection with international industry, the coal industry not only need domestic competitiveness, but also need to improve the international competitive advantage. As a developing country, whether to get rid of extensive expansion of the development model in economic growth, to change the past high growth, high consumption and high pollution, to achieve high growth, low consumption, low pollution, is the key to transformation of China's energy of economic growth and achieve sustainable development. In this context, to explore the influencing factors of the Chinese coal industry safety and evaluation of China's coal industry security has become a strong reality and urgency of the major theoretical and reality during China's energy sector meet the world economy globalization's opportunities and challenges. Therefore, how to scientifically evaluate the operational state of China's coal industry, and build the coal industrial evaluation index system under low-carbon economy has great significance.

Foreign study on the evaluation of industrial security is still in the exploration of the initial stage. With the advent of economic globalization and China's accession to the WTO, China's economic openness enhances unceasingly, some scholars started to pay attention to industrial security, and have proposed metrics and evaluation indicators on industrial security.

Professor He Weida has quite contribution in the field of Industrial Security, in his text, Professor He Weida (2002), decomposed industrial security evaluation index system into international competitiveness evaluation indicators of industry, external dependency evaluation indicators of industry, the controlling power evaluation indicators of industry, and, respectively, assessed the international competitiveness of industry with trade specialization coefficient indicators, evaluated the industry dependence on foreign industry with industrial export degree of dependence on foreign trade and industrial capital indicators, used the foreign equity control indicators to evaluate the controlling power of industry, and it can better give consideration systematic, testability and other requirements of the index system. Industrial Safety evaluation proposed by Professor Yang Gongpu, and Xia Dawei (2005), Professor Shi Zhongliang (2005) is extended to four evaluation indicators on the basis of Professor He Weida. Jing Yuqin (2006) proposed industrial domestic environmental evaluation, industrial competitiveness evaluation and industrial control evaluation, analyzed six major influencing factors-government regulation environment, market environment performance, structure, industry controlled situation and the country concentration, took the government performance indicators into them, and eliminated the industrial capital and industrial technology dependence on foreign which are highly related to the control rate of foreign ownership and foreign technology. The evaluation of industrial security status must be combined with qualitative research, in addition to quantitative analysis.

However, such studies deserve to be discussed further and improved. Under the environment of low-carbon economy, with resources and environmental problems becoming increasingly prominent, the constraints on economic development are increasingly influencing the stable and healthy development of the industry. Therefore, it requires taking resources and environmental factors into the evaluation index system of the industrial security.

61.2 Analysis on the Influence Factors of the Coal Industry Safety

The factors affecting the safety of the coal industry involved in internal and external aspects.

61.2.1 Internal Factors

Internal factors Industrial security involved in two major categories of the living environment of the domestic industry and competitive environment.

61.2.1.1 The Living Environment

There is a very direct relationship between the industrial safety and industrial environment. The domestic living environment of the industrial is the basis of the industrial survival. The industrial environment in the broad sense, include the status of the industry and the various influence factors such as industrial development, natural geographical factors, macroeconomic factors, political and legal factors, and socio-cultural factors. Industrial environment described in the context of industrial safety issues includes production environment, market demand, environmental and industrial policy environment which affect the industrial safety development. In addition, there is also the resources environment. Per capita recoverable reserves of China's coal is less, only two-thirds of the world level; the development scale is large, and reserve-production ratio is less than one-third of the world average; the rate of resource extraction is low and consumption is large, about 48 % of the world. Resources development and utilization are difficult to support long-term economic and social development.

61.2.1.2 The Competitive Environment

In the theories of industrial economics, excessive competition and monopoly are not the ideal market structure, and both are the deviation from the optimal allocation of resources. If a country's enterprises have lost control and influence on a reasonable competitive landscape, excessive competition is bound to affect the reasonable adjustment of industrial structure, thereby affect industrial safety.

61.2.2 External Factors

External factors affecting the industrial security include capital, technology and products factors from foreign due to global economic integration and market opening conditions.

61.2.2.1 Foreign Capital

In recent years, with the strong demand of the Chinese market for energy, and the consolidation trend in the Chinese coal mining industry, a growing number of international financial capital are looking for investment opportunities in China's coal mining industry. Suddenly foreign investment in the China's mining is active; the coal industry has become the hot pursuit of foreign capital, foreign mergers and acquisitions are in the ascendant, the scale and pace of mergers and acquisitions is rapidly rising. Large-scale mergers and acquisitions of foreign investors in the coal industry have a significant impact on the control of the industrial chain and state-owned assets of China, and it likely causes local monopoly problem.

61.2.2.2 Foreign Technology

Technology import is the important supply channels of a country's technological progress, any country only depending on their own invention alone is far to meet the needs of technical progress. Coal machinery products are often long-term work in a harsh and complex environment, the requirements of overall technique is safe, reliable, and able to adapt to the harsh operating conditions. Compared with foreign countries, coal equipment manufacturing industry in China has made some progress in recent years, but the overall technical and technological level is not high, and there is still a certain gap compared with the international advanced level in automation and control, service life and reliability.

61.2.2.3 Foreign Monopoly of Raw Materials and the Price of Resource Products

Coal machinery and most of the raw materials are made of steel processing; the cost is affected by the steel price fluctuations. Steel raw material prices are closely related to the macroeconomic situation, and the periodic is strong. In recent years, the fluctuation in steel prices is frequent; coal machinery enterprises are facing greater cost pressures.

Table 61.1 Coal industrial evaluation index system under low-carbon economy

Target	One-class index	Level 2 index	Level 3 index
Coal industrial evaluation index system	Living environment of industry	Production elements environment	Capital efficiency
			Cost of capital
			Overall labor productivity
			Unit labor cost
		Market demand environment	The growth rate of market demand
		Institutional environment	The degree of marketization
			The government regulation performance level
		Resources environment	Coal consumption strength unit GDP
			"Three-wastes" emissions unit production value
			"Three-wastes" comprehensive utilization rate
			The main pollutants
			Unit value added of industry water consumption
			One million tons mortality in the coal mine safety production
			The reserve-production ratio of Coal resources
	International competitiveness of industry	Trade competitiveness	Trade special coefficient
		Performance competitiveness	Capacity utilization
			The added value of the coal industry
			Proportion of products sold
		Structure competitiveness	Market concentration rate
		Technology competitiveness	R&D investment Proportion in the mining
			Professional technology personnel proportion
		Scale competitiveness	Domestic market share
			The international market share
	External dependency of industry	Industrial export level	Industrial export degree of dependence on foreign trade
		Industrial import level	Industrial import degree of dependence on foreign trade
	Industrial control	Foreign capital market scale	Industry foreign capital market control
		Foreign capital scale	Industry of foreign capital control
		Foreign technology scale	Foreign technology industry control

61.3 Establishment of Coal Industrial Evaluation Index System

Based on the coal industry security factors and principles of indicators system designing, the policy that states to promote energy conservation to achieve the strategic objectives of the low-carbon development, and characteristics of coal industrial development, establish a more systematic, comprehensive evaluation index system of low-carbon development on the coal industry. The indicator system consists of 4 one-class index of the living environment of industry, the international competitiveness of industry, external dependency of industry and industrial control, 14 level-2 indicators of production elements environment, market demand environment, institutional environment, resources environment, and 28 level-3 indicators of capital efficiency, cost of capital, personnel labor production rate, etc. in Table 61.1.

61.4 Conclusions

Bring the resources environment indicators of the coal industry into the industrial safety evaluation system is correction and perfect to the measure of industry safety standards. To safety assessment on the industry, problems which need to further consider or take attention, include: (1) The data in industrial safety evaluation needs to reprocess under the existing statistical standards; (2) The industrial safety evaluation indicators have relative characteristics, and the industrial safety assessment is a dynamic process; therefore, different industry types could cause change of certain factors and their weighting; (3) It requires a combination of qualitative analysis and quantitative evaluation of industrial safety.

References

Statistical Review of World Energy (2012) http://www.bp.com
He W, He C (2002) Preliminary estimate of the safety of the three major industries in China. China Ind Econ 2:25–31
Yang G, Xia D (2005) Modern industrial economics. Shanghai University of Finance and Economics Press, Shanghai
Shi Z (2005) Industrial Economics. Economic Management Press, Beijing
Jing Y (2006) Industrial safety evaluation index system. Economist 2:70–76

Chapter 62
China's Energy Economy from Low-Carbon Perspective

Xiaonan Qu

Abstract Energy security issues related to China's economic lifeline and livelihood of our people, also of great significance to maintaining world peace and stability and promoting common development. With China's rapid economic development of China's energy demand is also increasing year by year in this article mainly analyzes the current development of China's energy industry, and development issues, and made recommendations on China's energy development.

Keywords Low-carbon · Energy · Sustainable development

62.1 China's Energy Situation

China has a 9.6 million square kilometers' land area, and territorial sea, continental shelf, exclusive economic zone area of approximately 3 million square km. From Table, China's mainly fossil energy reserves are rankings at the forefront in the world. China's fossil energy reserves from the total amount.

Our country's energy is mainly petrochemical energy at which made up by oil, coal and natural gas mainly at present. However, fossil energy is non-renewable energy, the reserves are limited. With the rapid development of economy and society in recent years, the demand for energy is more and more robust. This limited petrochemical energy reserves are then declining, prices are rising. Coal is the main

X. Qu (✉)
School of Economics and Management, Beijing Jiaotong University, No.3 Shangyuancun, Xizhimenwai, Beijing, China
e-mail: 11120534@bjtu.edu.cn

Table 62.1 China's major energy reserves situation

Energy type	Recoverable reserves of the world	Recoverable reserves of China	Proportion of the world (%)	World ranking	Reserve-production ratio of the world	Reserve-production ratio of China
Oil	13.2832 trillion barrels	14.8 billion barrels	1.1	14	46.2	9.9
Natural gas	187.1 trillion cubic meters	2.8 trillion cubic meters	1.5	14	58.6	29
Coal	860.9 billion tons	114.5 billion tons	13.3	3	118	35

fossil energy in China, it takes 70 % of the whole energy consumption. BP statistics show the world's Reserve-production ratio of oil, coal and natural gas is 46.2 years, 118 and 58.6 years, however, the data in China are 9.9, 35 and 29 years.

At the same time, China's energy output and growth rate was significantly lower than the consumption and its growth rate. Oil, for example, between 2000 and 2010, the Chinese oil production increased by 24.85 % at the same time, oil consumption is simultaneous increase 91.2 %.

In 2007, crude oil apparent consumption of China reached 346 million tons, 159 million tons of net imports, the import dependence is as high as 46.1 %. According to the forecast, as China is increasingly dependent on overseas oil supplies, the oil imports from 350 million barrels/day in 2006 surge to 1,310 million barrels/day in 2030, the share of imports in its demand will rise from 50 to 80 %. Moreover, the world power demand annual rate will be 2.5 % by the year 2030, from a global perspective, by the end of 2030, the total amount of electric power production capacity will be 4.8 million megawatts. China will contribute 28 % of the total increasing amount. In China, the increasing demand for oil, gas and coal resources in electric power industry makes the energy security condition more severely.

The great consumption of fossil fuels leads to serious environmental disruption, the increasing carbon dioxide emissions will sharpen the greenhouse effect. Since 2007, China has became the world first emissions superpower. And because the U.S financial crisis, the energy consumption and greenhouse gas are continuous in negative growth emissions in recent years. However, China has 6–9 % high growth speed each year. China's per capital emissions in 2007 is more than the average level in the world, in many of the cities has surpassed many other developed countries, such as France, Sweden, and approached the European Union, Japan. In 2010, the calculated carbon dioxide emissions based on the Chinese National Bureau of Statistics is 7.693 billion tons. As the second place, carbon emissions in the U.S is 5.638 billion tons. Since the beginning of this century, China has increased its greenhouse gas emissions to account for about 40 % of the world. The greenhouse effect caused global warming, that brings increasing intensity and frequency of extreme weather or climate in recent years, and changes the law of natural disasters, which had a serious bad impact on global food production, human life and the natural environment. Gill Bates (2007) shows the climate of China occurred significant changes, it also greatly affect the ecosystem. This is an unavoidable problem in the sustainable economic and social development.

62.2 Some Problems for China's Energy Industry

62.2.1 Some Structural Problems for Energy Industry in China

The majority energy production and consumption rely on the coal. Compared to the production and consumption of the other two energies, namely the gas and the petroleum, those of the coal take up 86.1 and 76.5 % respectively. In a word, it is an outstanding figure of Chinese energy structural that the coal holds the dominant position. According to the figure published in 2004, 36.8 % petroleum, 23.7 % gas, 27.2 % coal, 6.2 % hydropower, and 6.1 % nuclear energy construct the consumption of world's energy. However, Chinese figure is quite different: 22.3 % petroleum, 2.5 % gas, 69.0 % coal, 5.4 % hydropower, and 0.8 % nuclear energy. After such a comparison, Chinese energy structure drags because of its lower gas and nuclear energy consumption and the much higher coal consumption. China holds 13.3 % of world's total coal mining that is the third largest figure in the world, just following the US and Russia while our country products 48.3 % and consumes 48.2 % of world's total figure. Wei et al. (2007) proposed that contrary to the fact that the exploitation rate and power conversion rate of coal is lower than the figures of petroleum based on prevailing technology, overuse of low level energies such as the coal brings high level emission of Carbon Dioxide. According to the survey in 2010, 81.07 % energy consumption Carbon Dioxide comes from the consumption of the coal. Furthermore, China will stand as a heavier player for Carbon Dioxide emission with its higher economic development and increasing coal consumption.

62.2.2 A Big Issue of Low Efficiency for Energy Consumption

Hu and Wang (2006) refer that recently high speed economic growth of China is at the cost of the huge consumption of fossil energy and the resulting countless damage to the environment. Such an extensive development model is harmful to China's long term economic growth and not sustainable.

As the survey has shown, China experienced annual 9 % energy consumption for ten years, but its GDP only took up 4 % of the world's figure while China consumed world's 31 % coal, 8 % petroleum and 10 % electricity with a 30 % mine recourse recycle rate that is 20 % lower than more advanced level of recycle. This trend strengthens after 2010 because high energy consumption and high emission industries boomed with the recovery of backward production capacity. Such enormous energy consumption not only exhausted natural resources capacity but also brought serious environmental problems.

Chinese government executed a series of energy saving solutions for the micro and macro level with exhausting energy and polluted environment. As a result, China set the restricted index of 20 % lower energy consumption per 10 thousand GDP than that of 2005 and 10 % lower emission of major pollutants. However, the situation is passive: on one hand, energy production keeps the track of "high energy consumption, high pollution and low production" that is traditional; on the other hand, energy demand is increasing because of higher living standard and high speed economic development. Currently, China ranks the first place for emission of the wastes, and thus faces more serious emission reduction pressure during "after Kyoto era".

62.2.3 *Some Problems with the Development of Net Energy*

In spite of the fact that New Energy Industry is developing fast, Chinese new energy industry is still lagging in respect to the conditions of developed countries. Developed countries are always walking on the front edge of new energy industry. Even though the new energy industry of China also booms recently, and the fact that China is the largest wind power machine and solar cells producer, but the insufficient investment of basic research and development, limits of core technology field restrict the long term development of this industry. As to the firms of this industry, they rely on the core technology provider abroad but also throw themselves to the fierce competition over the world, however, limited cash held and scarce financing resources are the major problems that limit their ability of creation.

China has no price advantage because of relative higher cost of generating new energy. New energy only takes a small percentage of Chinese electric power industry, less than 1 % of renewable energy excluding hydropower. In addition, new energy power's cost is much higher than that of coal power, namely, 150 % coal's cost for biomass power generation, 170 % for wind power generation, 1100–1800 % for Photovoltaic power generation, new energy is a small part of the traditional power generation method. What's more, China's net energy industry relies on no strong and powerful laws and regulations published by government and no strategic planning in the long run. Overcapacity problem is also very serious in China. Take the wind power as an example, in 2008, there were 12.15 million wind watt's power generation machines finished, but only 10 million watts power can be generated. But the fact is that only 73 % of the planned machines were completed with the real wind power generation capacity 8.94 million. The deeper reason is that the system has not been constructed with any synergy between net energy development and electric net construction.

62.3 Suggestions for the Energy Development in China

62.3.1 Adjust the Energy Industry Structure, Promote the Upgrading of the Industrial Structure Optimization

In recent years, although China continues to adjust and optimize industrial structure, industrial energy consumption accounts for the proportion of total energy consumption is still large, especially compared with developed countries, China's energy consumption per unit of output is still high in the secondary industry. Adjust the industrial structure, promote the optimization and upgrading of industrial structure are still the major issue. Adhere to the principle of energy conservation priority, promote ecological civilization and saving culture. Popularize technology achievements, strive to reduce energy consumption. Create economical models of development and consumption, improve energy utilization efficiency. Balance supply and demand reasonably. That are the ways for realizing the sustainable and healthy development of China's economy.

New energy industry development is inseparable from national policy support. Since 2009, our attitudes to the development of new energy industry has changed from "actively guide" to " high strategic valued". During the 12th Five-Year Plan, the overall goal of China's new energy development is to establish the initial adaptation of large-scale development of new energy grid and other infrastructure systems. Promoting the growth and upgrading of new energy equipment manufacturing industry, and the expansion of the new energy market. Raise the proportion of non-fossil energy sources in energy consumption to about 12 % in 2015. With the exception of hydropower, renewable energy accounted for primary energy consumption will rise to 6 % in 2020. The state will invest more than 3 trillion yuan to promote the development of new energy including solar energy, wind power, biomass.

62.3.2 Improve Energy Efficiency

Management and technology are the two aspects to improve energy efficiency. On the management side, energy saving standards, goals and policy measures should be developed in all sectors. The energy consumption in various sectors will be supervised, reward and punishment are also needed to encourage enterprises improving energy efficiency continuously. In technical terms, through the transformation of high technology, promotion of the traditional industry production, developing and popularizing new energy-saving technology, improving the utilization rate of raw coal and crude oil, strengthen the investment in the coal washing and processing, moulded coal, coal desulfurization and the use of clean energy.

62.3.3 Strengthen the Research and Development of New Energy Core Technology, Promote Independent Innovation Ability

Developed countries always lead the trend of new energy development. Despite the rapid development of China's new energy industry, as the world's largest installed wind power country, the largest producer of solar cell, the basic investments for research and development are obviously insufficient, the core technology is always the bottleneck. This restricts the long-term development of new energy industry. The shortage of funds and financing are the main setbacks for the development of new energy and the ability of independent innovation. An effective solution is a detailed set of financing strategy and the faithfully implement of it. The best way for China's new energy industry is following the world's trend, improving the ability of independent innovation.

Blind production should be guided to the rational and orderly production. Enhance project approval management and the access policy of new energy market, eliminate backward production capacity, promote mergers and acquisitions, motivate economies of scale and economy of scope, improve the efficiency of the industrial structure. Regulate the order of market competition, execute the "Anti-Unfair Competition Law" strictly to prevent abuse of market dominance and excessive competition. Improve the new energy enterprise management level and promote the continuous upgrading technology under the market competition mechanism. Expanding domestic demand, foster domestic market, increase the use of new energy products. Shifting the production subsidies to the domestic consumer subsidies to encourage enterprises excavating the consumption potential of domestic market. On one hand, achieve the original intention of new clean energy interiorly. On the other hand, reduce exports will reduce trade friction and create a favorable export environment. In fact, it is impossible to exchange the core technology for market. developed countries will never transfer the latest technology to other countries especially developing countries. To break through the bottlenecks of technology, developing countries should establish long-term developing consciousness, increase research and development investment, enhance the capability of independent innovation and technological breakthroughs.

References

Gill B (2007) Rising Star: China's new security diplomacy. Brookings Institution Press, Washington

Hu JL, Wang SC (2006) Total factor energy efficiency of regions in China. Energy Policy 34(17):3206–3217

Wei YM, Liao H, Fan Y (2007) An empirical analysis of energy efficiency in China's iron and steel sector. Energy 32(12):2262–2270

Chapter 63
Analysis for Transformation and Development of China PV Industry

Shengzhen Ma

Abstract Under the targets of low-carbon environment, China solar industry developed fairly fast during the last ten years. However, as an important component of China solar industry, PV industry is facing the most difficult time. This paper analyses the problems along the development of China PV industry and some suggestions for a better development of the industry.

Keywords Low-carbon economy · PV industry · Technological innovation · New energy sources

63.1 Domestic and Overseas Research Actuality

Zhao Wenyu (2001) did a research about the development strategy of China PV industry. He started his article with the gap of China PV industry and other countries, the challenges for joining in the WTO. His work showed us the world PV industry's development trend, speed, and the strategy China PV industry should apply. He made a preliminary developing forecast of the early twenty-first century.

National Development and Reform Commission, the Global Environment Facility (GEF), World Bank, China Renewable Energy Development Project Office set up an expert group for Development Study Project of the China PV industry in October, 2004. They did a research for the development of China PV industry, and composed "China PV industry development report". The report

S. Ma (✉)
School of Economics and Management, Beijing Jiaotong University,
No.3 Shangyuancun, Xizhimenwai, Beijing, China
e-mail: 11120545@bjtu.edu.cn

summarized the current situation of China's solar energy resources, technology, development and market prospects. It also raised policy and action plan for further promotion of China PV industry.

Ma Shenghong and her team members (2005) presented the future direction of photovoltaic power generation in China's energy structure. They believed that in the next 10–20 years, China's photovoltaic power generation will be mainly used in the following areas: off-grid power supply in rural areas, power distribution, large-scale desert power station and other commercial applications. The first step was to intensify the ability of photovoltaic generation, which includes: resources survey and evaluation, research, training system, quality control service system.

Michael Rogol's (2005) "The solar industry outlook report" was the most authoritative solar energy industry report throughout the world. This report was known as Lyons report. In this report, it was said that although the investment volatility of solar energy industry will increase, the current profit status would continue to increase several times over now, solar stock still had a strong rising power. The report predicted that the entire industry sales revenue in 2010 would reach $36 billion, pre-tax profit will be atleast $6.4 billion.

"Global PV industry development research report" (2009) summarized the development of global PV industry and surplus production problems. The report introduced the tendency of PV industry in Germany, United States, Spain, Japan and China. It pointed out that China was a potential market, the installed capacity of China PV industry will surpass Japan by 2020.

UBS Investment Research said Germany was the country that gives the best policy support in solar photovoltaic industry. German government proposed PV roof plan in 1991, provided low-interest loans for rooftop PV systems from 1999–2003, "The Renewable Energy Act" in 2004 formulated renewable energy generation targets which spurred the rapid development of photovoltaic industry, the amendment of Renewable Energy Act in 2010 significantly reduced photovoltaic electricity price, a decline of 16–25 %, the subsidies range expanded as well.

63.2 Present Situation of China PV Industry

63.2.1 Ten Years of Rapid Development

Since 2002, China's PV industry mushroomed thanks to the pull of the European market. Its rapid growth attracted international notice. In 2007, China has became the world's largest producer of solar cells, China's solar cell production reached 13 GW in 2010, battery components production increased to 10 GW, accounting for 45 % of the world total production, and the first solar cell producer in the world for five consecutive years. The installation of photovoltaic power generation reached 500 MW in 2010, grand total of 900 MW, ranking the world's top tens.

63.2.2 Encounter the "Winter"

China PV industry is experiencing more severe conditions. Internationally, accompanied by the debt crisis in Europe, Germany, the United Kingdom have reduced the solar energy plan. As the second largest market, Japan's market demand has reduced because of the tsunami. The United States raised anti-dumping and anti-subsidy probe for China PV products' low price. Some European companies also planed to initiate anti-dumping proceedings against the Chinese solar manufacturers in the mid-2012es. PV industry in China needs to be integrated, to get rid of the vicious competition in a disordered development trend, and further developing of domestic PV application market.

63.2.3 Looking Forward to Transition

In fact, out of the high-profit welfare development a few years ago, China PV industry is facing the history opportunity of transition. China Central Economic Work Conference made a resolution to cultivate and develop the emerging strategic industries actively at the end of 2011. Two directly related plans were made by "The Twelfth Five-Year Guideline". Due to the development of photovoltaic technology, the on-grid price will decrease to a reasonable level in the near future. Photovoltaic power generation should no longer dependent on state subsidy, it will participate in the fully competitive electricity market. China photovoltaic industry is facing a "shuffle", PV companies with capital, technology advantages will continuously promote the technology and scale of production along the routes of reduce costs. Outdated capacity in many SMES will be eliminated, forcing the China PV industry's transformation, upgrading, and better development.

63.3 Plight of the PV Industry Development

63.3.1 Overcapacity and Vicious Competition

For the blind optimism of the PV market, the temptation of high profits in the last few years, and local governments' pursuing of large projects to improve the GDP, the PV industry in China experienced an unprecedented investment boom in 2007. These investments helped to ease the shortage of silicon materials, but the blind capacity expansion led to domestic and global overcapacity, so the Chinese enterprises found themselves in a vicious price war competition. For example, in the polysilicon industry, excessive domestic polysilicon project planning leaded to higher investment and energy consumption. Unless the polysilicon enterprises

Table 63.1 2009–2012 Comparison of capacity and output in China PV companies

	2009	2010
Capacity of polysilicon	51,850	11,5250
Output of polysilicon	19,000	45,3900

upgrade technology, there will be idle capacity and heavy losses because of high manufacturing costs and brutal competition in the industry.

With the continued weakness of the domestic PV market, polysilicon price began to fall into the historic low. By contrast, in this sagging market environment, many silicon material companies did not reduce the yield and lower capacity planning, but improve the capacity and output of polysilicon through the following data in 2012, allowing us to query the "overcapacity" (Table 63.1).

63.3.2 Insufficient Domestic Market Demand, Excessive Dependent on International Markets

63.3.2.1 The Risk of "Two Heads Out"

High purity silicon is the raw material in China PV industry, however, its 95 % dependent on imports, the major markets are currently abroad, more than 95 % of productions are aiming at exports. The domestic demand for finished productions is very small, this makes the solar photovoltaic industry controlled by others, subject to the fluctuations of the international market of raw materials and finished goods prices. That leads no good to the stable development of the industry. Actually, Chinese PV industry's export was hit by the downturn of international photovoltaic market recently (Table 63.2, Fig. 63.1).

63.3.2.2 Facing the Risk of Anti-Dumping

During the post-crisis period, some countries launched anti-dumping proceedings for the low price of China PV companies according to politics or trade protectionism. These measures increased market risk, affected the development of the industry to some extent. The U.S. Commerce Department announced the preliminary results of anti-dumping duties on Chinese photovoltaic cells and modules range, the rate was from 31.14–249.96 %, which is the most severe anti-dumping consequence in the history. The anti-dumping duty of solar photovoltaic cells and photovoltaic components was 31.14 percent for the Suntech company. Yingli Solar was 31.18 %. Trina Solar was 31.22 %. All the other Chinese producers were 249.96 %. The German PV company Solar World also planed to initiate anti-dumping proceedings for Chinese solar manufacturers in Europe in May 2012.

Table 63.2 Global and China PV installed capacity of the year (China PV Industry Development Report 2011)

	2006	2007	2008	2009	2010
Total global	1,603	2,932	5,950	7,380	16,000
China	10	20	40	160	500
China's share of the world (%)	0.60	0.70	0.70	2.20	3.10

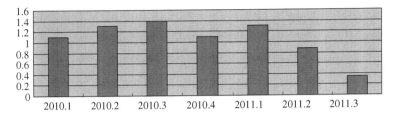

Fig. 63.1 Comparison of global PV equipment orders and shipments ratio (SEMI PV Group 2012)

63.3.3 Lack of Core Technology

Chinese solar energy companies are lack of independent research and development capacity, the key technology is basically resting in the hands of the foreign companies. Domestic enterprises are still in the "processing" stage, and making their living by low-cost, low energy and labor prices. This leads many companies to chase short-term returns, they are not willing to spend time on research and development, the capability of independent innovation is a serious deficiency.

At the same time, the matching technology for photovoltaic power generation is not mature in China. Such as grid-connected inverter could not produce independently, short life of the storage battery in autonomous system, lack of the core technology in high purity solar grade silicon.

63.4 Measures for the Healthy Development of Photovoltaic Industry

63.4.1 Explore Domestic Market Actively, Promote the Development of Industry Chain

63.4.1.1 Explore the Domestic Market Actively

Steps should be taken to explore and open up the domestic market, speeding up the industrialization and commercialization of photovoltaic technology. The government should support the procurement and construction of solar power plants, start lighting demonstration project. The support of the National Development and

Reform Commission is also significant, such as carrying out the pilot solar power plant construction, achieving grid-connected power generation, and promoting the application fields. Application in infrastructure and basic industries such as mobile communication base stations, gas stations, highway tunnels should be promoted.

63.4.1.2 Strengthen the Development of Industry Chain

Downstream enterprises in the PV industry include the production and installation of solar modules, application in power generation projects, combining use among solar PV cells, building materials and electromechanical industry, especially PV integrated buildings. In order to promote the development of related industries, revise and adjustment of the industrial technology roadmap is urgently needed. The extension of industrial chain will drive the complementation of related applications, while helping the enterprises pay attention to technology application and development according to the market. As an emerging energy industry with great developing potential, the solar energy industry's growth will also drive the boom of other application industries such as solar LED lighting.

63.4.2 Intensify Scientific and Technological Research and Development

63.4.2.1 Increasing Investment to Master the Key Technology

The government should encourage and support research institutions and enterprises to carry out research and development from basic theory to the production process. Looking for further breakthroughs in key technologies and processes such as polysilicon purification technology, the battery conversion rate, slice technology, systems integration. At the same time, deepening the research and development of thin film battery, concentrator cells, other new technologies and new products. Use economic measures to spur the enterprises' activity, guide the technical development in a comprehensive, balance and harmonious way.

63.4.2.2 Research Interaction within the Industry

In the long run, it is insufficient only rely on the support of government, we should rely on the intra-industry production, research and interaction to enhance the R & D capability and competitiveness, which will form a virtuous cycle, help the whole industry moving towards the peak level. The key enterprises should build a cooperation system with the world's best research institutions, to promote their independent innovation ability. Concentrate on breaking through the bottleneck of key technology, setting up technology innovation system of PV industry.

63.4.3 Strengthen the Self-Discipline, Formulate Industry Standards

There are some discordant voices over the past few years' rapid growth of PV industry, such as disordered competition within the industry, polluting the environment in order to reduce costs. Facing intense market competition in the future, PV companies must focus on self-growth and orderly development within the whole industry.

Photovoltaic industry has been called for relevant industry standards years ago. PV original, auxiliary materials specification, photovoltaic cells, the testing of products' performance, and application terminal of the power plant construction are in parts or even all lack of standards. Therefore, the state has set up a nationwide solar photovoltaic energy system standardization committee and the National Solar PV Products Quality Supervision Center. National Standards Commission also launched a photovoltaic research and construction work for the standard system, and introduced dozens of photovoltaic national standards. Within the PV industry chain, the PV companies should be encouraged to participate in the establishment of standards, regulate the industry's development, and make contribution to the quality and competitiveness of PV products in China.

63.4.4 Further Policy Support of the Government

63.4.4.1 Finance and Tax Policy Support

Preferential fiscal and taxation policies could encourage the development of photovoltaic industry. Special organization should be built as developed countries' fiscal and taxation policies in solar power industry, formulating specific policies to work with the overall planning for the solar power industry in China. Relevant national ministries should develop their policies, such as fiscal subsidies, credit support and tax incentives. Give financial discounts for PV industry infrastructure loans and provide scientific fees for technical research. Import productions and equipments needed for the PV industry will have tariff reductions provided by the state.

63.4.4.2 Form the National Strategy of PV Industry

Photovoltaic development plan should be promulgated to guide enterprises' decision-making. In recent years, European countries, the United States, Japan and other developed countries are vigorously promoting the development of solar photovoltaic industry. The introduction of supporting policies and substantial investments helped them to seize the cutting edge of technology. They also made it a bulwark against international financial crisis to revive the economy. Therefore, China should also seize the opportunity to develop photovoltaic industry, and publish a PV development plan over comprehensive survey and research. This is also a focal point to expand domestic demand, adjust the economic structure and maintain economic growth.

References

China PV Industry Development Report (2011) http://www.semi.org.cn/img/marketinfo/pdf/201104200951439250.pdf
Ma S (2005) PV industry's strategic status and development in China's energy structure. Ener China 7:24–32
Rogol M (2005) The solar energy industry outlook report. Credit lyonnais securities (Asia):22–34
SEMI PV Group (2012) China PV industry development report of 2012. China PV industry alliance
Zhao Y (2001) The development and trend of solar energy utilization in China. Rural Electrification 9:11–12

Chapter 64
Non-decomposable Minimax Optimization on Distribution Center Location Selected

Zhucui Jing, Menggang Li and Chuanlong Wang

Abstract The minimax optimization model proposed in this paper is an important model which has received some attention over the past years. In this paper, the application of minimax model on how to select of distribution center location was first introduced. Then a new algorithm using nonmonotone line search to solve the non-decomposable minimax optimization is proposed. Numerical results show the proposed algorithm is effective.

Keywords Non-decomposable · Minimax · Nonmonotone line search

64.1 Introduction

Since e-commerce developed prosperously, how to select the distribution center location plays a key role on a successful logistics company. Improving the operational performance, system cost reduction and other factors should be considered in this process. Suppose there are a number of locations for the logistics company to choose, the decision-making space is chosen as a set denoted by E^n as following:

$$E^n = \{x | x = (\text{transportation, facilities, system cost}...)\}. \tag{64.1}$$

Z. Jing (✉) · M. Li
CCISR, Beijing Jiaotong University, Beijing 100044, China
e-mail: jingzhucui@163.com

C. Wang
Department of Mathematics, Taiyuan Teachers' College, Taiyuan 030012, China

Suppose $q_i(x)$ represents the degree of benefit of the ith distribution center location caused under x, and $R(x)$ the set made up of all of the distribution center location which can get the most degree of benefit under x. Thus the set $R(x)$ can be expressed as

$$R(x) = \{i|q_i(x) = q(x),\ x \in E^n\},$$
$$q(x) = \max\{q_1(x), q_2(x), \cdots, q_m(x)\},$$

where m is the total number of the locations.

Denote by $f_i(x)$ the risk caused by the ith distribution center location under x. Therefore the selection of location model can be formulated as

$$\min_{x \in E^n} \max_{i \in R(x)} \{f_i(x)\}. \tag{64.2}$$

From the above example (Jiao et al. 2005), we can know minimax is an important problems which can be applied in various primary areas of the cerebral cortex, economic model, control model and voting problem (Rolls 1998; Penrose 1989; Wen et al. 2007). This kind of special nonlinear programming which is called non-decomposable minimax optimization has been researched by many authors from optimal theories (DiPillo et al. 1993; Dem'yanov et al. 1974; Grippo et al. 1986; Li 1997; Zhou and Tits 1993) to computing algorithms, such as the value function method, the weighting method, the goal programming method and the interactive method, etc (Gal et al. 1999; Wang and Xu 2004). This non-decomposable minimax optimization can be defined as follows,

$$\min_{x \in E^n} \max_{i \in R(x)} \{f_i(x)\}, \tag{64.3}$$

where

$$R(x) = \{i|q_i(x) = q(x),\ x \in E^n\},$$
$$q(x) = \max\{q_1(x), q_2(x), \cdots, q_m(x)\}.$$

Equation (64.3) is called unconstrained non-decomposable minimax optimization. The necessary and sufficient conditions of continuity and differentiability for Eq. (64.3) have been studied in (Dem'yanov et al. 1974; Grippo et al. 1986). In this paper, we focus on devising a new algorithm to solve Eq. (64.3) and we give the numerical results.

64.2 Algorithm

The sequence $\{x_k\}$ can be defined by

$$x_{k+1} = x_k - \alpha_k g_k, \alpha_k \in (0, 1],\ k = 0, 1, 2, \cdots, \tag{64.4}$$

Considering the drawback of line search which guarantees a monotonic decrease of the objective function, a nonmonotone line search technique has been proposed in (Rolls 1998; Li 1997). In this paper, we define the nonmonotone line search as follows,

$$f(x_k - \alpha_k g_k) \leq \max_{l=0,1,2} f(x_{k-l}) - \gamma \alpha_k \|g_k\|^2. \tag{64.5}$$

Now we give the algorithm using nonmonotone line search Eq. (64.5).

Algorithm 3.1 Parameters. $\gamma \in (0, \frac{1}{2})$, $\delta \in (0, \frac{1}{2})$.
  Data. $x_0 \in E^n$.

Step1.
i. Initialization. Set $k = 0$, and $x_{-2} = x_{-1} = x_0$.
ii. Compute the index set $R(x_k)$ and $f(x_k) = \max_{i \in R(x_k)} \{f_i(x_k)\}$.

Step2.
i. Set $\alpha_k = 1$.
ii. Compute g_k by solving the quadratic program

$$\min g^T g$$

s.t. $g \in G(x_k)$, if $\|g_k\| < \delta$, stop.

Step3.
i. Compute $R(x_k - \alpha_k g_k)$ and $f(x_k - \alpha_k g_k) = \max_{i \in R(x_k - \alpha_k g_k)} \{f_i(x_k - \alpha_k g_k)\}$.
ii. If $f(x_k - \alpha_k g_k) \leq \max_{l=0,1,2} f(x_{k-l}) - \gamma \alpha_k \|g_k\|^2$, go to step 4.
iii. Set $\alpha_k = \frac{1}{2}\alpha_k$, go to step 3.

Step4.
Set $x_{k+1} = x_k - \alpha_k g_k$, $k = k + 1$, go to step 2.

64.3 Numerical Results

In this section we report the numerical results obtained for a set of test functions, by means of the algorithm 3.1. Typical values for the parameters are: $\gamma = 0.1$, $\delta = 0.05$. This program is made in Matlab 6.5, running on the DELL OPTIPLEX GX270.

Functions 1.

$$f_1(x) = x^2, \; f_2(x) = (x - 5)^2 - 4;$$
$$q_1(x) = f_2(x), \; q_2(x) = f_1(x).$$

Table 64.1 Numerical results for a set of test functions when $\gamma = 0.1, \delta = 0.05$

Functions	x_0	k	$g*$	$x*$	$f(x*)$	Time
Functions1	10	4	0	0	0	<1 s
Functions2	(5, 5)	29	(0, 0)	(0, 0)	0	2 s

Functions 2.

$$f_1(x_1, x_2) = x_1^2 + x_2^2, f_2(x_1, x_2) = e^{(x_1^4 + x_2^2)} - 1, f_3(x_1, x_2) = e^{(x_1^2 + x_2^4)} - 1;$$
$$q_1(x_1, x_2) = -5x_1 + x_2, q_2(x_1, x_2) = x_1^2 + x_2^2 + 4x_2, q_3(x_1, x_2) = 5x_1 + x_2.$$

The Table 64.1 gives the results.

64.4 Conclusions

In this paper, a kind of generalized minimax optimization which is called the non-decomposable minimax optimization is discussed. A new algorithm using non-monotone line search to solve the unconstrained non-decomposable minimax optimization is given. Numerical results show the proposed algorithm is effective.

Acknowledgments The writing of this paper has been supported by Beijing Jiaotong University Project (No. B09c1100020), Beijing Philosophy and Social Science, Research Center for Beijing Industrial Security and Development, NSFC grant 11071184 and NSFC for Youth 11101401.

References

Dem'yanov VF, Malozemov VN (1974) Introduction to minimax. John Wiley & Sons, New York
DiPillo G, Grippo L, Lucidi S (1993) A smooth method for the finite minimax problem. Math Program 60:187–214
Gal T, Stewart TJ, Hanna T (1999) Multicriteria decision making: advances in MCDM models, algorithms, theory, and applications. Kluwer Academic Publishers, Boston pp 1–30
Grippo L, Lampariello F, Lucidi S (1986) A nonmonotone line search technique for Newton's method. SIAM J Numer Anal 23:707–716
Jiao YC, Leung Y, Xu ZB, Zhang JS (2005a) Variable programming a generalized minimax problem Part I: Odels and Theory. Comp Optim Appl, 30:229–261
Jiao YC, Leung Y, Xu ZB, Zhang JS (2005b) Variable programming a generalized minimax problem Part II: Algorithms, 30:263–295
Li X-S, Fang S-C (1997) On the entropic regularization method for solving minmax problems with applications. Math Methods Oper Res 46:119–130
Penrose R (1989) The Emperor's New mind: concerning computers, minds, and the Laws of Physics. Oxford University Press, New York
Rolls ET, Treves A (1998) Neural networks and brain function, Oxford University Press,USA

Wang C-L, Xu Z-B (2004) Necessary conditions and sufficient conditions for non-decomposable two-stage minimax optimizations. In: Yuan YX (ed) Numerical linear algebra and optimization. Science Press, Beijing, pp 100–109

Wen R-P, Ren F-J, Wang C-L (2007) Study on the nondecomposable minimax optimization. In: IEEE international conference on control and automation, Guangzhou pp 2510–2513

Zhou JL, Tits AL (1993) Nonmonotone line search for minimax problems. J Optim Theor Appl 76:455–476

Chapter 65
Green Finance and Development of Low Carbon Economy

Shuo Chen

Abstract Finance is the core of modern economy, and green finance is the key of low carbon economy. The development of low carbon economy can not be separated from the green finance. Low carbon economy is becoming the new trend of international economic development. This paper, based on the analysis of the 'low carbon economy' development and green finance foundation, suggests how to develop green finance to promote low carbon economy development.

Keywords Green finance · Low carbon economy · Development

65.1 The Definition of Green Finance

Green finance and low carbon economy are closely related concepts, and its essence is the financial sector to make the environmental protection as a basic policy, considering the potential environmental influence when we make the investment and financing decision. The environmental conditions are related to the potential return, the risk and cost is integrated into the daily business. In the financial management activities, we should pay attention to the ecological environmental protection and pollution control, through social environment and resource guide, to promote the sustainable development of society. Green finance specifically has two aspects: one is the financial industry how to promote environmental protection and sustainable development of economy and society; another refers to the financial industry's sustainable development. The former green finance's major role is to guide the flow of funds to save resources

S. Chen (✉)
College of Economics and Management, Beijing Jiaotong University, Beijing 100044, China
e-mail: Chenshuo6856@163.com

technology development and ecological environmental protection industry, and to guide enterprises to the production of green environmental protection, to guide consumer form the concept of green consumption. The latter makes clear that the financial industry is to maintain sustainable development, and pays attention to short-term interests to avoid excessive speculation.

Green finance was born in the United States in 1990s as the times require. People of the United States' financial circles, who pay attention to environment and global climate change, put the environment factor into financial innovation, and research how to evaluate the environmental risk, which in order to develop a successful environmental financial products, formed a suitable product structure, develop circular economy, and protection environment fund. After the concept of "Green finance" put forward, world governments, international organizations, financial institutions and non-governmental organizations in the field of environmental protection has tried a lot. Early in 1974, the former West Germany established the world's first environmental bank. In 1991, Poland has also set up environmental protection bank to support for promoting the environmental protection investment project. In 2003, the World Bank Group's International Finance Corporation in international banking industry has launched the "Equator Principles", and by Citibank, Barclays Bank, Bank of Holland and the West German State Bank of 7 countries of 10 leading international bank has announced the implementation of. The so-called "Equator Principle", i.e. on a voluntary basis, is a judgment, evaluation and management of project financing in environmental and social risks of the financial industry benchmark. It also provides a framework of project financing in environmental and social risk assessment, including different types of project risk classification, and lists with the environmental assessment process, monitoring and following-up guide related issues. In 2006, the equator principles financial institutions account for the global project financing more than 90 % of the market around the world. The financial product innovation and green finance business of "Green insurance", "green capital market" green areas develop very quickly in recent years in the United States, Britain and other developed countries.

65.2 Green Finance and Development of Low Carbon Economy

Low carbon economy is a new mode of economic development which different from the traditional pattern (high efficiency, low energy consumption, low pollution, low emission) in response to global climate change. It emphasizes less greenhouse gas emissions to obtain bigger economic output. Its essence is the energy efficiency and clean energy structure problem, and its core is the energy technology innovation and system innovation. The goal of low carbon economy is to slow climate change and to promote sustainable development. On the surface, low carbon economy is the result of hard work to reduce greenhouse gas emissions, but in essence, low carbon economy is a new change of the mode of economic development, energy consumption patterns, and the human way of life. It will be reform the modern industrial civilization which based on fossil fuels all-around the ground to ecological economy and ecological civilization.

The modern financial development is the driving force of the development of a modern market economy. The financial industry based on market mechanism, which in the pursuit of profit, can improve efficiency at the same time, and play actively fulfill social responsibility as a corporate citizen, to adjust the economic structure, to promote independent innovation, to conserve the energy, to protect the environment and to play a greater role in other aspects. In the short term, the transformation of the traditional financial industry to the green financial sector may experience improve challenges of customer threshold and increased the new mechanism cost, their interests will be temporary impact. But in the long run, green finance is not only beneficial to the coordination of economic development and environmental protection, maintain human long-term social interests and long-term development, alleviate the traditional banking business negative effect, but also to prevent the financial risk, put a environmental protection firewall for sound operation, and create more green business. The transform of traditional high carbon economy to a low carbon economy transformation calls for the traditional financial system to green financial restructuring and development.

China is the world's largest carbon emitter of the state which in rapid industrialization and industrialization development stages. The situation of the energy-saving emission reduction is increasingly serious. The environmental pollution and resources capacity have approached the limit state in many places, the government is increasingly strict bound to the enterprise environmental pollution liability, because the polluting enterprises' credit risk started to increase. The financial sector is also facing the potential crisis in the promoting energy-saving emission reduction, carrying out harmonious development of economic and natural, and realizing sustainable development. In the big background of the global attention to sustainable development of financial industry, economy and society, the implementation of the "green finance" is not only the implementation of scientific outlook on development the reality needs, but also conform to the international trend of China's financial industry, realize the inevitable choice that conforms with international.

65.3 Develop Green Finance, and the Policy Recommendations to Promote the Development of Low Carbon Economy

65.3.1 The Construction of Multi-Level Financial Support System

First is the support and the guide of policy finance. Policy financial institutions should play an important role in green financial system. The support for the development of low carbon economy's infrastructure investment, low carbon technology research, production and use of clean energy projects should be focused on. We should actively guide the social funds into the construction of low carbon economy. Second is the financial support of commercial banks. We can encourage commercial banks to actively participate in the CDM project's loan business, and carry out the "green credit" accountability system. We must make clear the "green credit" policy's request and market access's standards. The environmental protection standards and credit risk management requires organic combination. We can use the implement of "one ticket overruled make" for environmental protection, and carry out to the customer survey, loan marketing, credit, project evaluation, credit review, credit management and so on each link. To prevent credit risk from the source which brings from the changes of enterprises and project construction for environmental protection requirements change.

65.3.2 Innovation of Green Financial Products and Services

First is the creative product. Green credit as a starting point, we should promote carbon mortgage financing loan. We can explore the issue of green bonds, and absorb relatively stable long-term capital into low carbon economic projects of high demand capital and good comprehensive benefit. Second is to promote international financial cooperation. In the bank credit, we encourage commercial banks to study the equator principles, and put environmental factors into their loans, investment and risk assessment procedure. Through the cooperation with the International Finance Corporation, we can carry out a wide range of international project financing based on the new model of green credit. In carbon trading services, encourage the qualified financial institutions actively corporate with international investment banks and carbon fund, make a full grasp of the international carbon emission right transaction information, to enhance the intermediary service trade competence.

65.3.3 Make Preferential Policies to Encourage Low-Carbon Technology Innovation and Green Consumption

Financial institutions increase the financial support for carbon technology innovation, encourage the application of clean coal technology, the transformation of energy-saving technological, the development of new energy, renewable energy, alternative energy technology, and the comprehensive utilization of energy and other green low-carbon industry development. We should incent individual and units' "green consumption". On consumers' purchase of environment-friendly consumer goods, such as environmental protection, automobile, environmental protection appliances, banks can be joined together to launch business interest even interest-free loans to encourage consumption; conversely, if the purchase of serious environmental pollution in the consumer goods, such as luxury cars, high energy consumption, banks can refuse to loan to curb consumption loan effectively.

Part IV
Workshop on Green Supply Chain Management

Chapter 66
Research on Network Optimization of Green Supply Chain: A Low-Carbon Economy Perspective

Cuizhen Cao and Guohao Zhao

Abstract Based on the theory of low-carbon economy, this paper evaluates the influence of carbon emissions on supply chains' overall value using carbon footprint. It is showed that three objects, namely, profitability, service level and environmental protection should be coordinated, and to balance cost, response time and carbon footprint, the penalty function coefficient is introduced, which coverts a multi-objective optimization problem to a single objective one. Also, a network optimization model is formulated to offer a supplementary solution for optimization design of green supply chain network.

Keywords Low-carbon economy · Carbon footprint · Green supply Chain · Network optimization

66.1 Introduction

Since the beginning of twenty-one century, climate change has become a challenge mankind has to meet in the economic and social development. Carbon emission is one of the main causes of global warming. China, one of the top countries in terms of carbon dioxide (CO_2) emission, faces great pressure to reduce carbon emission.

C. Cao (✉)
College of Business Administration, Shanxi University of Finance and Economics,
Taiyuan 030031 Shanxi, China
e-mail: caocuizh@sina.com

G. Zhao
College of Management Science and Engineering, Shanxi University of Finance and Economics, Taiyuan 030006 Shanxi, China
e-mail: gzhao1958@126.com

As an important part of the Third Industry, green supply chain is an ecological industry system characterized by low carbon emissions, and so it is the government's possible policy choice to develop low-carbon economy. Green supply chain optimization drives at a balance between environmental protection and profitability, achieving systematic coordination in order to improve both environmental quality and competitive edge in a dynamic way and for a win–win effect. To use carbon footprint to measure carbon emissions as an environment index is a reliable way in dealing with environmental pollution, caused mainly by carbon emissions. Carbon emission has great bearing on the design of supply chain network, so to take the factor of carbon emissions into account is inevitably a tendency in the network design of supply chain (Fang and Xu 2012). Therefore, in the context of low carbon economy, to introduce the notion of carbon footprint to measure the influence of carbon emissions on the overall cost of supply chain is of great theoretical and pragmatic meaning to developing low carbon logistics, undertaking green supply chain management, and facilitating sustainable development strategy.

66.2 Literature Review

Green supply chain has been one of the hottest topics in recent years' supply chain management research (Zhang and Xu 2009), and many people pay their attentions to the green supply chain network optimization and design. Wang et al. (2011), analyzing environmental issues, build a related multi-objective model, and solve the model with heuristic algorithm. Sundarakani et al. (2010) formulates a model of carbon footprint across supply-chain under both static and dynamic situations, and checks it numerically, demonstrating the importance of carbon footprint in supply chain design. Chaabane et al. (2011) analyzes the green supply chain network design problem in carbon exchange market, and try to help decision makers find the efficient solution that balances the increase of supply chain cost and emission reduction. In China, thorough analysis of green supply chain network optimization is far from enough. Fang and Xu (2012), drawing on research findings in network designs of the green supply chain and sustainable supply chain, elaborates some important relevant issues in supply chain network design with carbon emission taken into consideration. However, up to now there has been no in depth, detailed design model for establishing a green supply chain network, overall there is no quantitative research in this respect, either.

Based on low-carbon economy theory, this paper evaluates the influence of carbon emissions on supply chains' overall value using carbon footprint, and seeks to balance the three targets,namely, supply chain profitability, service level and environmental protection, so as to achieve a green, low-carbon supply chain. Penalty function coefficient is introduced to balance cost, response time and carbon footprint, converting the multi-objective optimization problem to a single-objective one. An optimization model is developed to offer supplementary decision solutions for the optimization and design of the green supply chain network, aiming at pushing effective implementation of green supply chain management and realizing low-carbon development.

66.3 Thought and Assumptions of Network Optimization

66.3.1 Basic Idea of Green Supply Chain Network Optimization

The difference between general supply chain network design and the green supply chain lies in that the green supply chain not only needs to optimize cost and service level at the same time and reach the network structure solution at Pareto curve (He and Meng 2009), but also needs to consider the influence of carbon emission on the overall value of the supply chain and sustainable development. Therefore, the research on the green supply chain network optimization based on low-carbon economy's perspective should first clarify the internal mechanism of carbon emissions' influence on the overall cost of the supply chain network, measure the emission accurately on the supply chain level (Sundarakani et al. 2010) and analyze the cost influence. Then considering the effect of carbon emissions on strategic and tactical factors of the green supply chain network design, it should apply carbon emissions' limit to the selection of network nodes strategically, and network channels tactically, thus effectively building a green supply chain network.

66.3.2 Model Assumptions

In the paper, we use carbon footprint to measure the carbon emissions at all the nodes and channels of the supply chain, and the penalty function is applied to realize the balance of cost, response time and carbon footprint. Two coefficients of the penalty function are introduced: one is penalty coefficient of late delivery, which is defined as the unit increase in objective function because of late response to customer needs in the supply chain system; the other is penalty coefficient of carbon emission, which is the unit increase in objective function when the carbon emission exceeds the system's limit. Thus, the penalty coefficients can be set differently according to different regions, products, and phases to balance supply chains' cost, response time, and carbon footprint. Now the assumptions are as follows:

(i) All facilities to be built are only built at places that are given, and all these places can be scientifically simulated in systems' daily operation and satisfy needs of the green supply chains' management very well;

(ii) The capacity of all nodes, the building and operation cost, and carbon footprint can be tracked and acquired in real time via the supply chains' sharing system of information;
(iii) The unit transport cost, time, and carbon footprint between any source node and any destination node can be learned from system simulations and they have linear relations with transport quantity in the mean time.

66.4 Network Optimization Decision Model

66.4.1 Model Notations

For convenience, here we give some notations which are used in the sequel. P, D, R denote the set of nodes for manufacturing facilities, the set of nodes for distribution centers and the set of demand nodes, respectively. DC_e denotes the distribution center e and R_j denotes the demand quantity for demand node j. P_i, C_i^P, T_i^P, CF_i^P denote the manufacturing capacity, the construction and operation cost, the unit production time and the carbon footprint of unit production activity of plant i, respectively. D_e, C_e^D, T_e^D, CF_e^D denote the capacity, the construction and operation cost, the unit processing time, the carbon footprint of unit processing activity of DC_e, respectively. T_{GSN}, CF_{GSN} denote the target response time, the limit carbon footprint of the network, respectively. Let C_{ie}^D, T_{ie}^D, CF_{ie}^D denote respectively the unit transport cost, the unit transport time and the unit transport carbon footprint from plant i to DC_e, and let C_{ej}^R, T_{ej}^R, CF_{ej}^R denote respectively the unit transport cost, the unit transport time and the unit carbon footprint from DC_e to demand node j. CT_{GSN} denotes the penalty function coefficient for late delivery and CCF_{GSN} denotes the penalty function coefficient for carbon footprint.

In addition, let x_{ie} be the transport quantity from plant i to DC_e and y_{ej} be the transport quantity from DC_e to demand node j. Let u_i be the binary variable with 1 representing building plant i and 0 otherwise; let v_e be the binary variable with 1 representing building DC_e and 0 otherwise.

66.4.2 Math Model

Minimize TC
$$= \sum_{i \in P} \sum_{e \in D} C_{ie}^D x_{ie} + \sum_{e \in D} \sum_{j \in R} C_{ej}^R y_{ej} + \sum_{i \in P} C_i^P u_i + \sum_{e \in D} C_e^D v_e$$
$$+ CT_{GSN} \times \left(\sum_{i \in P} \sum_{e \in D} T_{ie}^D x_{ie} + \sum_{e \in D} \sum_{j \in R} T_{ej}^R y_{ej} + \sum_{i \in P} T_i^P u_i \sum_{e \in D} x_{ie} + \sum_{e \in D} T_e^D v_e \sum_{j \in R} y_{ej} - T_{GSN} \right)$$
$$+ CCR_{GSN} \times \left(\sum_{i \in P} \sum_{e \in D} CF_{ie}^D x_{ie} + \sum_{e \in D} \sum_{j \in R} CF_{ej}^R y_{ej} + \sum_{i \in P} CF_i^P u_i \sum_{e \in D} x_{ie} + \sum_{e \in D} CF_e^D v_e \sum_{j \in R} y_{ej} - CF_{GSN} \right)$$

Subject to

$$\sum_{e \in D} x_{ie} \leq P_i u_i, \quad i \in P \tag{66.1}$$

$$\sum_{j \in R} y_{ej} \leq D_e v_e, \quad e \in D \tag{66.2}$$

$$\sum_{e \in D} y_{ej} \geq R_j, \quad j \in R \tag{66.3}$$

$$\sum_{i \in P} x_{ie} = \sum_{j \in R} y_{ej}, \quad e \in D \tag{66.4}$$

$$u_i \in \{0, 1\}, \quad i \in P; \quad v_e \in \{0, 1\}, \quad e \in D \tag{66.5}$$

$$x_{ie} \geq 0, y_{ej} \geq 0, \quad i \in P, \quad e \in D, \quad j \in R \tag{66.6}$$

where the first part of the objective function is construction and transport cost, the second part is the penalty cost indicating the failure in responding to customer needs, and the third part is the penalty cost caused by carbon footprint which exceeds the system limit. Constraints (66.1) and (66.2) are the capacity limits; (66.3) is the limit of demand quantity; (66.4) is the limit of balancing node capacities; (66.5) and (66.6) limit the range of variables.

66.5 Analysis of Example

A multinational company TC needs to optimize supply chain network. Currently two high value-added innovative products have been developed to be launched in three markets. The response time to product orders is set to be 1 week. The whole supply chain's carbon footprint is 8,500 ton. There are three candidate plants and three distribution centers for these products. The detailed data is shown in Tables 66.1, 66.2, and 66.3.

Usually, innovative high value-added products need very agile and adaptable supply chains, and in the investment period, a higher response ability and service level for customers are required. So by combining market investigation with empirical analysis, and applying LINGO 11.0 to test the model repeatedly, we find that the penalty function coefficient CT_{GSN} should be set as the greater number 999, while CCF_{GSN} can be set as 650 by reference to the carbon taxes in different regions. After coding and solving the problem, the optimized solutions can be obtained: strategically, nodes selected are plants 1 and 3, DC 1 and 2; operationally, detailed distribution path can be optimized real time by using this model. This can guarantee the feasibility and effectiveness of the model.

Table 66.1 Unit transport cost, unit time, and unit carbon footprint between nodes

From	To	C ($)	T(h)	CF(t)	From	To	C ($)	T(h)	CF(t)
P1	DC1	2000	1	780	DC1	C1	4000	2	983
P1	DC2	4000	2	890	DC1	C2	1000	3	786
P1	DC3	3000	1	560	DC1	C3	3000	2	698
P2	DC1	2000	2	800	DC2	C1	6000	2	889
P2	DC2	1000	3	780	DC2	C2	4000	1	459
P2	DC3	4000	1	569	DC2	C3	2000	3	859
P3	DC1	8000	2	1120	DC3	C1	10000	3	963
P3	DC2	6000	4	790	DC3	C2	5000	6	569
P3	DC3	1000	2	799	DC3	C3	4000	4	698

Table 66.2 Capacities and construction cost of all nodes on the network

P	Capacity	Cost($)	DC	Cost ($)	Capacity	C	Demand
P1	700	45000	DC1	200000	400	C1	400
P2	600	65000	DC2	100000	800	C2	500
P3	500	90000	DC3	150000	1000	C3	300

Table 66.3 Unit processing time and carbon footprint of all nodes on the network

P	Time (h)	Carbon footprint (T)	DC	Time (h)	Carbon footprint (T)
P1	7	650	DC1	10	890
P2	9	890	DC2	13	987
P3	5	1000	DC3	12	780

66.6 Conclusion

The research on the green supply chain network optimization based on low-carbon economy's perspective should first clarify the internal mechanism of carbon emission's influence on the overall cost of the supply chain network, measure the emission accurately on the supply chain level and analyze the cost influence. Then in view of the influence of carbon emission on strategic and tactical factors of the green supply chain network design, penalty function coefficient is introduced to balance cost, response time and carbon footprint, which converts the multi-objective optimization problem to a single-objective one. By balancing the supply chain's profitability, service level, and environmental protection, a green, low-carbon supply chain network can then be built effectively.

Acknowledgments The work was supported by the National Natural Science Foundation of China (No. 71173141).

References

Chaabane A, Ramudhin A, Kharoune M, Paquet M (2011) Trade-off model for carbon market sensitive green supply chain network design. Int J Oper Res 10(4):416–441

Fang J, Xu L (2012) Green supply chain network design research: considering carbon emissions. Mod Manag Sci 1:72–91

He B, Meng W (2009) Research of reverse logistics network design: considering customer selection behavior. China Manag Sci 17(6):104–108

Sundarakani B, Souza R, Goh M, Wagner SM, Manikandan S (2010) Modeling carbon footprints across the supply chain. Int J Prod Econ 128(1):43–50

Wang F, Lai X, Shi N (2011) A multi-objective optimization for green supply chain network design. Decis Support Syst 51(2):262–269

Zhang X, Xu L (2009) Green supply chain research review. Econ Manag 2:169–173

Chapter 67
The Research on Evolutionary Game of Remanufacturing Closed-Loop Supply Chain Under Asymmetric Situation

Jian Li, Weihao Du, Fengmei Yang and Guowei Hua

Abstract Remanufacturing is an effective means to realize energy saving and emission reduction. The remanufacturing industry makes important contribution to achieve the goal of low-carbon economy. This paper develops an evolutionary game model with a two-echelon closed-loop supply chain to study evolutionary stable strategies (ESS) of manufacturers and retailers. Through analyzing evolutionary path of the game, we find that there are two possible evolutionary results affected by the profits of manufacturers.

Keywords Remanufacturing · Closed-loop supply chain · Evolutionary game · ESS

67.1 Introduction

Enterprises are the main carriers of energy consumption and carbon emission. How to combine enterprises' business decisions with low-carbon economy goals is important to drive the development of low-carbon economy. Because the quality and performance of remanufacturing products are not inferior to new products,

J. Li (✉) · W. Du
School of Economics and Management, Beijing University of Chemical Technology, Beijing 100029, China
e-mail: lijian@mail.buct.edu.cn

F. Yang
School of Science, Beijing University of Chemical Technology, Beijing 100029, China

G. Hua
School of Economics and Management, Beijing Jiaotong University, Beijing 100044, China

but energy consumption and materials consumption are both under half of original manufacturing (Ferrer 1997a, b; Zhang et al. 2009; Xu 2010).

A lot of articles related to remanufacturing closed-loop supply chain always consider economic and environmental benefits within short term, while there is a growing number of research papers studies remanufacturing closed-loop supply chain in consideration of long term. Many literatures study remanufacturing closed-loop supply chain mainly consider symmetric situation of small population difference (Xiao and Yu 2006; Dong 2012). But the research for asymmetric situation of big population difference is few. Therefore, this paper extends short term issue to long term issue on the base of Stackelberg model established in the paper of Du et al. (2012). This paper considers the dynamic game between manufacturers and retailers. We find the ESS of two populations by replicated dynamic equation and further analyze the possible factors which affect the ESS. The rest of the paper is organized as follows. Section 67.2 presents the research issue and builds the model. Section 67.3 tries to solve the model and find ESS of the model under market mechanism. Section 67.4 draws some conclusions and outlines some directions for future research.

67.2 The Evolutionary Game Model

The following assumptions are made in this paper:

Assumption 1. This paper considers a two-echelon closed-loop supply chain consists of the manufacturer population and the retailer population. Retailers engage in the work of recycling waste.

Assumption 2. The waste products recycled by retailers are all repurchased by manufacturers. And the returned products can be all used into remanufacturing.

Assumption 3. The quality of remanufacturing products is same to the new products. The market of remanufacturing products is in short supply.

Assumption 4. The price of remanufacturing products is lower than the price of new products. We only think about the revenue related to the activities of remanufacturing in the process of game.

Assumption 5. It needs to use some new raw materials in the process of remanufacturing.

The following notation is used in the model:
Q quantity of recycled waste products;
p_1 unit transfer price of manufacturer in the process of repurchasing waste products from retailers;
p_2 unit wholesale price of manufacturer in selling remanufacturing products;
p_3 unit wholesale price of manufacturer in selling new products;
C_f unit fixed cost of manufacturer in the remanufacturing process;
C_p unit cost of repairing of manufacturer in the remanufacturing process;

C_m unit cost of manufacturing of manufacturer in the original manufacturing process;

C_{b1} procurement cost of new raw materials required by unit product in the remanufacturing process. It follows a normal distribution that $C_{b1} \sim N(u,\delta)$;

C_{b2} procurement cost of raw materials required by unit product in the original manufacturing process;

What's more, these notations satisfy $C_{b2} > C_{b1}, p_3 > p_2 > p_1$

This paper studies the game between suppliers and retailers under asymmetric situation. The game between the two populations is a dynamic game. The manufacturer makes decisions after the retailer. The strategies and payoffs of two players are both different. In addition, the game is played in a space of uncertainty and bounded rationality. The strategies of two players influence each other. Under this circumstance, we think about two pure strategies of the manufacturer and the retailer respectively.

The retailer have two pure strategies:

S_{r1} recycle waste products
S_{r2} not recycle waste products

The manufacturer also have two pure strategies:

S_{m1} use the recycled waste products to remanufacture
S_{m2} purchase new raw materials to manufacture directly

The strategy faced by the retailer is whether to recycle or not to recycle. Recycle waste products needs to pay recycling cost. Due to the possible situation that manufacturers may not repurchase waste products from retailers, retailers have the risk of loss. Then the strategy faced by the manufacturer is whether to remanufacture or not to remanufacture. The buyback cost of remanufacturing is lower than the procurement cost of original manufacturing. Manufacturers need to bear fixed cost like management cost, disassembly cost, cleaning cost, inspection cost and so on after taking remanufacturing strategy. Assumed that x represents the individual ratio of retailer population who takes recovery strategy. Then 1-x represents the remaining individuals who do not take recovery strategy. x satisfies the condition of $0 \leq x \leq 1$. Similar to the meaning of x, y represents the individual ratio of manufacturer population who takes remanufacturing strategy. Then 1-y represents the remaining individuals who do not take recovery strategy. y satisfies the condition of $0 \leq y \leq 1$. The adjustment of strategies is a dynamic adjustment process. When expected benefits of the manufacturer or the retailer is lower than their average benefits respectively, the two players will change their strategies in next selection to pursue the higher benefit.

The profit functions of the retailer are different. Firstly, when the retailer takes recovery strategy and the manufacturer takes remanufacturing strategy, the profit of the retailer will be $\prod_{r1}$, $\prod_{r1} > 0$. Secondly, when the retailer takes recovery strategy but the manufacturer doesn't take remanufacturing strategy, the loss of the retailer will be $\prod_{r2}$, $\prod_{r2} < 0$. Lastly, when the retailer doesn't take recovery strategy, the profit of the retailer is 0 whatever strategy does the manufacturer choose.

$$\prod_{r1} = p_1 Q - C_r Q = (p_1 - C_r)Q, \tag{67.1}$$

$$\prod_{r2} = -C_r Q, \tag{67.2}$$

When faced same product demand, the profit functions of the manufacturer are different. $\prod m1$ is the profit when the manufacturer takes remanufacturing strategy. $\prod m2$ is the profit when the manufacturer takes original manufacturing strategy.

$$\prod_{m1} = p_2 Q - p_1 Q - C_{b1} Q - C_f Q - C_p Q = (p_2 - p_1 - C_{b1} - C_f - C_p)Q, \tag{67.3}$$

$$\prod_{m2} = p_3 Q - C_{b2} Q - C_m Q = (p_3 - C_{b2} - C_m)Q, \tag{67.4}$$

Let ES_{ri} and ES_{mj} denote expected benefits of the retailer and the manufacturer respectively when they take different strategy combinations. $i, j = 1, 2$.

$$ES_{r1} = y \prod_{r1} + (1-y) \prod_{r2}, \tag{67.5}$$

$$ES_{r2} = 0, \tag{67.6}$$

$$ES_{m1} = x \prod_{m1} + (1-x) \prod_{m2}, \tag{67.7}$$

$$ES_{m2} = x \prod_{m2} + (1-x) \prod_{m2} = \prod_{m2}, \tag{67.8}$$

Therefore, we define $\overline{ES}_r$ and $\overline{ES}_m$ as the average benefits of the retailer and the manufacturer respectively when they take different strategy combinations.

$$\overline{ES}_r = xES_{r1} + (1-x)ES_{r2}, \tag{67.9}$$

$$\overline{ES}_m = yES_{m1} + (1-y)ES_{m2}, \tag{67.10}$$

67.3 The Evolutionary Analysis of Game Model Under Market Mechanism

According to Malthusian equation, the quantity's growth rate of strategy selected by players should equal to its fitness minus its average fitness. Then the replicated dynamic equations of S_{r1} selected by the retailer and S_{m1} selected by the manufacturer are denoted as follows.

$$f_1(x,y) = x(ES_{r1} - \overline{ES_r}) \tag{67.11}$$

$$f_2(x,y) = y(ES_{m1} - \overline{ES_m}), \tag{67.12}$$

Substitute formula (67.5) and (67.9) into the replicated dynamic Eq. (67.11), we have

$$f_1(x,y) = \frac{dx}{dt} = x(1-x)[y\prod_{r1} + (1-y)\prod_{r2}] = x(1-x)(yp_1 - C_r)Q, \tag{67.13}$$

In the same way, we have

$$f_2(x,y) = \frac{dy}{dt} = y(1-y)x(\prod_{m1} - \prod_{m2})$$
$$= y(1-y)x(p_2 + C_{b2} + C_m - p_1 - p_3 - C_{b1} - C_r - C_p)Q, \tag{67.14}$$

67.3.1 The Evolutionary Game Analysis of Retailer Population

We can use the replicated dynamic equation to depict the evolutionary course of retailer population's strategy selection. Then we can get equilibrium point and ESS of the evolutionary game. According to the stability theorem of differential equation, the equilibrium point should satisfy $\frac{\partial f_1(x,y)}{\partial x} < 0$ if it's a ESS. To solve the partial derivative of $f_1(x,y)$, we have

$$\frac{\partial f_1(x,y)}{\partial x} = (1-2x)(yp_1 - C_r)Q, \tag{67.15}$$

According to formula (67.15), we can have Theorem 1 as follows.

Theorem 1

I. When $y = y = \frac{C_r}{p_1}$, there is no ESS.

II. When $y < \frac{C_r}{p_1}$, $x = 0$ is the ESS of retailer population's strategy selection.

III. When $y > \frac{C_r}{p_1}$, $x = 1$ is the ESS of retailer population's strategy selection.

67.3.2 The Evolutionary Game Analysis of Manufacturer Population

The same way to the content of former chapter, we can get equilibrium point and ESS of the evolutionary course of manufacturer population. It should satisfy $\frac{\partial f_2(x,y)}{\partial y} < 0$. To solve the partial derivative of $f_2(x, y)$, we have

$$\frac{\partial f_2(x, y)}{\partial y} = (1 - 2y)x(\prod_{m1} - \prod_{m2}), \quad (67.16)$$

According to formula (67.16), we can have Theorem 2 as follows.

Theorem 2

I. When $x = 0$, there is no ESS.
II. When $x > 0$ and $\prod_{m1} - \prod_{m2} < 0$, $y = 0$ is the ESS of manufacturer population's strategy selection.
III. When $x > 0$ and $\prod_{m1} - \prod_{m2} > 0$, $y = 1$ is the ESS of manufacturer population's strategy selection.

To show the retailer population's and manufacturer population's replicated dynamic trend in a same coordinate plane as Fig. 67.1 depicts:

We can find that when the profit of remanufacturing strategy selected by manufacturer is lower than the profit of original manufacturing strategy. This game will gradually converge to the equilibrium point (0, 0). It's said that the retailer population gradually tends not to select recovery strategy and the manufacturer population gradually tends to select original manufacturing strategy. (No recovery, Original manufacturing) will be the only choice of the two populations. In contrast, while the profit of remanufacturing strategy selected by manufacturer is higher than the profit of original manufacturing strategy, the game will gradually converge to the equilibrium point (1,1). It's said that the retailer population gradually tends to select recovery strategy and the manufacturer population tends to select

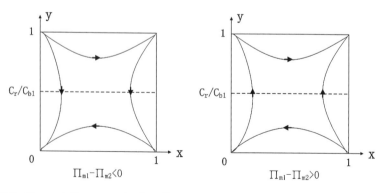

Fig. 67.1 The evolutionary graph of retailer group and manufacturer group

remanufacturing strategy. (Recovery, Remanufacturing) will be the only choice of the two populations. Through the above analysis, we can find that the ESS of the two populations are both determined by the profit of manufacturers.

67.4 Conclusions

Remanufacturing industry plays an important role in promoting the development of low-carbon economy as an emerging industry. But it's still immature. This paper analyzes the retailer population's and manufacturer population's long-term evolutionary course in the remanufacturing closed-loop supply chain. The analysis shows that the ESS of two populations is influenced by the profit of manufacturers achieved from remanufacturing activity. To strengthen the degree of technological innovation or improve employee skills is turned to be the main driving force of promoting the development of remanufacturing industry. This paper studies the evolutionary game of remanufacturing closed-loop supply chain from the perspective of two populations. The closed-loop supply chain in reality is more complicated. Therefore, to study the evolutionary game of remanufacturing closed-loop supply chain from the perspective of more populations is the direction for further research.

Acknowledgments This research was supported by the NSFC under grant number 70801003, 71171011 and 71071015.

References

Dong HZ, Song HL (2012) Research on duplication dynamics and evolutionary stable of reverse supply chain. Phys Procedia 24:705–709

Du WH, Li J, Yang FM, Hua GW (2012) The carbon subsidy analysis in remanufacturing closed-loop supply chain. Forthcoming in information—an international interdisciplinary journal

Ferrer G (1997a) The economics of personal computer remanufacturing. Resour Conserv Recycl 21:79–108

Ferrer G (1997b) The economics of tire remanufacturing. Resour Conserv Recycl 19:221–255

Xiao TJ, Yu G (2006) Supply chain disruption management and evolutionarily stable strategies of retailers in the quantity-setting duopoly situation with homogeneous goods. Eur J Oper Res 173:648–668

Xu BS (2010) Remanufacture engineering and its development in China. China Surf Eng 23:1–6

Zhang W, Xu BS, Zhang S, Liang XB, Shi PJ, Liu SC (2009) Status and development of remanufacturing research and application. J Acad Armored Force Eng 23:1–5

Chapter 68
A Sequencing Problem for a Mixed-Model Assembly Line on Supply Chain Management

Hugejile, Shusaku Hiraki, Zhuqi Xu and Shaolan Yang

Abstract The sequencing problem for mixed-model assembly has a direct influence on the parts supply chain, production logistics and mobile sale logistics; it is also one of the key factors to design and organize low-carbon model logistics. The model in this paper assumed that the parts is supplied in JIT and assembled into products in a mixed-model line. A part of the products are transported to the distributor directly by car-carrier after production. In this circumstance, the essay proposes a new sequencing method on mixed-model assembly line, which considers keeping a constant speed in consuming each part in the line, leveling the load on each process within the line, and the products transportation planning above-mentioned. At last an example is given to illustrate the characteristics and effectiveness of the method in the paper.

Keywords Mixed-model assembly line · Sequencing problem · Production loads · Transportation schedule

---

Hugejile (✉) · S. Yang
Ningbo University of Technology, Fenghua Road 201,
Ningbo City 315211 Zhejiang, China
e-mail: joxu@ehime-u.ac.jp

S. Hiraki
Faculty of Economic Sciences, Hiroshima Shudo University,
1-1-1, Otsukahigashi, Asaminami-ku, Hiroshima 731-3195, Japan
e-mail: hiraki@shudo-u.ac.jp

Z. Xu
Faculty of Law and Letters, Ehime University, Bunkyo 3, Matsuyama City,
Ehime 790-8577, Japan

68.1 Introduction

Nowadays the customer's need varies frequently and the life cycle of products becomes shorter. In order to survive in such fierce competition, many enterprises have introduced the advanced theory of Supply Chain Management (SCM).

In order to solve the sequencing problem for a mixed-model assembly line, the previous researches have covered the following fields such as the researches on the objective of keeping a constant speed in consuming each part in the line (Miltenburg 1989; Inman and Bulfin 1991; Cakir and Inman 1993; Kurashige et al. 2002), leveling the load on each process within the line (Thomopoulos 1967; Okamura and Yamashina 1979; Bard 1992; Tsai 1995; Xiaobo and Ohno 2000), or considering the two objectives above simultaneously (Miltenburg and Goldstein 1991; Bard et al. 1994; Kotani et al. 2004; Xu and Hiraki 1996; Huge et al. 2011). These researches seek to realize the optimized goal for logistics from part supplier to automobile manufacture enterprises, but fail to consider the logistics from automobile manufacturers to distributors, which may lead to the delay of production transportation planning.

The sequencing problem for the mixed-model assembly line proposed in this paper, considers the road transportation plan for a part of products, on the base of the research above.

68.2 To Decide the Production Volume of Each Product for Shorter Periods

The paper will take two stage-approach by Xu et al. (1996) to decide the sequencing problem. At the first stage, product period has been divided into several shorter periods by the time of the part withdrawn. We will attempt to determine the production volume of each product for each period, by keeping a constant speed in consuming each part and considering production transportation planning.

68.2.1 Assumptions

(1) The cycle time of the mixed-model mobile assembly line is fixed.
(2) Parts withdrawals are performed at fixed intervals.
(3) A part of the products in the schedule are transferred to the designated distributor by car-carriers.
(4) It is provided that both the destination to distributor of each car-carrier and the products loaded on the car-carriers.
(5) The car-carrier has set the start time according to the arrival time agreed by the designated distributor.
(6) The number of the car-carrier is ranked by the set start time.

68.2.2 Notation

I	the total quantity of product types $(i = 1, \cdots, I)$,
d_i	total production quantity of product i,
K	total production quantity of all products in schedule time,
J	total quantity of parts types $(j = 1, \cdots, J)$,
a_{ij}	necessary quantity of part j to be utilized for producing product i, (Table BOM)
R_j	total necessary quantity of part j to be consumed for producing all products,

$$R_j = \sum_{i=1,\cdots,I} d_i \cdot a_{ij} \ (j = 1, \cdots, J)$$

r_j	average necessary quantity of part j per unit product, $r_j = R_j/K$ $(j = 1, \cdots, J)$
sh^k	production start time of product with sequence number k,
fh^k	production finish time of product with sequence number k,
O_{jk}	it takes one unit time when j is withdrawal in position k, otherwise it takes zero,
ω	the set of k that one or more types of part are withdrawn in position k,

$$\omega = \{k | O_{jk} = 1 \ (j = 1, \cdots, J; k = 1, \cdots, K)\},$$

ω'	put the factor in set ω into order from small to big, $\omega'(b)$ means the factor b in set ω,
δ_k	the set of part j withdrawn in position k, $\delta_k = \{j | O_{jk} = 1 \ (j = 1, \cdots, J)\} \ (k \in \omega)$
X_{ik}	cumulative production volume of product i sequenced from position one to k,
Y_{jk}	total volume of part j required to assemble the products sequenced from position one to k,
L	total quantity of shorter periods, $L = |\omega|$,
T_{il}	production volume of product i in periods l,
K'	total quantity of production by road transportation,
N	total necessary quantity of car-carrier in road transportation $(n = 1, \cdots, N)$,
$V_{i,n}$	quantity of product i to be carried by car-carrier n,
E	the necessary transportation time from the finishing production to arriving at the loading area new car terminal,
g_n	set start time of car-carrier n, G_n: real start time of car-carrier n,
σ_k	the set car-carriers that has to start when the products sequenced from position one to k are loaded, $\sigma_k = \{n | fh^k + E \geq g_n \ (n = 1, \cdots, N)\} (k \in \omega)$.

68.2.3 Formulation

We decide to minimize the function (1) which is the sum of squares of deviations of the cumulative withdrawal quantity of all parts till the withdrawn time, under the constraints (2)–(9). The production volume of each product in each period could be determined by solving the mathematical programming problem.

Evaluation function

$$f_1 = \sum_{k \in \omega} \sum_{j \in \delta_k} (k \cdot r_j - Y_{jk})^2 \tag{68.1}$$

Constraint:

$$X_{iK} = d_i \quad (i = 1, \cdots, I) \tag{68.2}$$

$$\sum_{i=1,\cdots,I} X_{ik} = k \quad (k \in \omega) \tag{68.3}$$

$$X_{ik} \geq \sum_{n \in \sigma_k} V_{i,n \, (k \in \omega; i=1,\cdots,I)} \tag{68.4}$$

Therein,

$$\sum_{n \in \sigma_K} \sum_{i=1,\cdots,I} V_{i,n} = K' \tag{68.5}$$

$$Y_{jk} = \sum_{i=1,\cdots,I} X_{ik} \cdot a_{ij} \quad (k \in \omega; j \in \delta_k) \tag{68.6}$$

$$X_{i\omega'(b-1)} \leq X_{i\omega'(b)} \quad (i = 1, \cdots, I; b = 2, \cdots, |\omega|) \tag{68.7}$$

$$T_{il} = X_{i\omega'(l)} - X_{i\omega'(l-1)} \quad (i = 1, \cdots, I; l = 1, \cdots, L) \tag{68.8}$$

$$X_{i\omega'(0)} = 0 \quad (i = 1, \cdots, I)$$

$$X_{ik}, Y_{jk} (i = 1, \cdots, I; k \in \omega; j \in \delta_k) : \text{non} - \text{negative integer} \tag{68.9}$$

68.3 The Scheduling Order for Products in Shorter Period

In stage 2, considering the operation delay time and the set start time of car-carrier of products in each shorter period, we determined the products schedule. The model could be the same method as Huge et al. (2011), which is omitted in the paper.

68.4 Example

In the example, total quantity of product types $I = 4$, total quantity of parts types $J = 4$, total production quantity of each product $d_A = 30$, $d_B = 40$, $d_C = 30$, $d_D = 20$, total production quantity of all product $K = 120$. The products schedule needs to be determined. Cycle time $c = 60$ s, conveyor speed $v = 0.1$ min/s, sh^0 means 8 A.M., $E = 1$ h.

The production quantity of each product types, Table BOM, the total required quantity for each part types and withdrawal time is showed in Table 68.1. The withdrawal time for parts is four times and is 8:40, 9:00, 9:20 and 10:00 respectively. The scheduling time is divided into four shorter periods, $L = 4$.

During the 120 products in the scheduling time, 60 products have designated distributor (or customer) order and lead time. Each car-carrier could take 5 cars, and the whole transport needs 12 car-carriers. The specific transportation schedule is illustrated in Table 68.2.

Table 68.3 illustrates the operation time of each product in each process, and the initial position of operator of shorter period 1 when he starts operation.

Table 68.1 Production quantity of each product, BOM table, withdrawal time

Product type		A	B	C	D	Rj	8:40	9:00	9:20	10:00
d_i		30	40	30	20		40	60	80	120
Part Type	1	1	0	1	0	60	1	0	1	1
	2	0	1	1	1	90	1	0	1	1
	3	1	2	0	0	110	0	1	0	1
	4	1	0	0	1	50	0	1	0	1

Table 68.2 Car-carrier schedule

Number of car-carrier	1	2	3	4	5	6	7	8	9	10	11	12
Product type												
A	1	0	1	0	5	0	0	1	2	3	3	4
B	2	3	1	4	0	3	5	2	2	2	2	1
C	1	2	1	0	0	1	0	2	0	0	0	0
D	1	0	2	1	0	1	0	0	1	0	0	0
Set start time	16:40					17:00		17:10		18:00		

Table 68.3 The operation time of each process (second)

Number of process	1	2	3	4	5	6	7	8
Product types								
A	65	50	70	52	70	45	59	65
B	60	40	45	60	60	65	50	48
C	50	56	62	40	40	58	57	50
D	40	55	50	50	45	48	60	50

Table 68.4 Production schedule for each period

1	Production volume				Mixed ratio of product	Production schedule
1	10	15	10	5	2:3:2:1	DCBABCBA (5th time)
2	5	5	5	5	1:1:1:1	DABC (5th time)
3	5	10	5	0	1:2:1:0	BABC (5th time)
4	10	10	10	10	1:1:1:1	DABC (10th time)

Table 68.5 Model production volume and schedule based on (Xu and Hiraki 1996)

1	Production volume				Mixed ratio of product	Production schedule
1	10	10	10	10	1:1:1:1	DABC(10th time)
2	5	10	5	0	1:2:1:0	BABC(5th time)
3	5	0	5	10	1:0:1:2	DCDA(5th time)
4	10	20	10	0	1:2:1:0	BCBA(10th time)

Table 68.6 Real start time of car-carrier when that all products to be assembled

n	g_n	This model			Xu and Hiraki 1996)		
		$Z_n.fh^{Z_n}.G_n$			$Z_n.fh^{Z_n}.G_n$		
1	16:40	5	15:05	16:05	7	15:07	16:07
2		13	15:13	16:13	19	15:19	16:19
3		17	15:17	16:17	23	15:23	16:23
4		27	15:27	16:27	39	15:39	16:39
5		28	15:28	16:28	26	15:26	16:26
6	17:00	35	15:35	16:35	45	15:45	16:45
7		55	15:55	16:55	55	15:55	16:55
8	17:10	61	16:01	17:01	83	16:23	17:23
9		65	16:05	17:05	87	16:27	17:27
10	18:00	70	16:10	17:10	91	16:31	17:31
11		74	16:14	17:14	95	16:35	17:35
12		76	16:16	17:16	97	16:37	17:37

According to the known conditions, we could achieve the results below.

Table 68.4 illustrates this research result of production volume and Determine the scheduling order of each shorter period.

Table 68.5 showed production volume and production schedule in each shorter period, regardless of transportation planning.

Table 68.6 illustrates the real start time for car-carrier in the two models.

The result of this research illustrates that the set start time of all car-carrier have no delay. In the result of Xu et al. (1996), the set start time for numbers 8 and 9 car-carriers have delayed for 13 and 17 min respectively.

68.5 Conclusion

In this research, it is considered that only part of the scheduled products would be transported by car-carrier to the car distributor. Therefore it is proposed that a scheduling method for mixed-model assume objectives, the effectiveness of the research is testified. In order to solve the scheduling problem, this research has realized the objective of integrative optimization of the supply chain from part supplier to distributor.bly line would satisfy a constant speed in consuming each part, leveling the load on each process and production transportation plan. After compared with the research of Xu et al. considering two objectives, the effectiveness of the research is testified. In order to solve the scheduling problem, this research has realized the objective of integrative optimization of the supply chain from part supplier to distributor.

Acknowledgments This research was supported by Zhejiang provincial department of education scientific research project (Y201120349) and the scientific research fund project of Ningbo University of Technology.

References

Bard JF, Dar-el E, Shtub A (1992) An analytic. Framework for sequencing mixed model assembly. Int J Prod Res 30(1):35–48
Bard JF, Shtub A, Joshi SB (1994) Sequencing mixed-model assembly lines to level parts usage and minimizing line length. Int J Prod Res 32(10):2431–2454
Cakir A, Inman RR (1993) Modified goal chasing for products with non-zero/one bills of materials. Int J Prod Res 31(1):107–115
Huge J, Hiraki S, Jia C (2011) International conference on information technology. Serv Sci Eng Manag 3(26–28):1539–1542
Inman RR, Bulfin RL (1991) Manage. Science 37(7):901–904
Kotani S, Ito T, Ohno K (2004) Sequencing problem for a mixed-model assembly line in the Toyota production system. Int J Prod Res 42(23):4955–4974
Kurashige K, Yanagawa Y, Miyazaki S, Kameyama Y (2002) Production and operations management. Prod Plan Contr 13(8):735–745
Miltenburg J (1989) Manage. Science 35(2):192–207
Miltenburg J, Goldstein T (1991) Developing production schedules which balance part usage and smooth production loads in just-in-time production systems. Nav Res Log 38(6):893–910
Okamura K, Yamashina H (1979) Parametric appraisal of the JIT systems. Int J Prod Res 17(3):233–247
Thomopoulos NT (1967) Manage. Science 14(2):B59–B75
Tsai LH (1995) Manage. Science 41(3):485–495
Xiaobo Z, Ohno K (2000) Bibstring. Euro J Opera Res 124:560–570
Xu Z, Hiraki S (1996) Japan industrial management association, vol 46(6). pp 614–622

Chapter 69
Price Competition in Tourism Supply Chain with Hotels and Travel Agency

Yun Huang

Abstract Tourism was considered a "smokeless industry", depending on using and developing the natural and cultural resources of a country to attract visitors. This paper studies the impact of different power structures on the room rates decisions in a tourism supply chain with two hotels, one luxury and the other economic, and one travel agency. Stackelberg game and Nash game models are formulated to analyze the pricing decisions in the power structures. We conduct comparative study between the three structures and explore the effects of different parameters on equilibrium prices, demands and profits.

Keywords Tourism supply chain · Hotel · Travel agency · Pricing · Stackelberg game · Nash game

69.1 Introduction

Tourism is considered a "smokeless industry", largely dependent on using and developing the natural and cultural resources of a country as attractions for visitors. Increasing tourists' concerns on environmental issues also force companies to adopt sustainable supply chain management strategies. As the development of tourism supply chain, the relationships between different entities play important roles in business, tourism, academic exchange and so on (Zhang et al. 2009).

Y. Huang (✉)
Faculty of Management and Administration, Macau University of Science and Technology,
Taipa, Macau
e-mail: yuhuang@must.edu.mo

Different market power has profound influence on equilibrium prices and profits of the supply chain members (Choi 1991).

This paper is sparked by this conflict and coordination relationship between hotels and their channel partners (travel agencies or tour operators) of the tourism supply chain. A travel agency cooperates with hotels to provide tour packages of accommodation and sightseeing. The sightseeing routines are identical for all tourists; meanwhile the accommodation is divided into different levels to meet different demands of tourists. For example, the Hong Kong Student Travel agency (http://www.hkst.com.hk/) provides different Greece Romance travel tourism packages to the tourists (http://www.hkst.com/fileshow.asp?id=2991). The packages include a 4 days' visit to Athens and three nights' accommodation. The difference between them is the three night's accommodation, including economic and luxury accommodation provided by different hotels.

In order to study the pricing strategy of different hotels and travel agency under different market power, we consider a tourism supply chain with two types of hotels, one luxury and the other economic, and a travel agency. We mainly focus on the competition on room rates decisions between the hotels and travel agency. Three power structures are considered in this paper. The first two cases study the hotels and the travel agency take the chain leadership, respectively. Stackelberg game is used to model the two cases. In the third case, we study the situation that the hotels and the travel agency are of equal market status and a Nash game is played between them.

The rest of this paper is organized as follows. Section 69.2 reviews the literature on tourism supply chain, pricing strategy and game theory application. The subsequent section presents different game models in the tourism supply chain. Section 69.4 derives the optimal pricing decisions and profits in the tourism supply chain with hotels and travel agency under three different power structures. In Sect. 69.5, we conduct comparative studies on the effects of different power structures and different parameters. The paper concludes with some suggestions for further research.

69.2 Literature Review

There are two subject areas related to our work: tourism supply chain, pricing strategy and the application of game theory to supply chain management and tourism industry.

In general, the literature on tourism supply chain is very limited. Some studies have focused on the relationships between tour operators and the accommodation or hotels. Buhalis (2000) and Medina-Muñoz et al. (2003) examined the competitive and cooperative relationships between tour operators and hotels in a distribution channel. Zhang et al. (2009) systematically reviewed the literature of current tourism studies from the tourism supply chain management perspective and developed a framework for TSC management research.

Game theory is a powerful tool extensively used in manufacturing supply chains to study competitive and cooperative relationships. Choi (1991; Choi 1996) investigated a pricing competition problem for a channel structure consisting of two competing manufacturers and one common retailer who sells both manufacturers' products. Minakshi (1998) analyzed three channel structures dealing with two competing manufacturers and two retailers to examine the channel competition problem. Huang et al. (2011) studied supplier selection, pricing and inventory coordination problem using a dynamic game model. In recent years, applying game theory in tourism has attracted considerable attention. Aguiló et al. (2002) studied an oligopoly tourism market in which the tour operators have the market power to determine a higher price and keep their market share. Wachsman (2006) employed a Nash game model to study the interactions among hotels and airlines. Song et al. (2009) conducted investigation into pricing competition and coordination between Hong Kong Disneyland and a tour operator using game models. Many of these studies considered only one hotel and one travel agency, and the analysis of competition and coordination was confined to members in the same channel or echelon. When several different channel systems are considered, the interactions among channel members become more complicated. In this paper, we show that the different power structures play a critical role in determining the price decisions of the members in tourism supply chain.

69.3 Game Model

We consider the tourism supply chain with two hotels competing to provide two types of accommodation, luxury and economic respectively, to a travel agency. Then the travel agency sells the different hotels directly or combines with other tourism products as tourism packages to the tourists. The hotels and the travel agency are independent decision makers and optimize their own profits individually. Three cases are studied. One is the hotels take the leading role of the tourism supply chain. Suppose the hotels make the first move on room rate decisions and propagates the rooms to the travel agency. The travel agency as the follower then decides the optimal room rate for the tourism packages. The second one is the travel agency takes the chain leadership. That is the travel agency first prices the rooms and the hotels decide the optimal prices as the followers. The last one is the two hotels and the travel agency of the same market power. They make their pricing decisions simultaneously. No one dominates over others.

The revenue of the tourism hotels is as follows

$$\pi_i = D_i w_i \qquad (69.1)$$

where $i = L, E$, L stands for luxury hotel, and E for the economic one. w_i denotes the room rate set by the hotel i.

And the revenue of the travel agency T is

$$\pi_T = D_L(p_L - w_L) + D_E(p_E - w_E) \tag{69.2}$$

where p_i is the room rate decided by travel agency for rooms of hotel i.

The hotels and travel agency make pricing decisions non-cooperatively, and their objective is to maximize their own revenue. The capacity of hotel is neglected in this paper, which does not affect the discussion in this paper.

The tourist experience of luxury room is more desirable than that of the economic one. We denote tourist experience of the hotel i as s_i. Thus, we assume that $s_L > s_E$. $\Delta s = s_L - s_E$. Reflects the difference in tourism experience between the two different hotels.

We denote tourism preference of a hotel as θ. θ is a random variable following uniform distribution normalized to [0, 1]. The tourist utility is defined as a function of perceived experience and room rate to the tourists (Keane 1997).

$$u_i = v + \theta s_i - p_i, i = L, E \tag{69.3}$$

where v is a basic utility of a hotel room and homogeneous among all tourists to the two types of rooms. We should know that positive preference does not means positive utility. And if the utility is lower than zero, tourists will not purchase any of rooms. Given s_i and p_i, when $u_E = 0$, then $\theta = \hat{\theta}_E$, where $\hat{\theta}_E = \frac{p_E - v}{s_E}$. When $u_L = 0$, then $\theta = \hat{\theta}_L$, where $\hat{\theta}_L = \frac{p_L - v}{s_L}$. And when $u_L = u_E$, a tourist will be indifferent between the luxury and the economy rooms and we get $\theta = \theta^* = \frac{p_L - p_E}{s_L - s_E}$.

According to (Song et al. 2009), the demand of the luxury rooms is,

$$D_L = 1 - \theta^* = 1 - \frac{p_L - p_E}{s_L - s_E} \tag{69.4}$$

And the demand of the economic rooms is,

$$D_E = \theta^* - \hat{\theta}_E = \frac{p_L - p_E}{s_L - s_E} - \frac{p_E - v}{s_E} \tag{69.5}$$

69.4 Solution

Three power structures are considered in this paper, two leader–follower structures and one independent structure. We use Stackelberg game to model the first two structures and Nash game for the independent one.

69.4.1 Hotel Stackelberg

We use Stackelberg game to model the leader–follower power structure. For convenience, we call this game model as Hotel Stackelberg (HS). In this game, the hotels take the travel agency's reaction functions into consideration for their pricing decisions. The equilibrium is a set of pricing decisions in which both the hotels and the travel agency have no incentive to change their prices unilaterally.

69.4.2 TA Stackelberg

We then consider the structure that the travel agency takes the channel leadership. Stackelberg game is also employed to model this scenario. We call this game model as TA Stackelberg (short for TS). In this game, the travel agency takes the hotels' reaction functions into consideration for its pricing decision. We can obtain the Stackelberg equilibrium of the TS game as a solution for the travel agency Stackelberg model.

69.4.3 Vertical Nash

The third independent power structure is formulated as a Nash game. In this game, the hotels and the travel agency make pricing decisions simultaneously and non-cooperatively. Again for convenience, we call this game Vertical Nash (VN). In this game, the hotels choose their room rates conditional on the travel agency's pricing to maximize their profits. The travel agency chooses its optimal prices conditional on the hotels' pricing decisions.

69.5 Discussion

To understand the impacts of different power structures, we conduct a comparative study between the three cases.

Proposition 1 The basic utility v has positive impacts on the equilibrium prices, demands and profits.

This proposition shows us that the increase of v will increase the equilibrium room rates, demands and profits of all the chain members. High utility means high quality of the tourist experience, so the rooms would be more desirable. The proposition is very intuitive. A well-pleasing tour will attract more tourists and the tourists will accept a higher price easily.

Proposition 2 The tourism experience of economic tour package s_E has negative impacts on pricing, profits, but positive impact on demands of luxury tour package.

Proposition 2 indicates that the increase of tourist experience for economic room brings down the room rate of luxury hotel and its profit although their demand increases. Due to the competition between the hotels, the luxury hotel will reduce its hotels prices to attract more tourists to counter the impact of the increasing tourist experience from the economic hotel. The reduced price attracts more tourists, so their demand increases, but their profit still decreases.

69.6 Conclusion

Most previous studies have focused on tourism supply chain with one single member in each echelon. In the contemporary real market, coordination for the multiple members with different power structures is inevitable and necessary. This paper extends the growing literature of channel studies by analyzing pricing strategies in a two-level tourism supply chain with two hotels and one travel agency under three different power structures. Stackelberg and Nash games are used to model different structures. We investigate the effects of power structures, different parameters on the pricing decisions and profits for chain members.

This paper suffers from several limitations. The models in this paper only consider tourism supply chain with one or two members in each echelon. A more general model with multiple hotels, travel agencies could be developed. Besides, a range of distribution channels could be considered. For example, the hotels can sell their rooms to the travel agencies and customers directly. Such situations could also be studied. Thirdly, we consider only one strategic variable—pricing. A major direction for future research could involve some other strategic variables, such as quantity, advertising.

Acknowledgments The authors would like to acknowledge the financial support of Macau University of Science and Technology (Grant No. 0237).

References

Aguiló E, Alegre J, Sard M (2002) Analysis of package holiday prices in the Balearic Islands. Working paper, University of Balearic Islands, Palma de Mallorca
Buhalis D (2000) Relationships in the distribution channel of tourism: conflicts between hoteliers and tour operators in the Mediterranean region. Int J Hosp Tour Admin 1(1):113–139
Choi SC (1991) Price competition in a channel structure with a common retailer. Market Sci 10(4):271–296
Choi SC (1996) Price competition in a duopoly common retailer channel. J Retail 72(2):117–134
Huang Y, Huang GQ, Newman ST (2011) Coordinating pricing and inventory decisions in a multi-level supply chain: a game-theoretic approach. Transp Res Part E Logist Transp Rev 47(2):115–129
Keane MJ (1997) Quality and pricing in tourism destinations. Ann Tour Res 24:117–130
Medina-Muñoz RD, Medina-Muñoz DR, García-Falcón JM (2003) Understanding European tour operators' control on accommadation companies: an empirical evidence. Tour Manag 24(2):135–147

Minakshi T (1998) Distribution channels: an extension of exclusive relationship. Manage Sci 44(7):896–909

Song HY, Yang S, Huang GQ (2009) Price interactions between theme park and tour operator. Tour Econ 15(4):813–824

Wachsman Y (2006) Strategic interactions among firms in tourist destinations. Tour Econ 12(4):531–541

Zhang XY, Song HY, Huang GQ (2009) Tourism supply chain management: a new research agenda.". Tourism Manag 30:345–358

Chapter 70
Evaluation on Bus Rapid Transit in Macau Based on Congestion and Emission Reduction

Huajun Tang, Xinlong Xu and Bo Huang

Abstract With the fast development of the tourism industry, the land transportation system at present can barely sustain the overpopulated city. Based on the merits of the cheaper investment, massive capacity, less contamination, effectiveness and the short construction cycle, Bus Rapid Transit (BRT) is one of the best alternatives to enhance urban transportation service levels in a short period. This study takes an example of Macau and tries to identify the features of the public transportation system as well as analyzing the existing problems. Then the study conducts an assessment of the significance of launching a BRT in Macau with respect to the underlying economic revenue and exhaust reduction. Finally, it is supported that the BRT system can effectively reduce the traffic congestion and CO_2 emission.

Keywords Public transport priority · Bus rapid transit · Congestion reduction · Emission reduction

70.1 Introduction

With the rapid development of Macau economy, the number of vehicles in Macau increases dramatically. Particularly, since the year of 2003, in which the Sands and Wynn casinos opened, free shuttle buses have come into service, so as to provide convenient and quick travel for those who visit the casinos in Macau. Because of the quick development of the casinos and the lack of the local government's

H. Tang (✉) · X. Xu · B. Huang
Faculty of Management and Administration, Macau University of Science and Technology, Taipa, Macau
e-mail: hjtang@must.edu.mo

effective management, the population of vehicles (including public and private ones) grows remarkably, which not only leads to great traffic pressure, but also increases the vehicle emission in Macau. In addition, as a special region with famous tourism, it is fundamental for Macau to have an advanced, green and smooth transportation system. Therefore it is urgent and challenging to improve the transportation system to reduce its negative impacts and improve the environment in Macau.

70.2 Current Traffic Conditions

So far Macau has been the region with the highest vehicle-density. The streets in Macau are very narrow, while the roads available are very few, which leads to serious congestion, and many traffic accidents.

Vehicles in Macau can be classified into two categories: public and private motors. The public vehicles consist of taxis, buses and traveling buses. According to the statistic report from Macau Government, the population of taxies in 2011 reached to about 1,000, which is fewer than any other regions in Great China. However, they usually run in some busy streets so as to serve more passengers, which leads to much traffic congestion, since these streets have at most two lanes.

Private vehicles mainly consist of two-wheel and four-wheel motors. Since the return in 1999, the population of private motors increases rapidly. According to the report in Year 2012 from Macau Government, there are 81,684 four-wheel and 111,717 two-wheel vehicles, which is almost as twice as those (51,510 four-wheel and 57,292 two-wheel vehicles) in Year 2002. The population of the private vehicles in recent 11 years is given in Table 70.1.

Based on the current traffic conditions, it is urgent and challengeable to ease the traffic congestion and reduce its negative impact on the tourism in Macau.

Table 70.1 The population of private vehicles in macau in recent 11 years

Years	Four-wheel vehicle	Two-wheel vehicle
2002	51,510	5,7292
2003	52,379	5,8250
2004	55,809	62,164
2005	59,556	66,389
2006	63,916	72,528
2007	68,334	78,816
2008	71,726	85,368
2009	76,117	92,296
2010	78,753	97,724
2011	80,499	102,566
2012	81,684	111,717

70.3 Public Transport Priority

Considering that the current traffic conditions in Macau become more and more serious, Macau Government has set up an institution (i.e., Institution of Integrated Traffic and Transportation, IITT) to build a green, safe, and smooth integrated transportation system, so as to improve the traffic conditions.

In the literature there are several publications to study city transportation systems. For instance, Wang introduced several methods to improve public transport systems, according to different traffic conditions (Wang 2012). Zuo and Shao discussed some policies and strategies about public transport priority (Zuo and Shao 2012). Zuo et al. explored the difference between travel speed of buses and cars, and provided one model to express the shift relationship (Zuo et al. 2012). Furthermore, there also exist some publications on the construction and evaluation of public transit priority. For instance, Chen and Yan investigated on signal timing for urban intersections with genetic algorithms optimizing with the idea of bus priority (Chen and Yan 2005). Zhang et al. optimized the signal-planning method of intersections based on bus priority (Zhang 2004). However, to our best knowledge, there exists little research on the evaluation of public transport priority in Macau. Hence this study aims to focus on the application of BRT in Macau.

Based on the research in the literature and practical situations in Macau, it is emergent and fundamental to take public transport priority into account. This study will discuss the feasibility and positive impact of bus rapid transport (BRT), which is one of the best measures of public transport priority.

BRT system aims to use advanced buses to run along bus lanes, introduce the advantage of the rail transport, and keep the flexibility of buses. It is the combination of the rail transit and the public transit. According to its strength and real situations in Macau, IITT firstly provided one sample lane for BRT, the feasibility and effectiveness of which need to be evaluated. In the following this study will apply several models to prove the benefit of BRT with respect to congestion and emission reduction.

70.4 Evaluation of BRT

According to the spot investigation, one sample BRT may start from Mage, through River New Street, to Border Gate, so as to integrate the first-stage light railway, which will be completed in 2014. This BRT can provide rapid bus service for the residents at Cheongju, Fai Chi Kei, and Border Gate.

To analyze the benefit of the BRT sample road, taking into account practical distribution of the population, present bus service, and future blueprint from Transport Bureau in Macau, this study divides the sample road into two segments. The first segment starts from Mage to Si Dakou, and the second starts from Perfect the Road to Border Gate. In the following, the study will apply two models to conduct benefit analysis with respect to environment improvement and traffic congestion reduction.

70.4.1 Benefit Analysis of Traffic Congestion Reduction

New buses with environment protection (e.g., Dennis Type Buses) are used in the BRT sample road, which can reduce more noise and vehicle emission than those applied in common bus systems. In addition, the vehicle schedule adopts more frequent shifts and bigger bus capacity to reduce the population of vehicles and traffic congestion.

In the following the paper will use one model to evaluate the benefit on traffic congestion reduction. $X = (L/V_B - L/V_A) \times 1.25 \times M_t$, where X denotes the benefit (MOP$) of traffic congestion reduction from vehicles, V_B and V_A represent the running speed (km per hour) of vehicles before and after the implementation of the BRT system, respectively, L is the length (km) of the BRT road, and M_t stands for the value per unit time (MOP$ per hour), which can be computed as below:

$$M_t = \overline{GDP}/(365 \times 8).$$

According to the spot investigation, the distance from Mage to Border Gate is 4.3 km, and the reverse distance is 4.9 km, then its average distance $L = (4.3 + 4.9)/2 = 4.6$km. The current average speed of vehicles running along this road is 15 km/h, which is much lower than the limited speed 35 km/h. The average GDP per person in 2011 is MOP$ 398,073, which leads to M_t = MOP$136/h.

Once the BRT sample road is open, the average speed can reach to the limited value. The benefits corresponding to the different average speed are listed in Table 70.2.

According to the data in Table 70.2, the average loss per vehicle will be MOP$26.07 if the current traffic condition is kept until Year 2020. However, if the BRT road is open, then different running speed will lead to different benefit for each vehicle. When the running speed reaches to 35 km/h, then the average benefit for each vehicle will be MOP$ 29.79.

Table 70.2 The benefits corresponding to the different average speed under BRT

	Present	2020	BRT 1	BRT 2	BRT 3	BRT 4
V_B(km/h)	15	15	15	15	15	15
V_A(km/h)	15	10	20	25	30	35
M_t(MOPS/h)	136	136	136	136	136	136
X(MOP$)	0	−26.07	13.04	20.85	26.07	29.79

70 Evaluation on Bus Rapid Transit

Table 70.3 CO_2 emission corresponding to the different average speed under BRT

	Present	2020	BRT 1	BRT 2	BRT 3	BRT 4
V_A(km/h)	15	10	20	25	30	35
E(L/100 km)	42.49	44.33	40.81	39.31	37.97	36.79
E (L)	3.91	4.08	3.75	3.62	3.49	3.38
Eco2d(kg)	911.00	950.45	875.12	842.81	814.07	788.90
Eco2y(kg)	3197.26	3335.70	3071.34	2957.94	2857.07	2768.72

70.4.2 Benefit Analysis of Environment Improvement

In this subsection, the benefit of environment improvement will be analyzed. As it is known that BRT buses and the other buses use different roads, and they are independent. Once the BRT road is open, all the buses in the BRT road has less operational time, and less traffic congestion, which leads to less CO_2 emission. In the following this study adopts one model to evaluate the benefit of emission reduction.

Let E be total oil consumption on the sample road, M_v be the mileage of different vehicles running along the sample road, and E_v be the oil consumption for some vehicle along the road. In this study, E_v can be obtained through the equation $E_v = 0.001784842 \times V_A - 0.256157175 \times 35 + 17.94117582$. Hence, the vehicle emission can be evaluated as follows.

$$E = \sum (M_v \times E_v).$$

Then the emission of CO_2 can be computed through the following equation, $E_{CO_2} = f \times E$, where f is the coefficient of CO_2 emission, and is fixed as 2.241(kg CO_2 per liter oil), according to the D-type bus. Define E_{CO_d} and E_{CO_y} as the total CO_2 emission for the vehicle along the sample road in 1 day and in 1 year, respectively. Different vehicle speed along the road will lead to different CO_2 emission, which is shown in Table 70.3.

According to the data in Table 70.3, it is obvious that the CO_2 emission can be reduced through the increase of the vehicle speed.

70.5 Conclusions and Future Research

This study firstly presented the current traffic conditions in Macau, and suggested that it was urgent and fundamental to improve the traffic system with the use of BRT. Then the paper provided one sample road for BRT based on the spot investigation, and applied two models to evaluate the benefits with respect to traffic congestion and CO_2 emission reduction. Finally, it was supported that BRT system can effectively reduce the traffic congestion and CO_2 emission.

Since this study only takes BRT system into account, which is only the one of the measures to remit traffic problems in Macau, it will integrate BRT system with other measures (e.g., introducing electronic motor and light railway) in the new future.

References

Chen Q, Yan KF (2005) Research on signal timing for urban intersections with genetic algorithms optimizing with the idea of bus priority. Syst Eng Theory Pract 11:133–138

Wang W (2012) Insist on public transport priority and build smooth cities. Road Transp Manag 3:46–50

Zhang WH, Lu HP, Shi Q (2004) Optimal signal-planning method of intersections based on bus priority. J Traffic Transp Eng 4:49–53

Zuo ZY, Shao CF (2012) Study on the policy of public transport. Integr Transp 4:31–34

Zuo ZY, Yang GC, Shao CF (2012) Modeling modal shift of car travelers to buses based on public transport priority. J Transp Syst Eng Inf Technol 2:124–130

Chapter 71
The Analysis and Strategy Research on Green Degree of Enterprise in Green Supply Chain

Lijin Liu

Abstract This paper firstly compares the meaning and the general conceptual model of traditional supply chain with those of green supply chain and finds out the green supply chain's advantages; then it puts forward an evaluation index system of enterprises green degree and builds a multifactor fuzzy comprehensive evaluation model of enterprises green degree. Finally, it puts forward some measures to improve enterprises green degree, in order to enhance a whole green level of enterprises and supply chain.

Keywords Green supply chain · Green degree · Fuzzy evaluation · Strategy

71.1 Introduction

Since the second half of 2009 China's emission reduction targets published and the Copenhagen climate conference, environmental protection and green economy development has been concerned very extensively and profoundly, green supply chain management has become a very hot topic.

Supply chain is an overall functional network structural model, which connects manufacturers, distributors, suppliers, retailers and the final user (Ma Shihua et al. 2000). Its procurement, production, sales and logistics processes are of high investment, high consumption and high waste output. The collaboration between enterprises pays more attention to economic benefits, and ignores the environmental protection, which is opposite to the economy sustainable development.

L. Liu (✉)
ZhuHai College of JiLin University, Zhuhai Guangdong 519041, China
e-mail: Liulijin814@163.com

So Many scholars have carried out comprehensive research on green supply chain theory since 1996.

Green supply chain is not a uniform definition as same as traditional supply chain. The author of this paper consults massive literature and information and believes that the green supply chain pays more attention to environment protection in the implementation of supply chain management and emphasizes on the harmonious development of environment and economy. It is a model of modern enterprise management that calls on an ecological design of the whole supply chain following a thought of green purchasing, green manufacturing, green marketing, green consumption, green logistics, and tries to make the whole supply chain management achieve environment harmonious and unified through close cooperation of the enterprise in chain, and finally tries to realize the economic benefits, environmental benefits and social benefits optimization with higher resource efficiency. Green supply chain management is based on the sustainable development theory, ecological economics theory and ecological ethics theory, which can realize economy and environment harmonious coexistence, and achieve a win–win situation.

71.2 The Analysis and Evaluation Study on Enterprises Green Degree in Green Supply Chain

"Green" is closely related to the environmental impact and is also a symbol of "environmental protection". The green degree is usually relevant with environmental regulations and standards as the benchmark, when the supply of environmental effects meets the requirement, it is green. Enterprise's green degree in green supply chain can be defined as the green degree or the environmental friendly degree, namely environmental impact quantitative. Negative environmental impact is greater, the green degree is smaller, whereas the larger (Wang and Sun 2005).

Evaluation model usually comprises of the evaluation index system and evaluation method. This paper firstly builds the index system of green degree, and then uses the fuzzy comprehensive evaluation model to analyze it, in order to help enterprises judge its supply of environmental compliance with environmental standard and regulatory requirements.

71.2.1 A Design of Evaluation Index System on Enterprises Green Degree in Green Supply Chain

The establishment of any evaluation index system is not imaginary, but must be combined with the characteristics of being evaluated object. In general, the green degree evaluation must be based on environmental protection and resource conservation principle to select the most appropriate index.

Due to legal and public pressure, businesses are increasingly trying to improve environmental performance. The index Environmental performance U_1 can effectively reflect enterprises' environmental protection implement level in green supply chain, so we can design for the wastewater discharge compliance rate x_1, emissions compliance rate x_2, solid waste emissions compliance rate x_3, noise pollution x_4, cleaner production level x_5;

The effective use of energy and resources gradually attracted the attention of enterprises. The index energy properties U_2 can reflect the extent that enterprises in green supply chain use effective energy-saving, so we can be design for energy efficiency x_6, energy output ratio x_7, product energy consumption x_8; the index resource properties U_3 can reflect green supply chain for the degree of efficient use of raw materials, equipment, etc., so you can design materials recycling rate x_9, equipment effective utilization of x_{10}, eco-friendly materials usage x_{11}.

The enterprises in supply chain should focus on products with reusable value, the recovery of parts or materials used to conserve resources, and achieve the goal of waste reduction, while the forward logistics and recycling more efficient. Therefore the recovery levels U_4 indicators can be designed for product recovery x_{12}, product disassembly x_{13}, recycling rate x_{14}, and logistics equipment recycling rate x_{15}.

Assessment of Green degree could not be separated from the green management level U_5, which can reflect an enterprise's environmental awareness and the level of investment in environmental protection, and reflect an individual for sustainable development and environmental protection, the degree of support. So we can design staff's green awareness x_{16}, proportion of environmental management x_{17}, green consumer product acceptance x_{18}, and environmental protection investment ratio x_{19} Li and Li (2010).

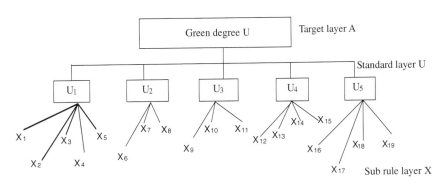

Fig. 71.1 The diagram of Multivariate hierarchical structure

In accordance with the principles of systematic, comprehensive and operability, this paper establishes a green degree evaluation index system of delivery access hierarchical model, as shown in Fig. 71.1.

71.2.2 The Fuzzy Comprehensive Evaluation Model of Enterprises' Green Degree in Green Supply Chain

In the evaluation index system, some indexes can obtain exact data, and others are very fuzzy. Considering the complexity and maneuverability and in order to reduce the arbitrary judgment subjectivity and to improve the objectivity and reliability of evaluation results, this paper uses a kind of two stage fuzzy comprehensive evaluation method combining the theory of fuzzy sets and analytic hierarchy process (AHP) (Chang and Zhang 1995).

(1) Determination of fuzzy set: Determining the evaluation object set: $U = \{u_1, u_2, u_3, u_4, u_3\}$, where u_i represents an ability of green degree evaluation of enterprises in the green supply chain; evaluation factors set: $X = \{x_1, x_2, x_3...x_{19}\}$. $u_1 = \{x_1, x_2, x_3, x_4, x_5\}$ $u_2 = \{x_6, x_7, x_8\}$ $u_3 = \{x_9, x_{10}, x_{11}\}$ $u_4 = \{x_{12}, x_{13}, x_{14}, x_{15}\}$ $u_5 = \{x_{16}, x_{17}, x_{18}, x_{19}\}$; comment set: $V = \{v_1, v_2, v_3, v_4\}$, v_1, v_2, v_3, v_4 were expressed as "very good", "good", "general", "poor".

(2) Do the first level fuzzy synthetic evaluation after determining the index weight: This article uses AHP (Xu et al. 1988) to determine the weight of each index. This approach can express a complex problem as a orderly hierarchical structure, and rank the decision scheme through people's judgment, which is practical, systematic and concise. Some indicators of Evaluation index system cannot be directly quantified, but it is enough to use the expert scoring method and Delphi method to obtain the extent of n_{ijn} that x_i belongs to reviews, and thus get fuzzy judgment matrix R_j.

The value method of n_{ijn} is to collect evaluation comments of expert group members, and get that there are v_{j1} comments on v_1, v_{j2} comments on v_2, v_{j3} comments on v_3, v_{j4} comments on v_4 for index x_j, and they are listed as follows:

$$n_{iji} = v_{ji} \bigg/ \sum v_{jn} (i = 1, 2, 3, 4) \qquad (71.1)$$

$$\sum v_{jn} = v_{j1} + v_{j2} + v_{j3} + v_{j4} \qquad (71.2)$$

So, fuzzy judgment matrix R_i can be obtained below.

$$R_1 = \begin{bmatrix} n_{111} & n_{112} & n_{113} & n_{114} \\ n_{121} & n_{122} & n_{123} & n_{124} \\ n_{131} & n_{132} & n_{133} & n_{134} \\ n_{141} & n_{142} & n_{143} & n_{144} \\ n_{151} & n_{152} & n_{153} & n_{154} \end{bmatrix} \quad R_2 = \begin{bmatrix} n_{211} & n_{212} & n_{213} & n_{214} \\ n_{221} & n_{222} & n_{223} & n_{224} \\ n_{231} & n_{232} & n_{233} & n_{234} \end{bmatrix}$$

$$R_3 = \begin{bmatrix} n_{311} & n_{312} & n_{313} & n_{314} \\ n_{321} & n_{322} & n_{323} & n_{324} \\ n_{331} & n_{332} & n_{333} & n_{334} \end{bmatrix} \quad R_4 = \begin{bmatrix} n_{411} & n_{412} & n_{413} & n_{414} \\ n_{421} & n_{422} & n_{423} & n_{424} \\ n_{431} & n_{432} & n_{433} & n_{434} \\ n_{441} & n_{442} & n_{443} & n_{444} \end{bmatrix}$$

$$R_5 = \begin{bmatrix} n_{511} & n_{512} & n_{513} & n_{514} \\ n_{521} & n_{522} & n_{523} & n_{524} \\ n_{531} & n_{532} & n_{533} & n_{534} \\ n_{541} & n_{542} & n_{543} & n_{544} \end{bmatrix}$$

The weights the X layer to the U layer Using AHP method is $N_i(i = 1, 2, 3, 4, 5), N_1 = (n_{11}, n_{12}, n_{13}, n_{14}, n_{15}), N_2 = (n_{21}, n_{22}, n_{23}), N_3 = (n_{31}, n_{32}, n_{33}), N_4 = (n_{41}, n_{42}, n_{43}, n_{44}), N_5 = (n_{51}, n_{52}, n_{53}, n_{54})$ among them.

Let the level evaluation vector of U_i be B_i, then

$$B_i = N_i \hbar R_i = \{b_{i1}, b_{i2}, b_{i3}, b_{i4}, b_{i5}\} \tag{71.6}$$

among them:

$$b_{i1} = n_{i1} \cdot n_{i1n} \oplus n_{i2} \cdot n_{i2n} \oplus \cdots \tag{71.7}$$

$$a \oplus b = \min\{1, a+b\} \tag{71.8}$$

(3) Do the second level fuzzy comprehensive evaluation.

Regard each $U_i(i = 1,2,3,4,5)$ as a factor, so that U is also a factor set, the single factor evaluation matrix for U:

$$R = \begin{bmatrix} B_1 \\ B_2 \\ B_3 \\ B_4 \\ B_5 \end{bmatrix} = \begin{bmatrix} b_{11} & b_{12} & b_{13} & b_{14} \\ b_{21} & b_{22} & b_{23} & b_{24} \\ b_{31} & b_{32} & b_{33} & b_{34} \\ b_{41} & b_{42} & b_{43} & b_{44} \\ b_{51} & b_{52} & b_{53} & b_{54} \end{bmatrix}$$

You can get the weights of U layer to the A layer according to the AHP method, $A = (a_1, a_2, a_3, a_4, a_5)$ and you can obtain two stage evaluation vector

$$B = A \hbar R = (b_1, b_2, b_3, b_4, b_5) \tag{71.9}$$

among them:

$$b_i = a_1 \cdot b_{1i} \oplus a_2 \cdot b_{2i} \oplus a_3 \cdot b_{3i} \oplus a_4 \cdot b_{4i} \oplus a_5 \cdot b_{5i} \tag{71.10}$$

(4) The evaluation results: Make $B = (b_1, b_2, b_3, b_4, b_5)$ normalize, and then add b_1 and b_2, the results can express the green supply chain green values better, which is an evaluation index value. If the result of b_1 and b_2 is more than 0.5, then enterprise's economic behavior is consistent with environmental standards.

71.3 Measures to Improve the Enterprises Green Degree in the Green Supply Chain

In general, it is a multi-objective program that improves the green degree value, which requires to achieve the emission reduction, energy saving, material saving, and to increase recycling efforts and enhance the green environmental protection level. Based on the results of the field research of green supply chain enterprises, this paper puts forward some measures to improve green degree, to enhance the green level of the enterprises and the whole supply chain.

(1) Focus on "three wastes" and other harmful emissions reduction.

Enterprises should strictly follow relevant environmental protection law and regulations, take the initiative to reduce emissions like waste water, waste, noise and harmful, toxic substances and others, and handle them inevitably caused by production process by using of physical, chemical and biological methods to discharge after purification or non-toxic harmless to reduce environmental pollution.

(2) Focus on energy saving and resource consumption

Energies like coal, electricity belong to non-renewable resources, so enterprises should use them safely, and improve equipment utilization and resource recycling rate as far as possible. Some materials such as steel, wood and other ones can be blanked to improve the utilization rate of raw materials by making full use of leftover materials. The production process of finished products should pay attention to the quality control and reduce scrap generated.

(3) Focus on Recycling

Enterprises should carefully do a good job of waste products and packaging recycling according to the "resource product recycling regulations," "packaging recycling management approach" and other regulations.

(4) Improve the level of green management

Enterprises should popularize environmental awareness, and allow employees to actively participate in the environment protection. Meanwhile it should dedicate environmental management workers, and increase investment to actively promote the environmental performance. Moreover it should actively produce green products, and develop consumer acceptance of green.

(5) Follow the Ideas of Green "Supply—Production—Sales" and Union the Business Strategy and Environmental Protection.

Green "Supply—Production—Sales" requires enterprises to do a good job of green procurement, green production, green marketing, green consumption and green logistics, and try to conserve resources and protect environments from the point of the green supply chain participants' view.

71.4 Conclusion

It's an inevitable trend of the research and implementation of international supply chain from traditional to green. This paper puts forward an evaluation index system and a model to evaluate enterprises green degree and gives several suggestions on improving enterprises green degree, hoping to promote the enterprise's sustainable development.

References

Chang D, Zhang L (1995) Method of fuzzy mathematics in economics management. BeiJing Economy Institute Press, BeiJing
Li JF, Li Y (2010) Research on greenness evaluation system and model of green supply chain. Proceedings of international conference on engineering and business management (EBM 2010). Scentific Research Press, USA, pp 348–349
Ma S, Lin Y, Chen Z (2000) Supply-chain management. China Machine Press, BeiJing
Wang N, Sun L (2005) Green supply chain management. Tsinghua University Press, BeiJing
Xu S (1988) The utility decision method-the principle of AHP. Tianjin University Press, Tianjin, pp 23–35

Chapter 72
The Ways for Improving the Operations of Hospital Industry: The Case in Macau

Yan Chen, Harry K. H. Chow and Ting Nie

Abstract Macau is one of the well-known tourism-dominated urban cities with high population density and limited natural resources. To maintain the reputation of tourism industry and cope with expanding public service demand, such as hospital service, hospital industry is now looking for solutions to enhance the service quality and efficiency. This paper proposes analysis and suggestions towards the improvement of operations activities in Macau's hospital industry. The analysis covers the impact of economic growth, employee distribution, population growth, current scale of hospital and infrastructure in Macau. Discussion about the suitability of adopting third party logistics at certain logistics activities is also conducted. Through the research contributions, it is expected that the globalized concept of "leisure and tourism oriented city" can be realized in Macau.

Keywords Macau · Hospital · Casino · Logistics · Supply chain management · Information system

Y. Chen (✉) · H. K. H. Chow · T. Nie
Faculty of Management and Administration,
Macau University of Science and Technology, Macau, China
e-mail: yachen@must.edu.mo

H. K. H. Chow
e-mail: khchow@must.edu.mo

T. Nie
e-mail: tnie@must.edu.mo

72.1 Introduction

As being a tourism and gambling oriented urban city, Macau is now suffering from the urban planning and public service demand problem. After the Macau government released the gambling market at 2002, more than 17 casinos coming from different countries have opened in Macau (The Statistics and Census Service 2012). These casinos not only provide gambling activities, but also offer multiple entertainment activities like international drama and exhibition infrastructures. Thus, the infrastructure developed by casinos does help the attraction of tourists. In recent years, the casino industry has provided significant contribution towards the growth of Macau's economic as well as tourism industry. Nevertheless, the economic growth does trigger number of environmental and public health problems, particularly the growing demand of hospital services. The increasing numbers of vehicle users and tourists, logistics movement of goods to casinos and public transport have further deteriorated the living environment and the health of public citizens. Power, industry and transport are the three major sectors responsible for fossil-fuel-related CO_2 emission in each country in the world (Timilsina and Shrestha 2009). The vehicles that are powered by gasoline and diesel fuel, emit the vast majority of pollutants. Tang and Wang (2007) have conducted a study about the traffic-induced air quality and noise problems into different urban areas in Macau. They found that the greater street canyon effects in the historical urban areas, the higher the carbon monoxide concentration is generated by the vehicles. Thus, air pollution problem does affect the development of a local tourism industry, and more importantly, create health implication problems of Macau's citizens, including increase of public health care costs and loss of productivity.

In views of the importance of economic growth and public health care concern, hospital industry presents a great demand for further enhancing the hospital services' quality and reducing the cost of service. The objectives of this paper include (i) Review current logistics practices in hospital, the way of cutting logistics costs, information system support and coordination; (ii) Identify the current situation of hospital and potential development of hospital industry; (iii) Address feasible solutions towards the operations improvement of hospital industry. This paper is organized as follows: A review of logistics activities in hospital is provided in Sect. 72.2. Section 72.3 conducts the background study about the hospital industry in Macau. The factor analysis about the operations improvement of hospital industry in Macau is conducted at Sect. 72.4. Finally, a conclusion is presented in Sect. 72.5.

72.2 Logistics Activities in Hospital

Similar to other industries, hospital operations such as material management, food and medical supply are relied on using the supply chain concept to manage. Liao and Chang (2011) identify certain factors that affect overall supply chain of the hospital's logistics system including (i) Safety stock. (ii) Lead time. (iii) Transportation capacity. The hospital management is required to keep reviewing these logistics functions, classifying value added and non-valued added activities, and identifying the costs associated with these activities thereby decreasing non-valued added activities (Aptel and Pourjalali 2001).

The logistics department of the hospital supports three internal logistics activities including (1) drug distribution, (2) food service and laundry, and (3) supply and processing of sterilized items in hospital. At first, with regard to the drug distribution, (Fineman and Kapadia 1978) identify three models of drug distribution, Model 1: direct delivery to medical department through central warehouse. Model 2 is the semi-direct delivery via the warehouses of the medical department. The final model is direct delivery via daily replenishment of small medical department storage facilities.

Food service and laundry is the second type of logistics activities, in general, these activities are run internally so as to better control the quality of food and products. If these activities are subcontracted, hospital usually awards the activities to service providers who have certification. The job duty of logistics department is to control and evaluate the service providers (Aptel and Pourjalali 2001). The assessing criteria include the food quality, hygiene control and cost efficiency. Supply and processing sterilized items in hospital is the final type of logistics activities. These logistics activities are referred to handling and storing sterilized items. The sterilization processes include decontamination, washing, rinsing and packaging. Sterilization of hospital surgical and medical treatment supplies is to ensure instruments and equipment to be clean and with acceptably low level of microbial and viral infectious agents (Fineman and Kapadia 1978). These sterilized items are required special logistics handling and storage in order to avoid contamination. The level of special logistics handling and storage is subject to three categorized items including (Fineman and Kapadia 1978) 1. Critical items-item is introduced into the body such as hypodermic syringes. 2. Semi-critical items-item is introduced into body openings such as anaesthesia equipment, cystoscopes, thermometers,..., etc. 3. Non-critical items- item only contacts with intact skin, such as water bottles and ice bags. Given the special logistics handling and storage may require additional manpower and equipment cost, the way how to minimize the inventory and replacement stock of sterile items as well as the storage method is the major practice for this kind of logistics activities.

72.3 The Hospital Industry in Macau

Since 2012, Macau Government has opened its gaming industry which awarded six gaming concession and sub-concessions to companies based in Las Vegas (Wynn Resorts, Las Vegas Sands, and MGM Mirage) and Hong Kong (Galaxy Casino and Melco Crown), the economic is dramatically reformed to gaming-led tourism and recorded a substantially growth in recent year (Tang and Sheng 2009).

There are around 44,806 citizens who are working in the gaming industry of Macau which is equivalent to 11 % of the total employed population (The Statistics and Census Service 2012). The numbers of employees working in the gaming industry have been increased more than 47 % as compared with the figure recorded in 2004. The casino gaming industry is one of the economic pillars to support the local economy. The gaming industry contributes $18.6 billion, which is accounted for 77 % of government revenues DSEC: Employed Population by Industry, Retrieved April 20 2012). Despite the gaming industry sector brings the huge economic benefit to Macau, many casino employees have suffered from potential health problems in which they are exposed to second hand smoke (SHS) at work (Chan et al. 2012). Most of the casinos allow smoking, it is known that smoking can cause cancer, the types of cancer include lungs, larynx, esophagus, mouth, kidney and pancreas (U.S. Department of Health and Human Services: The Health Consequences of Smoking: A Report of the Surgeon General. Centers for disease control and prevention, National center for chronic disease prevention and health promotion, Atlanta, GA 2004). The second-hand smoking has the same impact to health problems and can also cause cancer of lungs as direct smoking (Cormany and Baloglu 2011). Currently, there are two public hospitals, Hospital Conde S. Januário Hospital Centre (CHCSJ) and Macau University of Science and Technology Hospital, and one private hospital, Hospital Kiang Wu providing a total of 1,172 beds in 2010. It is believed that the increasing number of new admissions will further increase the operating cost as well as medical material cost of the hospital operations budgets. Thus, to allow focus on medical treatment and improve the operation efficiency, the re-engineering and streamlining of some non-medical treatment activities such as supply chain and logistics are therefore taken as the major priority of the top management.

72.4 Suggestions of Improving Hospital Operations

72.4.1 Outsource of Logistics Activities

To cope with the expansion of logistics services demand and concentrate on the core competence, many industries have already outsourced their logistics functions to third party logistics providers. Lieb and Bentz (2005) address the growth of third party logistics (3PL) industry due to increased globalization, pressure to reduce cost and enhance the performance achievement. Koh and Tan (2005) state that the annual growth in 3PL industry in China has been increased 25 % on average, leading the U.S. (10–15 % annual 3PL growth). The distinctive advantages of 3PL providers include improve customer service, respond to competition and asset elimination (Handfield and Nichols 1999). Despite outsourcing logistics to 3PL shows the numerous potential, the research done by Sahay and Mohan

(2006) argues that nearly 55 % of the companies terminated the relationship with 3PL after 3–5 years. The reasons are due to the perception of the 3PL users are uncertain about the service levels and unrealistic expectation (Lambert et al. 1999). Further, Zhang et al. (2005) state that 3PL providers have to respond to changing customer needs. In order to do so, the establishment of performance measurement to evaluate 3PL providers is necessary.

Refer to the logistics outsourcing in hospital, research studies conducted by Aptel and Pourjalali (2001) show that food services and laundry services in certain U.S. and France hospitals have been outsourced to service providers which owned relevant certification. Apart from outsourcing the food and laundry services, outsourcing the medical related logistics activities shows the potential for the cost reduction of inventory. In fact, the medical material cost nearly consumes about 30 % of total hospital operations cost (Tung et al. 2008). Another advantage of outsourcing medical logistics activities is to help improve the usage of space. Due to limited supply of land use in Macau, outsourcing the medical storage areas allows better utilization of land use such as increasing the number of beds. Hospital management is therefore suggested to cooperate with 3PL providers to develop a long term partner relationship by establishing a series of performance indicators as well as standard operations procedures in order to better align the service level and expectation.

72.4.2 Development of E-health Information Management System

To control the hospital resources utilization and coordinate the information flow between supply chain parties in a hospital is another direction to cut cost and achieve competitive advantage in the medical sector. Many research studies proved the contributions of information systems towards the cost minimization and operational efficiency improvement. Gilbert (2001) addresses the value of an e-health system for reducing the procurement cost in hospital. More and Mcgrath (2002) clarify the e-health system is a kind of information and communication technology to help hospital management in decision-making, record and storage of relevant data of various supply chain participants including suppliers, hospitals and patients. Merode et al. (2004) try to review the potential of using the enterprise resource planning system (ERP) to support hospital management. In order to respond to the non-deterministic processes such as stochastic demand of front line patient services, visualizing the resources status and short term planning are essential in hospital. Anoraganingrum and Eymann (2009) advocate the application of radio frequency identification system (RFID) to improve the hospital efficiency. They propose using a RFID system to improve the performance of sterilization and equipment monitoring.

72.5 Conclusion

The substantial growth of economy and the development of gambling market in Macau has brought side effects of health problem to workers in casinos. In views of the importance of economic growth and public health care concern, the development of hospital industry shows a great potential value for the general public. Currently, there are three hospitals in Macau, and it is expected that the increased demand of the public health care will trigger the concerns of hospital management regarding operation efficiency enhancement and cost reduction of medical materials. Therefore, streamlining the supply chain and improving logistics activities in hospital industry is essential. The feasible suggestions for the development of hospital industry and improvement of the operations are summarized as follows:

Cooperate with external service providers in order to better concentrate on the core competence of hospital management. A long term partner relationship is suggested to develop between hospital and service providers through establishing a series of performance indicators as well as standard operations procedures in order to better align the service level and expectation.

Develop an e-health information system to improve the operation efficiency and inventory control of medical materials. The selection of information systems includes e-health, ERP and RFID systems. Different information systems show the potential value for improving the hospital efficiency and reducing costs. Nevertheless, the cost and benefit analysis, investment payback of information technology (IT) and influences of IT implementation towards the hospital industry are other important topics that the hospital management should not omit when adopting these technologies and systems.

References

Anoraganingrum D, Eymann T (2009) A conceptual framework to define indicators of information technology in the hospital. Int J Healthc Technol Manag 10(1/2):2–14

Aptel O, Pourjalali H (2001) Improving activities and decreasing costs of logistics in hospitals, A comparison of U.S. and French hospitals. Int J Account 36:65–90

Chan SH, Pilkington P, Wan YKP (2012) Policies on smoking in the casino workplace and their impact on smoking behaviour among employees: case study of casino workers in Macao. Int J Hospital Manag 31:728–734

Cormany D, Baloglu S (2011) Medical travel facilitator websites: an exploratory study of web page contents and services offered to the prospective medical tourist. Tour Manag 32:709–716

Fineman SJ, Kapadia AS (1978) An analysis of the logistics of supplying and processing sterilized items in hospitals. Comput Oper Res 5:47–54

Gilbert SB (2001) New millennium strategic initiatives for health care purchasing. Hosp Mater Manag 22(3):71–78

Handfield RB, Nichols EL Jr (1999) Introduction to supply chain management. Prentice Hall, Upper Saddle River

Koh SCL, Tan Z (2005) Using e-commerce to gain a competitive advantage in 3PL enterprises in China. Int J Logist Syst Manag 1(2/3):187–210

Lambert DM, Emmelhainz MA, Gardner JT (1999) Building successful logistics partnerships. J Busin Logist 20(1):165–181

Liao HC, Chang HH (2011) The optimal approach for parameter settings based on adjustable contracting capacity for the hospital supply chain logistics system. Expert Syst Appl 38:4790–4797

Lieb R, Bentz BA (2005) The North American third party logistics industry in 2004: the provider CEO perspective. Int J Phys Distrib Logist Manag 35(8):595–611

Merode GG, Groothuis S, Hasman A (2004) Enterprise resource planning for hospital. Int J Med Inf 73:493–501

More E, Mcgrath M (2002) An Australian case in e-health communication and change. J Manag Dev 21(7/8):621–632

Sahay BS, Mohan R (2006) Managing 3PL relationships. Int J Integr Supply Manag 2(1/2):69–90

Tang UW, Sheng N (2009) Macao. Cities 2:220–231

Tang UW, Wang ZS (2007) Influences of urban forms on traffic-induced noise and air pollution: results from a modeling system. Environ Model Softw 22:1750–1764

The Statistics and Census Service (DSEC): Employed Population by Industry (2005). http://www.dsec.gov.mo/ PredefinedReport.aspx? Report (2005). Accessed 20 April 2012

Timilsina G, Shrestha A (2009) Transport sector CO_2 emissions growth in Asia: underlying factors and policy options. Energy Policy 37:4523–4539

Tung FC, Chang SC, Chou CM (2008) An extension of trust and TAM model with IDT in the adoption of the electronic logistics information system in HIS in the medical industry. Int J Med Inf 77:324–335

U.S. Department of Health and Human Services: The Health Consequences of Smoking: A Report of the Surgeon General. Centers for disease control and prevention, National center for chronic disease prevention and health promotion, Atlanta (2004)

Zhang Q, Vonderembse MA, Lim J-S (2005) Logistics flexibility and its impact on customer satisfaction. Int J Logist Manag 16(1):71–95

Chapter 73
The Social Costs of Rent-Seeking in the Regulation of Vehicle Exhaust Emission

Yan Pu and Xia Liu

Abstract The regulation of vehicle exhaust emission requires the regulators to establish the vehicle exhaust emission level desirable for the society and then select inspection agencies to check every registered vehicle periodically. Both these decisions create opportunities for rent seeking. In this paper, we present the incentives of rent-seekers for being selected as inspection agencies and analyze the consequences for social welfare. We find differences in firms' rent-seeking choices compared to a traditional rent-seeking model. We see that a fundamental aspect of firms' incentives to seek rent depends on the number of incumbent inspection agencies and the present value of every successful rent-seeker's rent income, which mainly depends on the distortion degree of inspection process and which is inversely related to social welfare.

Keywords Rent seeking · Regulation · Vehicle exhaust emission · Social costs

73.1 Introduction

Substantial resources are devoted to altering government policies in the form of rent-seeking in China. It is unanimous among economists that rent-seeking is socially wasteful, if rational from the rent-seeker's perspective. The entire

Y. Pu (✉)
Economics and Management College, Sichuan Normal University, Chengdu, China
e-mail: puswallow@yahoo.com.cn

X. Liu
Business School, Zhengzhou University, Zhengzhou, China
e-mail: liuxia432@zzu.edu.cn

literature on rent-seeking that has developed over the past 40 years has focused on (a) building rent-seeking contest models from different perspectives, such as under varying cost structure, under uncertainty, and with the assumption of rent-seekers being symmetric or asymmetric, risk-neutral or risk averse, and so on (Stein 2002); (b) discussing the scope of the social costs of rent-seeking activities; (c) studying how the following factors influence rent-seekers' expenditures at equilibrium: the marginal return of rent-seeking, the winning chance of rent-seeking, that is, the probability of being successful, and entry costs, etc. (Ritz 2008; Anderson and Freeborn 2010). So far, economists have almost clarified the scope of the social costs of rent-seeking, which, at first, were considered to consist of "Harberger Triangle" and "Tullock Rectangle", then include the social cost of improper selection of high-cost producers in the rent-seeking contest, the social cost of policy-making in the rent-seeking contest and the social cost that the whole economy incurred because of rent-seeking (Sobel and Garrett 2002; Ihori 2011; Gao 2011).

Rent-seeking is quite rampant and prevalent in the transition of China from the planning economy to the market economy with the Chinese characteristics and it has led to serious distortion of government policies. The regulation of vehicle exhaust emission is no exception. Rent-seeking in the regulation of vehicle exhaust emission has caused large social loss to the sustainable development of the Chinese economy, so the development of low-carbon transportation and logistics has attracted increasing attention in recent years in China. The purpose of this paper is to report our first-stab efforts in this regard. We build a slightly different model from Tullock's classic Efficient Rent-seeking model to analyze the social costs of rent-seeking in the regulation of vehicle exhaust emission. Our findings, based on the regulation of vehicle exhaust emission in China, verify the existence of the social costs to rent-seeking.

This paper is organized as follows. Some realistic rent-seeking cases in the regulation of vehicle exhaust emission in China are presented in Sect. 73.2. Section 73.3 contains a model slightly different from Tullock's efficient rent-seeking model to analyze the rent-seeking contest and a presentation of other social costs in addition to the direct rent-seeking expenditure in the regulation of vehicle exhaust emission in China. Section 73.4 offers conclusions and suggestions for further research.

73.2 Some Relevant Cases

In January 1998, Liu Gongchao, one car owner in Beijing, China, installed one automobile exhaust purifier produced in Korea, which was testified to overpass the vehicle exhaust emission standard stipulated by the relevant authority in Beijing. However, in July of the next year, the Environmental Protection Bureau of Beijing, the Transportation Bureau of Beijing, and the Traffic Management Bureau of Beijing jointly issued an announcement, which stated that all the minibuses fixed with carburetor that were registered after January 1, 1995 shall be equipped with an electronic air supply device and a three-element purifier; otherwise, they cannot be accepted for annual inspection. Liu's car was rejected for annual inspection. Liu thought that the three bureaus had abused administrative power and restricted car owners to purchase the designated goods, so he filed a case to the People's Court of Haidian District of Beijing against the Environmental Protection Bureau of Beijing.

The court held that because the announcement jointly issued by the Environmental Protection Bureau, the Transportation Bureau, and the Traffic Management Bureau of Beijing was not aimed at specific vehicles and could be used repeatedly, it belonged to abstract administrative behavior which is generally binding on all vehicles, and it should not be considered as administrative monopoly. The court turned down Liu's claim thereafter.

Some similar problems took place in other cities. In Nanchang, the capital of Jiangxi Province of China, in order to regulate the vehicle exhaust emission, the environmental protection authority firstly required vehicle owners to install filters, then carbon monoxide canisters, and then clean the cylinders during the past 10 years. These measures did not improve the environmental quality of Nanchang, only helping the relevant authority reap a large amount of easy money. Further, it was reported that in Xiangfan, one city of Hubei Province of China, Lanzhou, the capital of Gansu Province of China, and other cities, as long as the car owner purchased designated purifier, their car can pass annual inspection no matter whether the car owner install the purifier on the car or not. It was even reported that as long as the car owner paid money, he would get a certificate for his car.

73.3 Social Costs of Rent-Seeking in the Regulation of Vehicle Exhaust Emission in China

The regulation of vehicle exhaust emission has created a lot of opportunities for rent-seeking, which reallocates resources away for productive, positive-sum, activities into unproductive, zero-sum, activities. Therefore, it is unanimous among economists that rent-seeking is a socially wasteful activity. Theory tells us that the real social costs of rent-seeking are not only the expenditures of rent-seekers, but the opportunity cost of these resources in terms of forgone production, which are hard to observe and notice in reality. It can be seen from the above cases that, in addition to the traditional opportunity cost of rent-seeking, they included the social cost of higher efficient device being replaced by lower efficient device, the social cost of vehicles with high level of exhaust emission passing annual inspection and the social cost of various kinds of artificially created barriers which help breed rent-seeking activities.

(a) The traditional cost of rent-seeking

Rent-seeking in the regulation of vehicle exhaust emission in China is slightly different from other rent-seeking activities.

The purpose of rent-seeking in the regulation of vehicle exhaust emission in China is to obtain the inspection license of vehicle exhaust emission. At present, there already exist some inspection agencies in nearly every city. Rent-seekers can only obtain part of the rent if rent-seeking is successful. In the future, part of their rent will be absorbed away by the later successful rent-seekers. Therefore, the number of incumbent inspection agencies and the duration of the possible rent have a more significant effect upon the equilibrium expenditure of rent-seekers than in the other rent-seeking contest models.

Assume there are two identical risk-neutral players, A and B, where a stands for A's rent-seeking expenditures, b for B's expenditures, n for the number of incumbent inspection agencies, t for the average number of days that a new comer needs to enter the market, r for the interest rate for the time period of t and R is the total rent created within one day. Then the successful rent-seeker can get $\frac{tR}{(n+1)(1+r)}$ for the current term, and when the next inspection agency enters into the market, his rent will change to $\frac{tR}{(n+2)(1+r)^2}$. Similarly, we can get the present value of the total rent for a new comer when there already exist n inspection agencies as follows:

The present value (PV)

$$= \sum_{N=1} \frac{tR}{(n+N)(1+r)^N} \qquad (73.1)$$

Because of entry barriers, such as fixed entry costs, close relationships with the relevant authority and so on, the number of new inspection agencies which can enter into the market, N, is limited.

Player A maximizes the expected value

$$E(a) = \frac{a}{a+b}(PV-a) + \frac{b}{a+b}(-a) \qquad (73.2)$$

which is reduced to

$$E(a) = \frac{a}{a+b}PV - a \qquad (73.3)$$

It can be seen that A's optimal investment depends on B's effort and the present value of the rent at stake.

B faces an identical choice and generates a similar reaction function. If both players behave according to their strategy of maximizing the expected value, a simply Cournot-Nash equilibrium appears, each player bidding $\frac{1}{4}$ of the present value of the rent at stake ($\frac{1}{4}PV$). With m identical players the equilibrium investment by A is $\frac{m-1}{m^2}PV$ and the total expenditures by all the players reach $\frac{m-1}{m}PV$. If m is big enough, then complete dissipation of the rent will be realized.

The social loss is quite large because of a very high PV, which can be seen from the PV's equation. PV depends on five variables: the number of incumbent inspection agencies, the number of new inspection agencies, the interest rate for t days, the total rent R in the market and the average days that a new comer takes to enter into the market. The bigger t and R are, the smaller n is, the larger the present value is, the more social resources will be exhausted.

(b) The social cost of higher efficient device being replaced by lower efficient device

The second category of social cost caused by rent-seeking in the regulation of vehicle exhaust emission in China is the social cost of higher standard device being replaced by lower standard device. In order to control vehicle exhaust emission, the relevant authority often limit car owners to purchase their designated equipment by issuing announcements or notices. So long as the announcement or regulation is not aimed at specific objects and can be used repeatedly, it is legal and does not belong to administrative monopoly, which is subject to the Chinese Anti-Monopoly Law.

(c) The social cost of vehicles with high level of exhaust emission passing annual inspection

It has been mentioned above that in many cities of China, as long as car owners purchased the designated vehicle exhaust purifying equipment, they can pass annual inspection even if they do not install the purchased equipment on the car. In some places, as long as car owners make the payment to agencies or individuals, which have close relationships with the regulator and which have already developed into a huge industry in China, they can bet the certificate for their cars. On the basis of "hazard risk", cars with higher level of exhaust emission are more willing to turn to those agencies and escape real annual inspection, causing serious damage to the environment. Even in some cities where car owners must drive their cars to the inspection site and accept real annual inspection, they can also pass annual inspection even if their exhaust emission level fails to meet the standard, as long as they pay some "lubrication money" to the relevant "helpers".

(d) The social cost of various kinds of artificially created barriers

The next but not the last measure for inspection agencies to reap easy money is to set different kinds of inspection items as barriers. These measures have two purposes; one is to reap money directly and the other is to create opportunities for rent-seeking. The watchdog of environment issues standards, announcements or regulations on a frequent basis. Sometimes, it is only a short period of time between the implementation dates of the old and new announcements or standards. Although the old equipment is still useful and efficient, it should be discarded and replaced with the new purifying one, according to the new regulation, which constitutes a social loss of resources.

As for obscure obstacles, such as waiting time and psychological pressure while waiting to get cars inspected, car owners find it hard to overcome them except for rent-seeking. It is a well-known fact that car owners have to queue and wait at several places before getting inspection finished. The forgone productive activities constitute a social cost, too.

73.4 Conclusions

In this paper, we have explored several aspects of the social costs of rent-seeking activity in the regulation of vehicle exhaust emission. It seems that rent is under dissipated in the regulation of vehicle exhaust emission; however, the resources exhausted for rent seeking only account for a small share of the social costs and they constitute the direct costs of rent seeking. In addition to these costs, higher efficient device being replaced by lower efficient device, vehicles with high level of exhaust emission passing annual inspection by rent-seeking and various kinds of artificially created difficulties and obstacles which help breed rent-seeking activities constitute much higher social costs. Vehicle owners are easily exposed to rent-seeking in China. Even if he follows laws and regulations issued by the state, or purchase the most expensive and best exhaust purifier at the time, he will not necessarily pass annual inspection, if he does not observe the announcement or notice issued by the local authorities.

Therefore, how to regulate the regulator of vehicle exhaust emission may be the key issue to eliminate rent-seeking and reduce waste of social resources. According to the Chinese Administration Law, all abstract administrative behaviors, usually in the form of announcements and notices issued by industry regulators or local authorities, belong to legal behaviors and thereafter have legal effect. Only concrete administrative behaviors which are targeted as specific objects and cannot be used repeatedly may be taken as illegal. Such provisions create lots of rent-seeking opportunities for regulators and authorities. Therefore, the first step to control vehicle exhaust emission in China shall be the rigorous regulation of the regulator.

The second measure lies in the establishment of a transparent and uniform emission standard, which can eliminate confusion among vehicle owners. The next but not the last step is to rigidly combat any collusion between regulators and inspection agencies, inspection agencies and vehicle owners, and inspection intermediaries and vehicle owners. Only in this way can the blue sky return to China.

References

Anderson LR, Freeborn BA (2010) Varying the intensity of competition in a multiple prize rent seeking experiment. Public Choice 143:237–254

Epstein and Nitzan (2002) Stakes and welfare in rent-seeking contests. Public Choice 112:137–142

Gao Y (2011) Government intervention, perceived benefit, and bribery of firms in transitional China. J Bus Ethics 104:175–184

Ihori T (2011) Overlapping tax revenue, soft budget, and rent seeking. Int Tax Public Finance 18:36–55

Ritz RA (2008) Influencing rent-seeking contests. Public Choice 135:291–300

Sobel Russell S, Thomas A, Garrett (2002) On the Measurement of Rent Seeking and its Social Opportunity Cost. Public Choice 112(1-2):115–136

Printed by Publishers' Graphics LLC